PHOSPHORUS CHEMISTRY
DIRECTED TOWARDS BIOLOGY

Honorary President

Lord Todd, PRS

Organizing Committee

Chairman: J. Michalski
Deputy Chairmen: M. Wiewiórowski
W. J. Stec
Secretary: A. Okruszek
Members: K. Golankiewicz
P. Mastalerz
M. Mikołajczyk
W. Ostrowski
D. Shugar
A. Skowrońska
K. L. Wierzchowski
J. Wróbel

INTERNATIONAL UNION OF PURE AND APPLIED CHEMISTRY
IUPAC Secretariat: Bank Court Chambers, 2-3 Pound Way
Cowley Centre, Oxford OX4 3YF, UK

INTERNATIONAL UNION OF PURE AND APPLIED CHEMISTRY
(Organic Chemistry Division)

in conjunction with

Polish Academy of Sciences
(National Committee of Chemistry)
(Centre of Molecular and Macromolecular Studies)

PHOSPHORUS CHEMISTRY DIRECTED TOWARDS BIOLOGY

Lectures Presented at the
International Symposium on Phosphorus Chemistry
Directed Towards Biology, Burzenin, Poland
25-28 September 1979

Editor

W. J. STEC

*Polish Academy of Sciences
Centre of Molecular and Macromolecular Studies
Lódź, Poland*

PERGAMON PRESS

OXFORD · NEW YORK · TORONTO · SYDNEY · PARIS · FRANKFURT

U.K.	Pergamon Press Ltd., Headington Hill Hall, Oxford OX3 0BW, England
U.S.A.	Pergamon Press Inc., Maxwell House, Fairview Park, Elmsford, New York 10523, U.S.A.
CANADA	Pergamon of Canada, Suite 104, 150 Consumers Road, Willowdale, Ontario M2J 1P9, Canada
AUSTRALIA	Pergamon Press (Aust.) Pty. Ltd., P.O. Box 544, Potts Point, N.S.W. 2011, Australia
FRANCE	Pergamon Press SARL, 24 rue des Ecoles, 75240 Paris, Cedex 05, France
FEDERAL REPUBLIC OF GERMANY	**Pergamon Press GmbH, 6242 Kronberg-Taunus, Hammerweg 6, Federal Republic of Germany**

Copyright © 1980 International Union of Pure and Applied Chemistry

All Rights Reserved. No part of this publication may be reproduced, stored in a retrieval system or transmitted in any form or by any means: electronic, electrostatic, magnetic tape, mechanical, photocopying, recording or otherwise, without permission in writing from the copyright holders.

First edition 1980

British Library Cataloguing in Publication Data
International Symposium on Phosphorus Chemistry Directed Towards Biology, Burzenin, 1979
Phosphorus chemistry directed towards biology.
(International Union of Pure and Applied Chemistry. IUPAC symposium series).
1. Phosphorus in the body - Congresses
I. Title II. Stec, W. J. III. International Union of Pure and Applied Chemistry. *Organic Chemistry Division* IV. Polska Akademia Nauk. *National Committee for Chemistry* V. Series
574.1'921   QP535.P1   80-40294
ISBN 0 08 023969 2

In order to make this volume available as economically and as rapidly as possible the author's typescript has been reproduced in its original form. This method has its typographical limitations but it is hoped that they in no way distract the reader.

Printed in Great Britain by A. Wheaton & Co. Ltd., Exeter

CONTENTS

Foreword	vii
Polymers Containing Phosphorus in Bacterial Cell Walls by J. BADDILEY	1

Studies on Oligoribonucleotides Synthesis
  by R.W. ADAMIAK, E. BIAŁA, K. GRZEŚKOWIAK, R. KIERZEK, A. KRASZEWSKI,
  W.T. MARKIEWICZ, J. OKUPNIAK, J. STAWINSKI and M. WIEWIÓROWSKI
	9

Reactive Phosphorylating Intermediates in the Nucleic Acids Chemistry by D.G. KNORRE and V.F. ZARYTOVA	13

Synthesis of $tRNA_f^{Met}$ from E. coli
  by M. IKEHARA, E. OHTSUKA, S. TANAKA, S. NISHIKAWA, T. TANAKA,
  T. MIYAKE, E. NAKAGAWA, T. WAKABAYASHI, R. FUKUMOTO, H. UEMURA,
  Y. TANIYAMA, T. DOI, A.F. MARKHAM and J. ANTKOWIAK
	33

The Synthesis of a Potentially Metabolically-Labile Derivative of 2'-Deoxy-5-Fluorouridine 5'-Phosphate by R.N. HUNSTON, A.S. JONES and R.T. WALKER	47
Synthesis and Biological Activity of Some Analogues of Nucleic Acids Components by A. HOLÝ	53
The Synthesis of Biologically Relevant Molecules via Phosphorus Ylids by H.J. BESTMANN	65
The Phosphate Transfer Process by A.J. KIRBY	79
Phosphate Synthesis, Exchange, and Interaction in Nucleotides and Related Compounds by N.J. LEONARD	85
Chiral Anticancer Oxazaphosphorinanes - Stereospecific Synthesis, Configurational Assignments and Biological Evaluations by W.J. STEC	95
Synthetic Polynucleotides as Interferon Inducers by D.W. HUTCHINSON	111
The Case for Monomeric Metaphosphate by A. SATTERTHWAIT and F.H. WESTHEIMER	117
Nucleoside Phosphorothioates - Tools for the Investigation of Enzyme Mechanisms by F. ECKSTEIN	125
The Stereochemical Course of Several Enzyme Catalyzed Reactions at the Phosphodiester Level by F.R. BRYANT, J.F. MARLIER and S.J. BENKOVIC	129
Models of Biopolymers by Ring-Opening Polymerization of Cyclic Phosphorus Containing Compounds by S. PENCZEK	133
Some Aspects of Oligonucleotide Synthesis by C.B. REESE	145

Applications of ^{31}P NMR to Biological Systems with Emphasis on Intact Tissue Determinations 157
  by *T. GLONEK*

Oligodeoxyribonucleotide High Speed Solid Phase Synthesis as a Molecular Biology Research Instrument 175
  by *V.K. POTAPOV, V.P. VEIKO, O.N. KOROLEVA, S.I. TURKIN and Z.A. SHABAROVA*

Chemical Synthesis of DNA - Possibilities and Applications 181
  by *H. KÖSTER*

Silyl- and Stannyl-Esters of Phosphorus Oxyacids - Intermediates for the Synthesis of Phosphate Derivatives of Biological Interest 197
  by *T. HATA and M. SEKINE*

The Use of Diethyl Azodicarboxylate-Triphenylphosphine System in the Synthesis and Transformation of Natural Products 213
  by *O. MITSUNOBU*

Nucleoside Phosphates as Probes of the Mechanism of Protein Biosynthesis 219
  by *S.M. HECHT*

FOREWORD

  The chemistry of phosphorus in its fundamental and applied aspects has grown rapidly in the past decade. Progress has been made in nearly every scientific discipline dealing with this key element. The understanding of reaction mechanisms of organophosphorus compounds and the development of analytical tools for their studies promoted research in nucleic acid chemistry, investigations on metabolism and biosynthesis of biologically relevant phosphorus compounds and elucidation of the mode of action of enzymes responsible for their biological transformations.

  Although the achievements in the fields of phosphorus chemistry connected with Life Sciences are significant, still phosphorus chemistry hesitates to open broadly towards biochemistry. International Symposia on Phosphorus Chemistry (Paris 1969, Gdańsk 1974, Halle 1979) to a very low extent covered topics dealing with biological aspects of phosphorus chemistry.

  Remembering, that promotion of knowledge comes through interdisciplinary approaches we undertook an effort to organize an International Symposium on Phosphorus Chemistry Directed Towards Biology, which was held in Burzenin n/Łódź, Poland, Sept. 25-28, 1979. Organization of this meeting was inspired by the International Conference "Synthesis, Structure and Chemistry of Transfer Rybonucleic Acids and their Components", held in Dymaczewo n/Poznań, Poland, on Sept. 13-17, 1976. I would like to acknowledge the advice that we received from Professor M.Wiewiórowski from his experience in arranging successful Nucleic Acid Seminars. It was, however, the determination and the dedication of the Chairman of the Organizing Committee Professor Jan Michalski, which made the Symposium possible.

  We owe a special debt of gratitude to Lord Todd, PRS, who kindly accepted our invitation to participate in this meeting. A number of prominent scientists, and it would be impractical to list here all of them, also responded favourably to our request to deliver lectures.

  The major part of this first International Symposium on Phosphorus Chemistry Directed Towards Biology consisted of lectures on phosphorus-bearing nucleic acid components. Macromolecular chemistry of phosphorus was reflected not only in the field of oligonucleotides, but also in papers discussing cell-wall components. Part of the lectures were devoted to the mechanism of biological phosphorylation and the mode of action of enzymes. Art in Synthesis has been demonstrated not only in the construction of biologically relevant phosphorus compounds but also in applications of organic derivatives of phosphorus in the synthesis of other biologically important molecules. Due to time limitation we were not able to cover all aspects of Phosphorus Chemistry Directed Towards Biology. We believe that this first Symposium helped many "pure" phosphorus chemists to cross disciplinary lines and thus further development in biologically oriented phosphorus chemistry will be stimulated by this type of approach.

Symposium Editor

Wojciech J. Stec

POLYMERS CONTAINING PHOSPHORUS IN BACTERIAL CELL WALLS

J. Baddiley

Microbiological Chemistry Research Laboratory, University of Newcastle,
Newcastle upon Tyne NE1 7RU, U.K.

The bacterial cell wall is of major importance in maintaining the integrity of the cell in widely different environments. It is the site of group- and type- specific antigens and its biosynthesis is the primary site of attack by many antibiotics, including the penicillins and cephalosporins. In Gram-positive bacteria the wall is composed of peptidoglycan and usually teichoic acid or, less frequently, teichuronic acid. The two main polymers are covalently attached to each other.

Peptidoglycan is always present and comprises about half of the dry weight of most walls. It contains glycan chains in which the sugars N-acetylglucosamine and N-acetylmuramic acid alternate regularly. Attached to the muramic acid residues are peptide chains which cross-link adjacent glycans. Although the glycan chains are common to all bacteria, the peptides vary in their amino acid composition in different organisms. The structure of a unit of peptidoglycan from *Staphylococcus aureus* is shown in Fig.1.

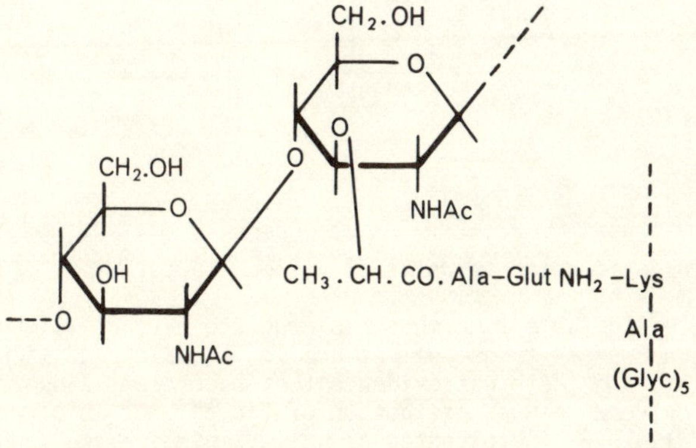

Fig.1. Peptidoglycan in walls of *Staphylococcus aureus*

The other major wall components in Gram-positive bacteria are the teichoic acids. These polymers vary considerably in structure, but all are rich in phosphodiester linkages. Many are polymers of glycerol phosphate (Fig.2a) to which sugar residues and D-alanine in ester linkage are usually attached. Others are similar polymers containing ribitol phosphate (Fig.2b). In some organisms the sugar residues form a part of the main polymer chain (Fig. 2c), and in others sugar 1-phosphate residues are present either alone or together with glycerol phosphate. In addition to teichoic acids the walls sometimes contain smaller amounts of phosphate-free acidic polysaccharides (teichuronic acids), and the relative amounts of teichoic or teichuronic acid can vary according to the amount of phosphate in the growth medium.

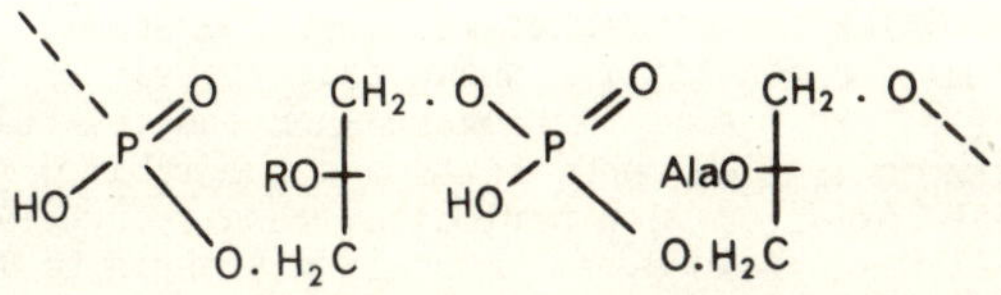

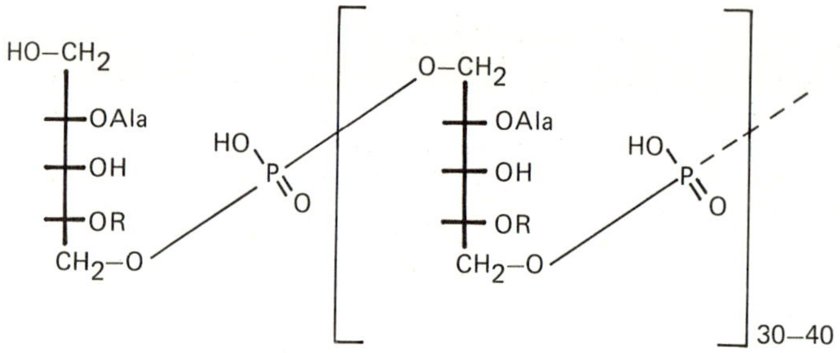

Staph. aureus, R = N-acetylglucosaminyl
Bacillus subtilis W23, R = β-glucosyl

b

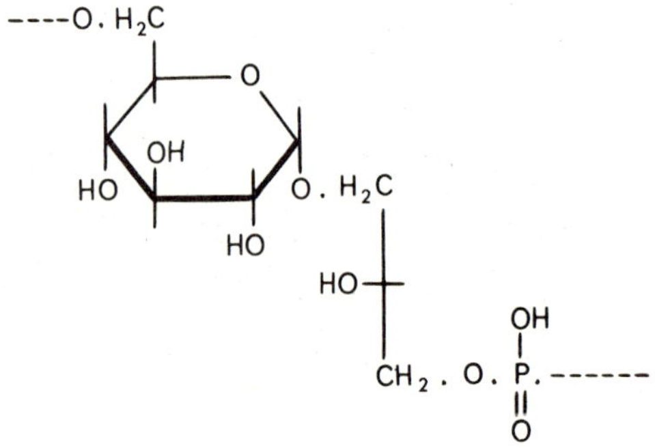

c

Fig.2a,b,c. Teichoic acids in walls of various bacteria

It has been known for a considerable time that, in order to extract teichoic acids from wall preparations, it is necessary to use either acidic or basic conditions. The extraction is not immediate and there was evidence that a covalent linkage occurs between teichoic acid and peptidoglycan. Thus, extraction of the polymer involves hydrolysis of at least one covalent linkage. Autolysates of walls, in which the linkage between the two polymers was still intact, did not contain phosphomonoester residues, and so it was concluded that the linkage between the polymers involves the terminal phosphate of the teichoic acid chain and, as muramic acid phosphate was formed on vigorous hydrolysis of walls, it was assumed that linkage involves unidentified muramic acid residues in the glycan chains (reviewed in Ref.1).

Although it seemed possible that linkage between the polymers was direct through a phosphodiester involving the terminal phosphate of the teichoic acid chain, it was not clear why such a linkage should be labile towards both acid and alkali. A detailed chemical analysis of walls of *S.aureus* has established the nature of this linkage. In order to carry out this study it was necessary to use walls of very high purity. It was particularly important to remove protein and phospholipid contaminants which are usually present in appreciable amounts in washed wall preparations. Analysis of the purified walls showed that about 90% of the phosphorus could be accounted for as ribitol phosphate in the main teichoic acid chain. However, about 8% of the total phosphate in an acid hydrolysate was glycerol phosphate.

Using a mutant of *S.aureus* which lacked *N*-acetylglucosamine substituents on the ribitol phosphate residues in the main chain, it was shown that the glycerol phosphate is bound to the wall in the form of a trimer (2). Moreover, when the teichoic acid was extracted with alkali under carefully controlled conditions, and the ribitol phosphate units in the main chain were subsequently oxidized with periodate, then reduced with borohydride, the fragment shown in Fig.3 was obtained. It contained a chain of three glycerol phosphate residues and one ethylene glycol phosphate. The ethylene glycol was derived from the terminal ribitol in the teichoic acid. This was confirmed by a similar degradation of walls from the wild type organism, where the presence of N-acetylglucosamine on the terminal ribitol residue restricted its degradation to ethylene glycol and gave instead a substituted glycerol (3). It follows that in this organism, and in the mutant, the main ribitol phosphate chain is attached directly through a terminal phosphate to a chain of three glycerol phosphate residues. Thus the teichoic acid is attached to the peptidoglycan through a linkage unit comprising three glycerol phosphate residues and possibly other unidentified components.

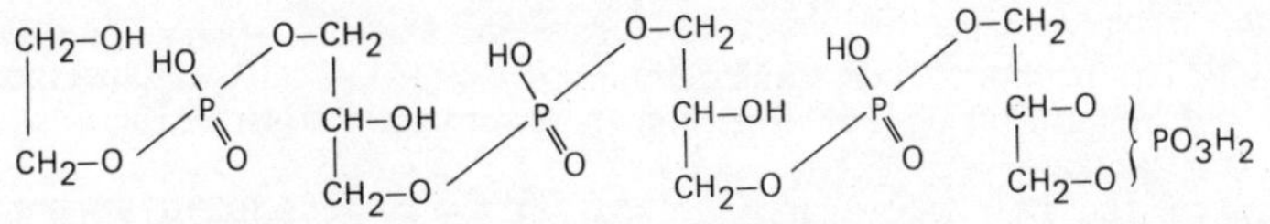

Fig.3. Degradation product from walls of *Staphylococcus aureus* mutant obtained by extraction with alkali, oxidation with periodate and reduction with borohydride.

The analysis described above was supported by biosynthetic studies. It was shown that membrane preparations from a number of organisms could synthesize polymers containing linkage unit from appropriate nucleotide precursors. However, the incorporation of glycerol phosphate residues into polymer from the precursor CDP-glycerol required the presence of both CDP-ribitol (Fig.4) and UDP-N-acetylglucosamine (4,5). The requirement for this last nucleotide suggested that N-acetylglucosamine or its phosphate might be an unidentified component in the linkage unit. The chemical degradations would not have revealed such a component, partly because of the considerable amount of the amino sugar in the peptidoglycan, but also because later work showed that the hydrolysis conditions used in the degradation would result in the amino sugar remaining with the peptidoglycan.

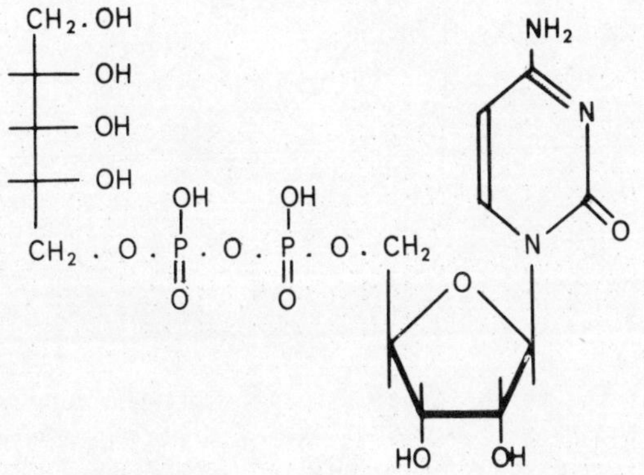

Fig.4. Cytidine diphosphate ribitol

In a further chemical study the teichoic acid was extracted from the wall under very gentle conditions, sufficient to hydrolyse sugar 1-phosphate linkages but insufficient to hydrolyse the main chain or the phosphodiesters in the linkage unit. The extracted polymer possessed a reducing end which was shown to be N-acetylglucosamine. Oxidation of the wall with periodate, followed by reduction with borohydride and then acid extraction gave the product shown in Fig.5 thereby confirming the presence of an N-acetylglucosamine residue in the linkage unit (6). The ready fission of the chain at a reducing centre suggests that the glucosamine was attached originally to the peptidoglycan through a phosphodiester linkage.

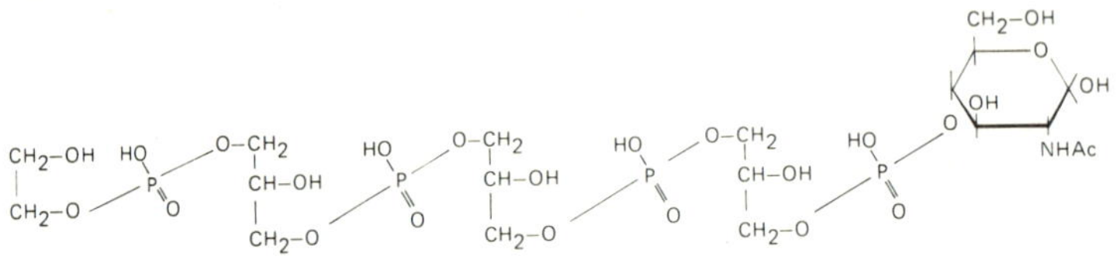

Fig.5. Degradation product from walls of *Staphylococcus aureus* obtained by oxidation and reduction as for Fig.3 then extraction with acid

The probable structure of the teichoic acid, linkage unit and attachment to peptidoglycan for the *S.aureus* mutant is shown in Fig.6. This structure explains the behaviour of walls towards various hydrolytic conditions. Under very gentle acid conditions the sugar 1-phosphate linkage is hydrolysed and the teichoic acid is thus released with a reducing end. Vigorous acid hydrolysis would give the expected products from the teichoic acid and linkage unit, together with a small amount of muramic acid 6-phosphate. Under gentle alkaline conditions the most labile phosphodiester linkage is that between the glycerol and N-acetylglucosamine in the linkage unit. The products would be a teichoic acid possessing three glycerol phosphate residues and an N-acetylglucosamine 1-phosphate still attached to the peptidoglycan. It was already known that this phosphodiester linkage was especially labile towards alkali, as the teichoic acid from *Micrococcus* I3 had been shown to possess a main chain with a repeating structure in which glycerol phosphate is attached to the 4-hydroxyl of N-acetylglucosamine 1-phosphate (7). The conditions of alkaline hydrolysis of that teichoic acid, and the nature of the products, correspond very closely with those observed for release of teichoic acid of *S.aureus* from the wall.

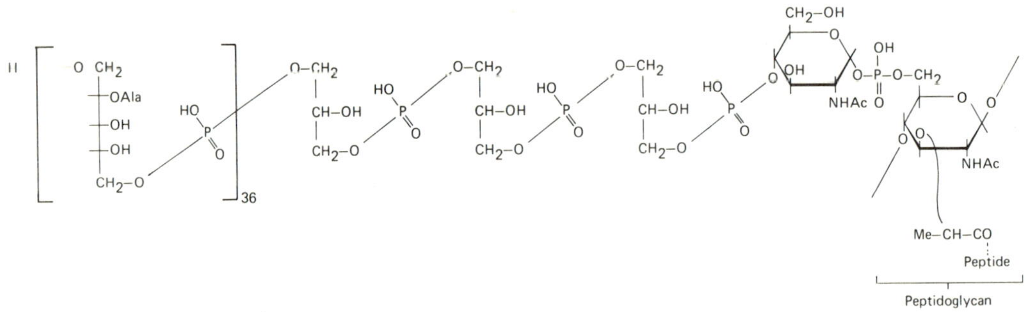

Fig.6. Structure of the teichoic acid-linkage unit-peptidoglycan comprising most of the wall of a mutant of *Staphylococcus aureus*

The evidence presented above was based largely upon chemical degradation, and the structural conclusions have been fully confirmed in biosynthetic studies. The biosynthesized polymers containing both main chain and linkage unit were bound to the membrane fragments by virtue of their amphiphilic properties. This suggests that a lipid forms a part of the complex. Moreover, using labelled nucleotide precursors, it was found that during the biosynthesis smaller lipids became labelled (4). These could be extracted with butanol, and pulse-labelling studies indicated that butanol-soluble intermediates are formed and these are utilized subsequently in building the polymeric product.

An indication of the nature of these intermediates was obtained from the demonstration that tunicamycin powerfully inhibited their formation (8,9). This antibiotic specifically inhibits the formation of prenyl phosphate intermediates in the biosynthesis of peptidoglycan, and also inhibits the synthesis of similar intermediates in glycoprotein synthesis. Thus, it appears that prenyl phosphate intermediates are formed during the teichoic acid synthesis. Fractionation of butanol extracts yielded three lipids; one of these (lipid I) is prenyl pyrophosphate N-acetylglucosamine and the others are derived from it by the addition of respectively one and two glycerol phosphate residues. The structure of lipid II, containing one glycerol phosphate, is shown in Fig.7. In all cases these structures have been established by chemical degradation, although the exact nature of the prenyl residue has been assumed in all cases to be the same as that which participates in peptidoglycan synthesis, namely undecaprenyl (10). Earlier work (11) had indicated that prenyl phosphate intermediates play a part in the synthesis of teichoic acids, and experiments in which both teichoic acid and peptidoglycan were synthesized simultaneously suggested that they share undecaprenyl phosphate from a common pool (12).

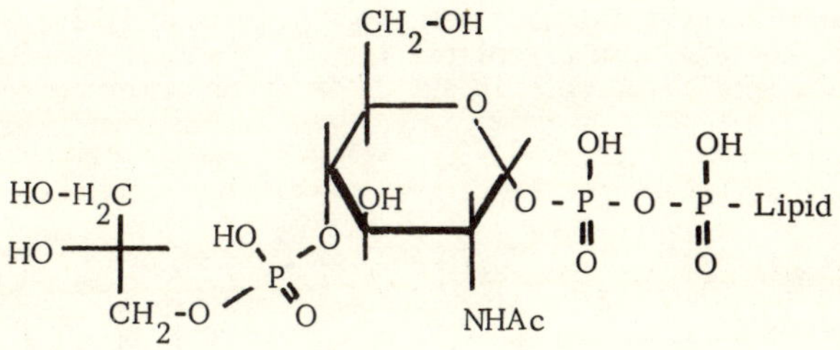

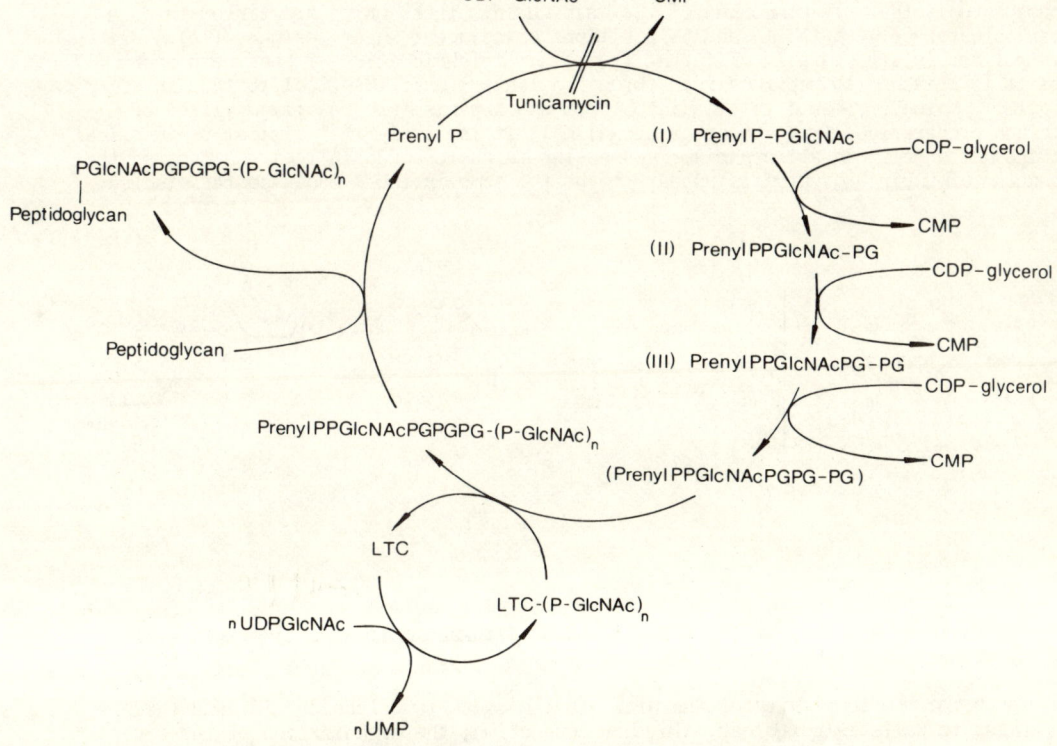

Fig.7. Lipid II, an intermediate in linkage unit synthesis

Isolated intermediate lipids I, II and III have been shown to be utilized in that order when added to membrane preparations synthesizing linkage unit and polymers. However, as the complete linkage unit in all preparations possesses three glycerol phosphate residues, a fourth lipid intermediate containing the complete linkage unit would be expected. Evidence for this has been difficult to obtain, and before lipid IV was characterized it was shown by double labelling experiments that all three glycerol phosphate residues in a complete teichoic acid - linkage unit - prenol originate from CDP-glycerol (10). More recently, experimental procedures have been developed which have made possible the demonstration that lipid IV is in fact formed. Work is incomplete but preliminary studies are in agreement with a structure in which the complete linkage unit is attached through a pyrophosphate to a prenyl residues (H.I.A. McArthur, I.C. Hancock and J.Baddiley, unpublished).

Fig.8. Scheme for biosynthesis of teichoic acid

The general scheme for the biosynthetic route to teichoic acid in *S.aureus* is shown in Fig.8. In this scheme the main chain of ribitol phosphate residues arises through successive transfer of ribitol phosphate residues from CDP-ribitol to an unidentified lipid called lipoteichoic acid carrier (LTC). The formation of such a teichoic acid - LTC complex is known to occur when membrane preparations form polymer from CDP-ribitol alone (13,14), but it is still uncertain whether this is an obligatory step in the living organism, or whether the ribitol phosphate chain is formed by direct transfer from nucleotide to the terminal glycerol of the linkage unit lipid. Although the nature of LTC is not known with certainty, it seems possible that it is related to a glycerol phosphate polymer containing a glycolipid residue which is present in the membranes of nearly all Gram-positive bacteria and is known as membrane teichoic acid (Fig.9) (reviewed in Ref.15).

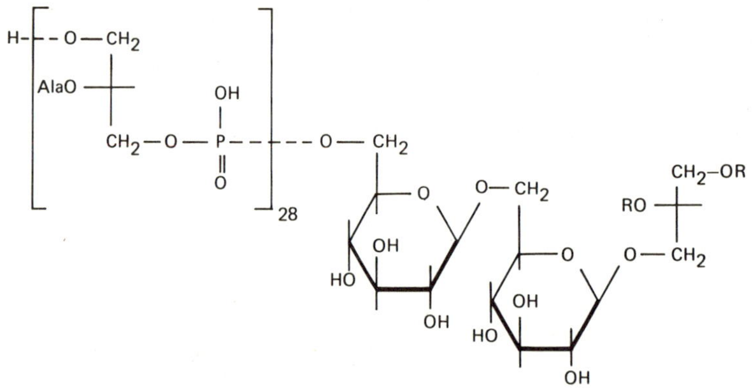

Fig.9. Membrane teichoic acid of *Staphylococcus aureus*

The scheme outlined above for *S.aureus* was in fact elucidated in part using another organism. In *Micrococcus varians* the wall contains a polymer in which N-acetylglucosamine 1-phosphate is the only component of the main chain; these units are linked through phosphodiester between the 1- and 6- positions on adjacent sugar residues (16). Although this polymer is structurally different from other teichoic acids, it has been shown (17) that it is attached to peptidoglycan through a linkage unit identical to that in *S.aureus* (Fig.10). Moreover, lipid intermediates from one organism are readily utilized by membrane preparations from the other organisms. It is likely that similar or identical linkage units occur in all organisms possessing teichoic acids in their walls, and the mechanism of their biosynthesis appears to be the same in those cases so far studied.

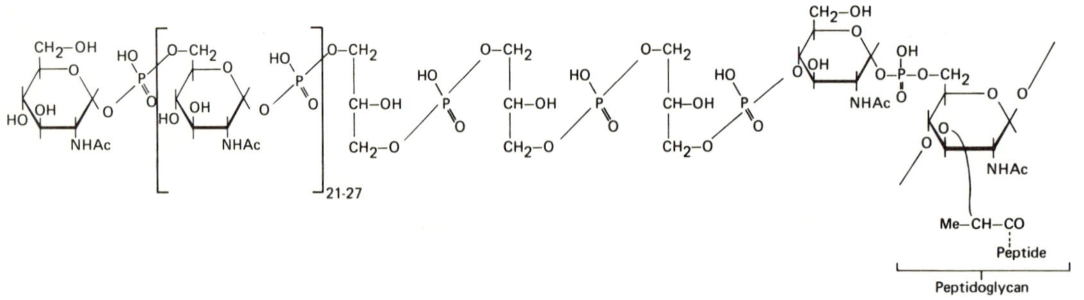

Fig.10. Structure of the teichoic acid - linkage unit - peptidoglycan of *Micrococcus varians*

The reason for the complexity of the biosynthetic route to teichoic acids and their attachment to walls is not clear. In this connection, the organization of the several participating enzymes is perhaps relevant. Metabolic experiments suggest that these enzymes form a complex on the membrane, and this is supported by the observation that the complete system is released from the membrane during freezing and thawing. When they are solubilized in the presence of a non-ionic detergent, the enzymes responsible for the linkage unit synthesis can be separated from those responsible for main chain synthesis, but even after density gradient centrifugation and ion-exchange chromatography the prenyl phosphate and enzyme complex do not separate from each other. This indicates a powerful affinity, not only between the enzymes themselves, but also between enzymes and prenyl phosphate (J. Leaver, I.C. Hancock and J. Baddiley, unpublished). It is possible then to envisage a complex, bound to the membrane, comprising the enzymes and prenyl phosphate. The prenyl phosphate would thus move from one enzyme to the next, accepting polymer or linkage unit components.

Protoplasts of *Bacillus subtilis* W23, unlike the intact organism itself, are able to utilize externally supplied nucleotides for the synthesis of teichoic acid - linkage unit - prenol. In fact, they are able to do this about 100 times more effectively on a membrane weight basis than do fragmented membrane preparations (K. Bertram, I.C. Hancock and J. Baddiley, unpublished). Similarly, intact organisms which have been allowed to autolyse their walls for a very short time, insufficient to cause noticeable loss of wall material, also utilize externally supplied nucleotides with great ease. Experiments are incomplete, but they suggest that the enzyme complex for teichoic acid synthesis is actually located on the outer surface of the cytoplasmic membrane. If this is its location in the intact cell then it would seem that nucleotide precursors normally synthesized inside the cell would be transported selectively through the membrane in order to participate in teichoic acid synthesis at its outer surface. Further work should enlighten this fascinating problem.

REFERENCES

1. J. Coley, E. Tarelli, A.R. Archibald and J. Baddiley, *FEBS Letters 88*, 1-9 (1978).
2. J.E. Heckels, A.R. Archibald and J. Baddiley, *Biochem.J. 149*, 637-647 (1975).
3. J. Coley, A.R. Archibald and J. Baddiley, *FEBS Letters 61*, 240-242 (1976).
4. I.C. Hancock and J. Baddiley, *J.Bacteriol. 125*, 880-886 (1976).
5. R. Bracha and L. Glaser, *J.Bacteriol. 125*, 872-879 (1976).
6. J. Coley, A.R. Archibald and J. Baddiley, *FEBS Letters 80*, 405-407 (1977).
7. A.R. Archibald, J. Baddiley, J.E. Heckels and S. Heptinstall, *Biochem.J. 125*, 353-359 (1971).
8. I.C. Hancock, G. Wiseman and J. Baddiley, *FEBS Letters 69*, 75-80 (1976).
9. R. Bracha and L. Glaser, *Biochem.Biophys.Res.Commun. 72*, 1091-1098 (1976).
10. H.A.I. McArthur, F.M. Roberts, I.C. Hancock and J. Baddiley, *FEBS Letters 86*, 193-200 (1978).
11. D. Brooks, and J. Baddiley, *Biochem.J. 113*, 635-641 (1969).
12. R.G. Anderson, H. Hussey and J. Baddiley, *Biochem.J. 127*, 11-15 (1972).
13. M.M. Burger and L. Glaser, *J.Biol.Chem. 239*, 3168-3177 (1964).
14. F. Fiedler and L. Glaser, *J.Biol.Chem. 249*, 2684-2689 (1974).
15. P.A. Kambert, I.C. Hancock and J. Baddiley, *Biochim.Biophys.Acta 472*, 1-12 (1977).
16. A.R. Archibald and G.H. Stafford, *Biochem.J. 130*, 681-690 (1972).
17. J. Heptinstall, J. Coley, P.J. Ward, A.R. Archibald and J. Baddiley, *Biochem.J. 169*, 329-336 (1978).

STUDIES ON OLIGORIBONUCLEOTIDES SYNTHESIS

R. W. Adamiak, E. Biała, K. Grześkowiak, R. Kierzek,
A. Kraszewski, W. T. Markiewicz, J. Okupniak, J. Stawiński
and M. Wiewiórowski

*Department of Stereochemistry of Natural Products, Institute of Organic Chemistry,
Polish Academy of Sciences, Noskowskiego 12/14, Poznań, Poland*

Abstract - Some chemical problems concerning hypermodification and deprotection steps during the synthesis of the anticodon loop of $tRNA_i^{Met}$ from the yellow lupine are disscused.

INTRODUCTION

Since 1972, when we started the chemical synthesis of the anticodon loop $CCCAUt^6AA$, interested by its chemical challenge and potential utility for structural studies, significant progress has been made in the elucidation of the structure of the whole tRNA molecule, mainly by X-ray analysis (1,2).

The function of hypermodified nucleosides, adjacent to anticodon triplets is still unclear, but on the basis of X-ray analysis it was possible to propose a hypothesis (3) that these nucleosides play an important role in the conformational changes of the anticodon loop occuring in the translation process, and in this way they may influence the correct reading of codes in the mRNA during protein biosynthesis.

In our opinion, the synthetic anticodon arm may serves as a simple, but reliabe model for studing codon - anticodon interaction. To achieve this, the synthesis of the anticodon loop as a first step on the way to anticodon arm synthesis was necessary.

We soon found that this was a hard way, and the problems we have faced during hypermodification and the final deprotection steps will be presented in this article.

THE INTRODUCTION OF THE HYPERMODIFIED NUCLEOSIDE N^6/THREONYLCARBONYL/ - - ADENOSINE /t^6A/ INTO THE OLIGONUCLEOTIDIC CHAIN

One can imagine two chemical routes serving to introduce hypermodified nucleoside into oligomer : /1/ the synthesis of suitably protected t^6A and its use as a incoming monomer for the coupling reaction and /2/ the transformation of an adenosine residue carrying a free exo-NH_2 group into t^6A on the protected oligomer level. After preliminary experiments it soon appeared to us that the second route would be much more economic and efficient. Moreover, the second approach can lead directly to the unmodified and hypermodified oligomer. Both of them would be very useful in elucidating the influence of hypermodified nucleoside on the oligomer conformation.

In Figure 1 a general scheme for the synthesis of the anticodon loop, according to route 2 is shown. This strategy was based on two assumptions.
1. The introduction of an adenosine unit carring a free exo-NH_2 group would be possible without extensive N-phosphorylation during the coupling reaction.
2. The transformation of the free exo-NH_2 group of this residue into the N,N´-disubstituted urea system of t^6A would be selective and efficient.
In fact we have found that under the general coupling procedure with the use of TPS-tetrazole (4) as an activating reagent, no N-phosphorylation occured (5). The second problem was more difficult to solve. We have found previously (6) that phenyl carbamate derived from adenosine is reactive enough to undergo aminolysis with threonine ester to form the desired N,N´- disubstituted urea system. The problem was, however, to transform selectively a free exo-NH_2 group to phenyl carbamate on the oligomer level in the presence of N-protected nucleosides also possesing reactive N-H bonds. Phenyl chloroformate, previously used on the monomer level showed no such selectivly.

Searching for a new selective reagent, crystalline phenoxycarbonyltetrazole was synthezised (7) and proved to be a reagent suitable for our purpose.

In Figure 2 the two-step hypermodification reaction is shown. As was mentioned above the aminolysis of the carbamate derivative by means of the p-nitrobenzyl ester of threonine also went undirectionally in mild conditions, therefore both steps: carbamate formation and aminolysis could be carried out one after another in one flask. The practical yield of dimer and tetramer were higher than 90% and heptamer with a hypermodofied t^6A unit was isolated in over 85% of the yield (5).

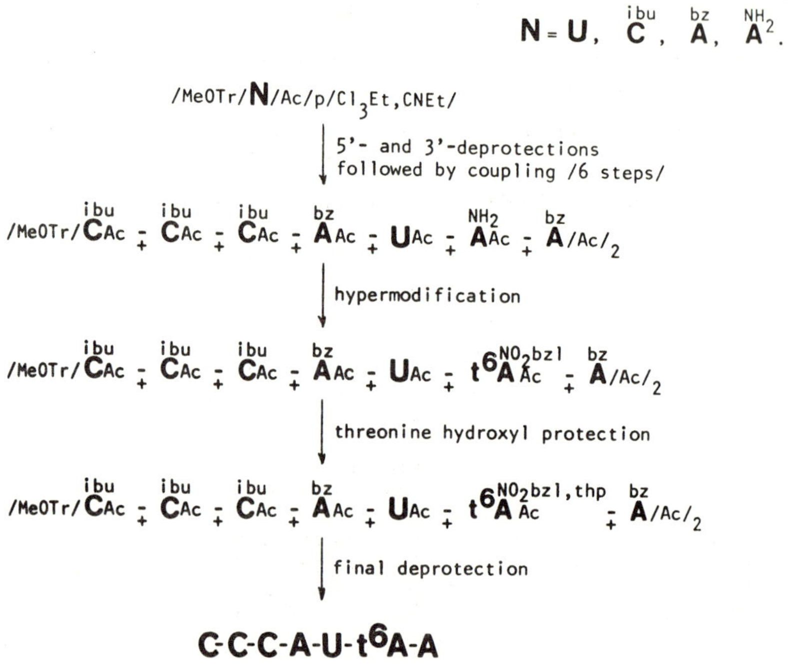

$$N = U, \ C^{ibu}, \ A^{bz}, \ A^{NH_2}.$$

/MeOTr/N/Ac/p/Cl$_3$Et,CNEt/

↓ 5'- and 3'-deprotections
followed by coupling /6 steps/

/MeOTr/C^{ibu}_{Ac} ÷ C^{ibu}_{Ac} ÷ C^{ibu}_{Ac} ÷ A^{bz}_{Ac} ÷ U_{Ac} ÷ $A^{NH_2}_{Ac}$ ÷ A^{bz}/Ac/$_2$

↓ hypermodification

/MeOTr/C^{ibu}_{Ac} ÷ C^{ibu}_{Ac} ÷ C^{ibu}_{Ac} ÷ A^{bz}_{Ac} ÷ U_{Ac} ÷ $t^6A^{NO_2bzl}_{Ac}$ ÷ A^{bz}/Ac/$_2$

↓ threonine hydroxyl protection

/MeOTr/C^{ibu}_{Ac} ÷ C^{ibu}_{Ac} ÷ C^{ibu}_{Ac} ÷ A^{bz}_{Ac} ÷ U_{Ac} ÷ $t^6A^{NO_2bzl,thp}_{Ac}$ ÷ A^{bz}/Ac/$_2$

↓ final deprotection

C-C-C-A-U-t^6A-A

Figure 1. The strategy of anticodon loop synthesis.

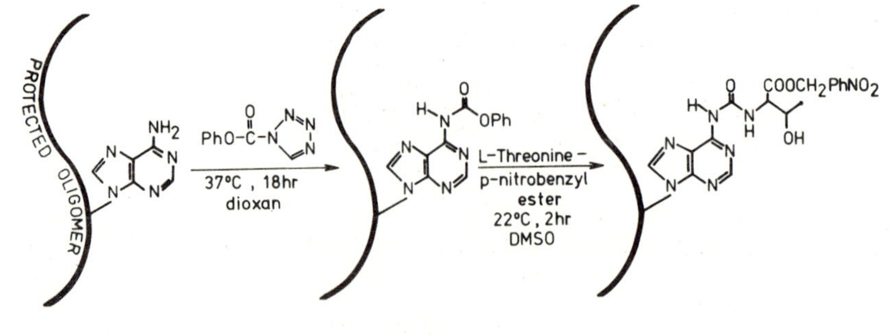

Figure 2. Transformation of adenosine residue into N^6/threonylcarbonyl/--adenosine on oligomer level /hypermodification/.

The high yield of hypermodification on the heptamer level seems to suggest that this reaction can probably also be applied to longer oligonucleotides. We were very satisfied with this results, but of course we still do not know a lot of things about this reaction. In our case adenosine with a free exo-NH$_2$ group was situated near the 3'-end of heptamer, but it is hard to predict if the efficiency of the hypermodification would also be as high if this adenosine residue was in different places or in a longer oligonucleotidic chain. Also the influence of the sequence of the oligomer on this reaction is unknown.

FINAL DEPROTECTION

The total yield of natural oligonucleotides depends not only on the efficiency of the chemical synthesis of the protected oligomer, but also on the efficiency of the deprotection steps. The crucial step in the phosphotriester approach is the complete removal of all phosphate protecting groups, since their presence in the oligomer during the subsequent deprotection of the base - labile group leads to very serious degradation of the oligonucleotidic chain.

At the very beginning we tried to use the Zn /acetylaceton/ pyridine method developed earlier by us (8) for this purpose.

This method is very efficient, but during the next step i.e. when isobutylamine was used to remove the base - labile groups, we have observed serious degradation, probably due to the presence of traces of Zn cations in the reaction mixture. In the case of the hypermodified heptamer, besides this kind of degradation we noticed additional side reactions, probably connected with the presence of the p-nitrobenzyl ester group of a threonine fragment. To avoid this, we tried to remove the p-nitrobenzyl at first by hydrogenolysis over PdO in ethanol.

Surprisingly, we noticed a simultaneous deprotection of 2,2,2-trichloroethyl groups and also in consequence of the generation of hydrogen chloride the removal of the monometoxytrityl group. When we replaced ethanol by pyridine the monometoxytrityl group was untouched, and after extending the time of reaction to 22 hours, we applied this method for the removal of 2,2,2-trichloroethyl groups during the synthesis of CCCAUt^{6}AA. The yield of the total deprotection was about 30%.

In order to improve this yield, further studies on the hydrogenolysis process were undertaken. Since the hydrogenolitic cleavage of the carbon - halogen bond depends on several factors typical to heterogeneous processes, we found during these studies that without the standardization of reaction conditions, e.g. the purity of all substrates, stoichiometry, the mode of addition of the catalyst and the mode of its transformation into its active form, the mode of stirring or shaking, the temperature, the time or the kind of solvent and more important, the kind of base, the efficiency of hydrogenolysis varied very much.

Under optimal conditions /i.e. pyridine as a base and solvent, 15-20 eq. of PdO per 2,2,2-trichloroethyl group, room temperature, H_2 under atmospheric pressure, 22 hours/ the reaction is highly reproducible, but the yield of deprotection drops dramatically when the oligoribonucleotide chain is increased in length. For example: UpA 99%, UpApA 90%, CpCpCpApUpApA 30% but simple heptamer in the deoxy series, /Tp/$_6$T over 90%. Our studies on the optimalization of the removal of 2,2,2-trichloroethyl groups by hydrogenolysis on PdO led to the following conclusions.

The method could be reproducible and efficient for short oligonucleotides. Starting from pentamer in the ribo series, a significant decrease in rate of hydrogenolysis was observed, and in the case of heptamer the rate dropped down to the 1/3 of the rate for the dimer. Recently we have found that to achieve a better yield of removal of 2,2,2-trichloroethyl groups, a period of time longer than 22 hours is necessary. We are currently studing whether the elongation of the time of hydrogenolysis affects pyrimidine 5,6 - double bonds. On the other hand, we think that this process, commonly occuring in acidic and neutral media (9), is strongly supressed in pyridine solution.

In the deoxy series, the hydrogenolysis goes in right direction faster than in the ribo series, and thus this method could also be applied to longer oligodeoxynucleotides.

Having in mind the better yield of removal of 2,2,2-trichloroethyl groups by the Zn /acetylacetone/ pyridine system, we tried to improve this method further. Since TL chromatography indicated that the removal of 2,2,2-trichloroethyl groups went smoothly also on the heptamer level and that the degradation started not earlier than during the next step i.e. in the alkaline medium, we decided to find a better method for the removal of traces of Zn cations before that step.

The following procedure gave us the best results. The reaction mixture obtained after the Zn /acetylacetone/ pyridine treatment was partitioned between a 0.02 M TEAB buffer, /pH 7.5/ and a 2% chloroformic solution of 8-hydroxyquinoline to remove Zn cations. Nevertheless, when the aqueous phase after evaporation was treated with isobutylamine in methanol we observed almost total degradation.

This indicated that Zn cations were still present in the reaction mixture and even this low concentration was enough to cause complete degradation. Remarkable progress was made when the reaction mixture obtained after 8-hydroxyquinoline treatment was subjected to LH-20 Sephadex column chromatography. Sephadex LH-20 has both hydrophobic and hydrophylic properties and contains a small percentage of carboxylic functions which cause the retardation of metal cations during the chromatographic process. We took advantage of these properties to separate the deprotected oligomer from side products and other impurities. The main product /over 90% of all nucleotidic materials/ obtained after LH-20 Sephadex chromatography was treated with 10% isobutylamine in methanol for 30 hours. Chromatography on PEI - cellulose plates indicated the presence of the unprotected heptamer CCCAUAA as a main product /80%/. The overall yield of deprotection of this heptamer using the modified Zn /acetylacetone/ pyridine method was over 70%.

The deprotected heptamers checked by DEAE - Sephadex chromatography, TLC on PEI plates and by enzymatic digestion, were homogeneous, had correct 3´- 5´internucleotic bonds and displayed a correct nucleoside composition. The sequence analysis of both oligomers is in progress.

To summarize the main problems faced during the chemical synthesis of the anticodon loop of tRNA$_1^{Met}$ were solved satisfactorally. Some of the methods which we developed or improved e.g. coupling, 5´ and 3´ deprotection or hypermodification, work very well and others, for example final deprotection still need further improvement.

We hope that the experience we have gained during this chemical work will allow us to synthesize in the near future a whole anticodon arm as a suitable model for studing codon-anticodon interactions.
Before that comparative structural studies will be carried out on both heptomers /with and without modification/ together with experiments on their interactions with oligo-ribonucleotides containing the AUG codon triplet.

REFERENCES

1. S.H.Kim, Progr.Nucleic Acid Res. Mol. Biol., 17, 181 /1976/.
2. A.Rich, and U.L.RajBhandary, Ann.Rev.Biochem., 45, 805 /1976/.
3. S.H.Kim, in "Advances in Enzymology and Related Area of Molecular Biology", Ed.A.Meister, John Wiley and Sons, Inc. 1978, p. 279.
4. J.Stawiński, T.Hozumi, and S.A.Narang, Can.J.Chem., 54, 670 /1976/
5. R.W.Adamiak, E.Biała, K.Grześkowiak, R. Kierzek, A. Kraszewski, W.T.Markiewicz, J.Okupniak, J. Stawiński and M. Wiewiórowski, Nucl. Acids Res., 5, 1889 /1978/.
6. R.W.Adamiak and M.Wiewiórowski, Bull.Acad.Polon.Sci.ser.sci.chim., 23, 241 /1975/.
7. R.W.Adamiak and J.Stawiński, Tetrahedron Lett., 1935 /1977/.
8. R.W.Adamiak, E.Biała, K.Grześkowiak, R.Kierzek, A.Kraszewski, W.T.Markiewicz, J.Stawiński and M.Wiewiórowski, Nucl.Acids Res., 4, 2321 /1977/.
9. M.Green, S.S.Cohen, J.Biol.Chem., 225 397 /1957/.

REACTIVE PHOSPHORYLATING INTERMEDIATES IN THE NUCLEIC ACIDS CHEMISTRY

D. G. Knorre and V. F. Zarytova

Institute of Organic Chemistry, Siberian Division of the Academy of Sciences of the U.S.S.R., Novosibirsk 630090, U.S.S.R.

<u>Abstract</u> - The last achievements in the investigation of the reactive phosphorylating intermediates used in the oligonucleotide chemistry are considered. Phosphomonoesters when treated with arenesulfonyl chlorides in pyridine solution are converted to reactive derivatives containing phosphorylpyridinium residue as revealed by ^{31}P and ^{1}H NMR data. This step may be carried out prior to addition of nucleophile to be phosphorylated. The use of cross-linked polystyrenesulfonyl chloride permit to perform subsequent phosphorylation in the absence of a condensing reagent. The kinetic prove is presented of the role of the phosphorylpyridinium derivatives as the main reactive intermediates in the ternary system containing triisopropylbenzenesulfonyl chloride (TPS) and both nucleoside and nucleotide components.
The possible role of phosphodiester triazolides and tetrazolides in the phosphotriester formation with arenesulfonyltriazolides and tetrazolides is discussed. Tetrazole derivative of diphenylphosphate is described and shown to be powerful phosphorylating reagent. Tetrazole efficiently catalyzes the reaction of alcohols with pyrophosphate tetraesters derived by treatment of phosphodiesters with condensing reagents. Therefore, the liberation of tetrazole in the first step of the reaction of phosphodiesters with arenesulfonyl tetrazolides may explain the powerful condensing action of the latters in the phosphotriester formation. Dimethylaminopyridinium derivative of diphenylphosphate is described and shown to phosphorylate efficiently alcohols. The catalysis by DMAP of the reaction of phosphodiesters with alcohols in the presence of TPS is demonstrated.

INTRODUCTION

Phosphorylation is one of the most important reactions in the nucleic acids chemistry. It is the main method of the formation of the internucleotide bonds in the oligonucleotide synthesis. Phosphorylation is one of the convenient approaches to prepare the oligonucleotide derivatives bearing reactive groups for affinity labelling of the oligonucleotide specific systems (single-stranded nucleic acids (Refs. 1 & 2), mRNA binding center of ribosomes (Ref.3) etc). The conversion of oligonucleotides to triesters by phosphorylation of ethanol was recently proposed to enhance the ability of the formers to cross the outer cell membrane and stabilize them against intracellular nucleases with partial preservation of the ability of oligonucleotides to base pairing with single-stranded nucleic acids (Refs. 4-7).
Phosphorylation is performed either with phosphorylating reagents or via transient phosphorylating intermediates. In connection with the chemistry of nucleic acids and their components the problem is discussed in details in the famous monographs of Khorana (Ref.8) and Michelson (Ref.9). The modern state of the problem is described in the recently appeared manual of Shabarova and Bogdanov (Ref.10). Concerning the mechanism of the formation of internucleotide bonds in the oligonucleotide synthesis the question was reviewed by the authors (Refs. 11 & 12).
The present paper deals with the achievements reached in the last two years in the investigation of the reactive phosphorylating intermediates mainly in connection with the oligonucleotide synthesis and preparation of the oligonucleotide derivatives.

THE MAIN TYPES OF THE PHOSPHORYLATION PROCEDURES

All numerous chemical processes used to perform phosphorylation in the nucleic acids chemistry may be divided into three main groups.

Nucleoside and oligonucleotide lacking end phosphomonoester group may be converted to reactive entities by treatment with some bifunctional phosphorylating derivatives of orthophosphate.

As an example of this approach we may present the method recently proposed by Agarwal et al (Ref.13) for the triester version of the oligonucleotide synthesis. The 3'-OH group of the growing oligonucleotide chain is treated with p-chlorophenyl-phosphoditriazolide. The phosphodiester triazolide formed is used to phosphorylate the next nucleoside

$$XOH + ClC_6H_4O-\overset{\overset{O}{\|}}{\underset{\underset{N}{|}}{P}}-N\overset{N=}{\underset{N}{\diagdown}} \longrightarrow XO-\overset{\overset{O}{\|}}{\underset{\underset{OC_6H_4Cl}{|}}{P}}-N\overset{N=}{\underset{N}{\diagdown}} \xrightarrow{+ HOX'} XO-\overset{\overset{O}{\|}}{\underset{\underset{OC_6H_4Cl}{|}}{P}}-OX' \quad (1)$$

X,X' - nucleoside or oligonucleotide moiety.

Two other groups of methods both involve the conversion of the preexisting phosphate residue to reactive phosphorylating moiety. In both cases we deal with three main components of the overall process: phosphate derivative, the reagent converting the former to phosphorylating derivative (activating reagent) and nucleophile (alcohol, amine, phosphate) to be phosphorylated.

According to one methodology phosphate is activated in the absence of the other nucleophile. After completion of the reaction due to exhaustion of either phosphate or activating reagent, or both, nucleophile is added. Thus phosphomonoesters may be converted to the mixed anhydrides with diphenylphosphoric acid by treatment with diphenylphosphochloridate (Ref.14)

$$\underset{PhO}{\overset{PhO}{\diagdown}}\overset{O}{\underset{Cl}{\overset{\|}{P}\diagdown}} + \,\,^-O-\overset{\overset{O}{\|}}{\underset{\underset{O^-}{|}}{P}}-OR \longrightarrow \underset{PhO}{\overset{PhO}{\diagdown}}\overset{O}{\underset{}{\overset{\|}{P}}}-O-\overset{\overset{O}{\|}}{\underset{\underset{O^-}{|}}{P}}-OR$$

(I)

I readily reacts with amines and phosphomonoesters giving correspondingly phosphoramidates and pyrophosphate diesters. By this method the first reactive derivatives of oligonucleotides bearing N-2-chloroethyl-N-methylamino group at 5-end were obtained (Ref.1)

$$\underset{PhO}{\overset{PhO}{\diagdown}}\overset{O}{\overset{\|}{P}}-O-\overset{\overset{O}{\|}}{\underset{\underset{O^-}{|}}{P}}-O-X-(O-\overset{\overset{O}{\|}}{\underset{\underset{O^-}{|}}{P}}-O-X)_n + \underset{CH_3}{\overset{ClCH_2CH_2}{\diagdown}}N-\underset{}{\bigcirc}-CH_2\,NH_2 \longrightarrow$$

$$\underset{CH_3}{\overset{ClCH_2CH_2}{\diagdown}}N-\underset{}{\bigcirc}-CH_2NH-\overset{\overset{O}{\|}}{\underset{\underset{O^-}{|}}{P}}-O-X-(O-\overset{\overset{O}{\|}}{\underset{\underset{O^-}{|}}{P}}-O-X)_n$$

X = dT, dA, A   n = 1, 2

The other similar method extensively elaborated by Z.A. Shabarova and coworkers is the conversion of the phosphate group to the mixed anhydride with mesithylene carbonic acid (Refs. 15 & 16)

$$CH_3-\underset{CH_3}{\overset{CH_3}{\bigcirc}}-COCl + \,\,^-O-\overset{\overset{O}{\|}}{\underset{\underset{O^-}{|}}{P}}-OR \longrightarrow CH_3-\underset{CH_3}{\overset{CH_3}{\bigcirc}}-CO-O-\overset{\overset{O}{\|}}{\underset{\underset{O^-}{|}}{P}}-OR$$

These anhydrides are rather mild phosphorylating reagents towards amines. They too were recently used to produce reactive derivatives of oligonucleotides for the affinity labelling (Refs. 17 & 18).

We shall further refer this method as the two-step procedure in the binary systems phosphate + activating reagent and activated phosphate + nucleophile. Certainly the real number of the direct participants in these steps may exceed two. In some cases solvent (especially pyridine) participate in the reaction. Moreover the activa-

ting reagent may consist of two components, e.g. triphenylphosphine + dipyridyldisulfide in the oxidative-reductive condensation (Ref.19).
The other methodology based on the phosphate activation is commonly known as the use of the condensing reagents. In this case the activating (condensing) reagent is added to the mixture of phosphate and nucleophile to be phosphorylated. We shall further refer this methodology as the performance of the reaction in the ternary system (phosphate, phosphorylated nucleophile and condensing reagent).
The classical example of a condensing reagent is dicyclohexylcarbodiimide (DCC). The latter is supposed to convert phosphates to highly reactive O-phosphoryldicyclohexylisoureas (Ref. 20). This type of the reactive phosphorylating intermediates cannot be accumulated in the reaction mixture due to rather fast isomerisation to phosphorylureido derivatives and may be used only in situ. Thus it was demonstrated that the treatment of phosphodiester (Tr)dTpdT(Ac) with DCC in pyridine solution results in the nearly quantitative formation of the ureido derivative of phosphodiester (Ref. 21).

$$(Tr)dT-O-\overset{O}{\underset{-O}{P}}-O-dT(Ac) + \bigcirc-N=C=N-\bigcirc \xrightarrow{(H^+)}$$

$$\left[\begin{array}{c} (Tr)dT-O-\overset{O}{\underset{O}{P}}-O\ dT(Ac) \\ \bigcirc-NH-C=N-\bigcirc \end{array} \right] \longrightarrow \begin{array}{c} (Tr)dT-O-\overset{O}{\underset{}{P}}-O-dT(Ac) \\ \bigcirc-NH-CO-N-\bigcirc \end{array}$$

Recently the highly selective method of the attachment of amines to phosphomonoester residues of oligonucleotides was proposed (Ref.22) based on the reaction between oligonucleotides and highly basic amines in dimethylsulfoxide or dimethylformamide (DMFA) with the mixture of triphenylphosphine and dipyridyldisulfide as a condensing reagent. The reaction scheme may be presented as follows (Ref.19).

$$Ph_3P + \bigcirc_N-S-S-\bigcirc_N \rightarrow \left[Ph_3P^+-S-\bigcirc_N \right] \bigcirc_N-S^-$$

$$RO-\overset{O}{\underset{O^-}{P}}-O^- \longrightarrow \left[Ph_3P^+-O-\overset{O}{\underset{O^-}{P}}-OR \right] + 2 \bigcirc_N-S^- \xrightarrow{R'-NH_2} Ph_3PO + RO-\overset{O}{\underset{O^-}{P}}-NHR'$$

Neither of the intermediates is accumulated in the extent permitting to observe it in the ^{31}P NMR spectrum. In the absence of amine pyrophosphate diester forms as a final product presumably due to phosphorylation of the second phosphomonoester molecule. This pyrophosphate is completely unreactive towards amine. Therefore the overall process can not be divided into activation and phosphorylation steps. Phosphodiester groups are not activated in these conditions as well as there is no further activation of pyrophosphate diesters and phosphoroamidates. It is especially essential that the reagent does not touch internucleotide bonds of ribooligonucleotides. All other commonly used condensing reagents even as mild as mesithoyl chloride were found to convert internucleotide phosphate groups of oligoribonucleotides to cyclic phosphotriesters. This may result in the subsequent cleavage or isomerisation of the internucleotide bond (Refs. 23-25). Hence, the method may be used to produce oligoribonucleotide derivatives bearing some amine attached to either 5' or 3' phosphate without any damage of oligoribonucleotide chain.
When the two-step procedure is possible it looks preferential to use this methodology. It is especially essential with electrophilic condensing reagents (DCC, arenesulfonyl chlorides) as it permits to escape the consumption of the condensing reagent in the reaction with nucleophile to be phosphorylated and the accumulation of the undesirable by-products.
Thus the first attempt to obtain γ-ATP amides was performed by treatment of the mixture of ATP and amine (methyl ester of phenylalanine) with DCC. The yield of the desired amide was 20% (Ref. 26). Later it was demonstrated that the treatment of ATP with DCC in the absence of other nucleophile results in the near quantitative formation of adenosine-5'-trimetaphosphate II (Refs. 27 & 28).

$$\text{O-P(=O)(O}^-\text{)-O-P(=O)(O}^-\text{)-O-P(=O)(O}^-\text{)-OAdo} \xrightarrow{\text{DCC}} \text{(cyclic trimetaphosphate)-OAdo} \qquad \text{(II)}$$

The latter may be quantitatively converted to amide by subsequent addition of the corresponding amine. Any complications due to contact of electrophilic DCC with amine are completely avoided.

Moreover the separation of the activation and phosphorylation steps permits to perform each of them in its optimal conditions. E.g. N-cyclohexyl N-[β-(4-N-methylmorpholinium)ethyl] carbodiimide p-toluenesulfonate (CME-carbodiimide) converts ATP in aqueous solution in high yield to II. The reaction proceeds readily at pH 5-6. In these conditions II does not react in the significant extent with such strong amines as benzylamine as the latter is nearly completely protonated. Therefore, in the ternary system (ATP, benzylamine, CME-carbodiimide) only small amounts of γ-ATP benzylamide may be obtained. However when the first step is performed at pH 5-6 up to maximal accumulation of II and then benzylamine is added at pH 9 II preformed in the first step is converted to desired amide in high yield (Ref. 29).

The two-step procedure may be easily performed with arenesulfonyl chlorides - condensing reagents most popular in the oligonucleotide synthesis by the diester methodology. This will be discussed in details in the next section.

PHOSPHORYLPYRIDINIUM DERIVATIVES OF NUCLEOSIDES AND OLIGONUCLEOTIDES

In 1959 Todd have proposed that metaphosphate ester is formed as reactive phosphorylating intermediate when phosphomonoester is treated with some condensing reagents (Ref. 30). In 1963 Michelson have suggested that in the pyridine solution metaphosphate is converted to still preserving high reactivity phosphorylpyridinium derivative III (Ref. 9)

$$\text{RO-P(=O)(O}^-\text{)-N}^+\text{(pyridine)}$$

(III) a) R = dT(Ac)
       b) R = phenyl

The real proof of the existence of the latter derivative has come from ^{31}P NMR investigations of the mechanism of the oligonucleotide synthesis.

It should be emphasized that pulsed ^{31}P NMR spectroscopy with Fourier transform is an extremely powerful method to follow the kinetics of phosphorylation. Due to significant differences in the position and structure of ^{31}P NMR signals of different phosphate derivatives the kinetics of accumulation and consumption of several P-containing reagents, intermediates and products, may be followed simultaneously directly in the reaction mixture by measuring the integral intensities of the respective signals. All experimental kinetic data presented in the following sections are obtained by this method using Bruker HX-90 pulse spectrometer operating at 36.43 MHz. Approximately 2 min is necessary to record one spectrum resulting from Fourier transform of ~ 50 accumulations (For details see Ref. 12).

Using this method we have demonstrated that nucleotides (pdT(Ac)) as well as arylphosphates when treated with triisopropylbenzenesulfonyl chloride (TPS) in the pyridine solution form highly reactive derivative with a singlet ^{31}P NMR signal several ppm units upfield as compared with the starting phosphomonoester (Refs. 31 & 32). In these experiments we have failed to observe spin-spin splitting of ^{31}P NMR signal at α-proton of III expected to be of the order of 10Hz and, therefore, could not surely assign structure III to the derivative under consideration. Although pyridine was found to be absolutely necessary for the formation of the reactive derivative we were not able to exclude completely some other ways of the stabilising effect of pyridine on the metaphosphate.

Recently (Ref. 33) we have succeeded to prepare the similar derivative of phenylphosphate in the methylene chloride solution using the reaction proposed by prof. M.N.Kolossov (Shemyakin's Institute of Bioorganic Chemistry, Moscow)

$$\text{Ph-O-P(=O)(OCH}_3\text{)-Cl} + \text{N-pyridyl-CH}_3 \longrightarrow$$

$$\text{Ph-O-P(O)(O}^-\text{)-N}^+\text{(C}_6\text{H}_4\text{)-CH}_3 + \begin{cases} CH_3Cl \\ CH_3-N^+(C_6H_4)-CH_3 \cdot Cl^- \end{cases} \quad (2)$$

(IIIb)

At -50°C the signals of α-protons as well as γ-CH$_3$ protons of both γ-picolinium moiety of IIIb and of the excess γ-picoline are seen in the ^{1}H NMR spectrum (Fig. 1)

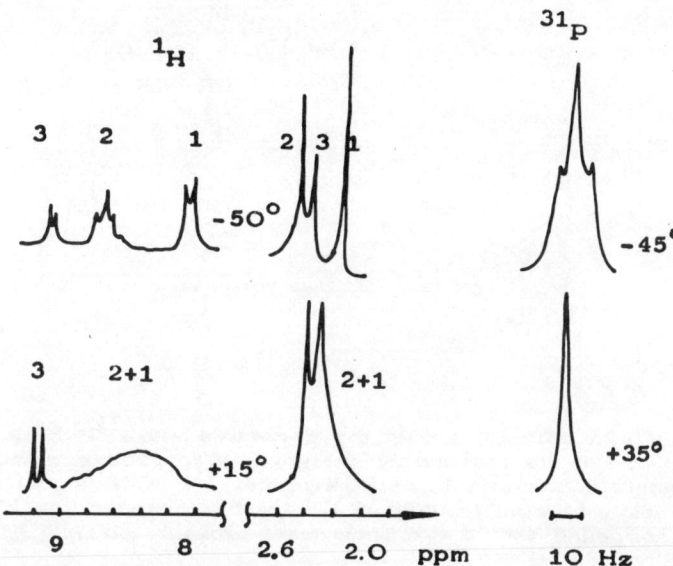

Fig. 1. ^{1}H and ^{31}P NMR spectra of IIIb : 1) signals of γ-picoline; 2) signals of γ-picoline moiety of IIIb; 3) signals of N-methyl-γ-picolinium formed in the reaction (2).

The signals of the former are shifted downfield as compared with those of γ-picoline as it should be due to the appearance of the positive charge at the γ-picolinium residue. The signals of α-protons besides being splitted at vicinal β-protons are additionally splitted with J $\sim$ 6 Hz as expected for P-N-C-H system. In the ^{31}P NMR spectrum recorded without ^{31}P-{^{1}H} spin-spin decoupling the ex-pected triplet (splitting at two vicinal α-protons) is seen with J $\sim$ 6 Hz.
At room temperature the ^{1}H NMR signals of both free γ-picoline and γ-picolinium moiety of IIIb are collapsed to broad signals. The splitting of ^{31}P NMR signal disappear. This means that the picolinium residue of IIIb is in the state of fast exchange with free γ-picoline. Due to this exchange we could not observe any ^{31}P-^{1}H spin-spin coupling in the ^{31}P NMR spectrum of phenylphosphate treated with TPS at room temperature in the pyridine solution (Ref. 32).
The above data entirely prove the structure III for the reactive phosphorylating derivative of phosphomonoesters formed in the reaction with arenesulfonyl chlorides. Therefore, the overall process of the activation of phosphomonoester with arenesulfonyl chloride in pyridine may be presented as follows

$$RO-P(O)(O^-)-O^- + ArSO_2Cl + N(C_5H_5) \rightarrow RO-P(O)(O^-)-N^+(C_5H_5) + ArSO_3^- + Cl^- \quad (3)$$

III was also demonstrated to be the final product of the treatment of phosphomonoesters with the mixture of triphenylphosphine and dipyridyldisulfide in pyridine in the absence of strong bases (Ref. 34). In contrast to arenesulfonyl chlorides the reaction is completely supressed in the presence of highly basic amines (e.g. triethylamine) pyrophosphate diester being the final product of the reaction.
Compounds III were demonstrated to react readily with primary and secondary amines, phosphomono- and phosphodiesters, pyrophosphate diesters, the reaction being accomplished within few minutes (Refs. 32 & 35).

$$III + R'NH_2 \longrightarrow RO-\underset{\underset{O^-}{|}}{\overset{\overset{O}{\|}}{P}}-NHR'$$

$$III + HN\diagup\!\!\diagdown \longrightarrow RO-\underset{\underset{O^-}{|}}{\overset{\overset{O}{\|}}{P}}-N\diagup\!\!\diagdown$$

$$III + RO-\underset{\underset{O^-}{|}}{\overset{\overset{O}{\|}}{P}}-O^- \longrightarrow {}^-O-\underset{\underset{OR}{|}}{\overset{\overset{O}{\|}}{P}}-O-\underset{\underset{OR}{|}}{\overset{\overset{O}{\|}}{P}}-O^-$$

$$III + {}^-O-\underset{\underset{OR}{|}}{\overset{\overset{O}{\|}}{P}}-O-\underset{\underset{OR}{|}}{\overset{\overset{O}{\|}}{P}}-O^- \longrightarrow {}^-O-\underset{\underset{OR}{|}}{\overset{\overset{O}{\|}}{P}}-O-\underset{\underset{OR}{|}}{\overset{\overset{O}{\|}}{P}}-O-\underset{\underset{OR}{|}}{\overset{\overset{O}{\|}}{P}}-O^-$$

$$III + R'O-\underset{\underset{O^-}{|}}{\overset{\overset{O}{\|}}{P}}-OR'' \longrightarrow \begin{array}{c} R'O-\overset{\overset{O}{\|}}{P}-OR'' \\ | \\ O \\ | \\ {}^-O-\overset{\overset{O}{\|}}{P}-OR \\ \| \\ O \end{array}$$

IIIa reacts readily with triazole forming the derivative with ^{31}P NMR signal at δ = 10.4 ppm which may be reasonably assigned to respective triazolide. When treated with tetrazole it converts to derivative with ^{31}P NMR signal at δ = 12.1 ppm presumably respective tetrazolide. With γ-N-dimethylaminopyridine (DMAP) it is coverted to the compound with δ = 6.8 ppm most probably phosphoryldimethylaminopyridinium derivative.

$$IIIa + HN\diagup\!\!\!\!\!{}^{N=\!\!=}\diagdown_N \longrightarrow {}^{N=\!\!=}\diagdown_N\!\!\!\!\!\diagup N-\underset{\underset{O^-}{|}}{\overset{\overset{O}{\|}}{P}}-O-dT(Ac)$$

$$IIIa + HN\diagup\!\!\!\!\!{}^{N=\!\!=}\diagdown_{N=N} \longrightarrow {}^{N=\!\!=}\diagdown_{N=N}\!\!\!\!\!\diagup N-\underset{\underset{O^-}{|}}{\overset{\overset{O}{\|}}{P}}-O-dT(Ac)$$

$$IIIa + N\diagup\!\!\diagdown-N(CH_3)_2 \longrightarrow (CH_3)_2N-\diagup\!\!\diagdown\overset{+}{N}-\underset{\underset{O^-}{|}}{\overset{\overset{O}{\|}}{P}}-O-dT(Ac)$$

The reaction with tetrazole and DMAP proceeds nearly quantitatively only in the presence of some excess of these compounds. In the same time the compound III exists as a single final P-containg product of the reaction (3). This means that no significant reaction of III with chloride and arenesulfonate takes place the equilibrium being displaced towards III.

Using cross-linked polystyrenesulfonyl chloride III may be separated from the excess of the condensing reagent and obtained in the pyridine solution pyridinium chloride being the single additional component of the reaction mixture (Ref. 36). It was used to phosphorylate nucleosides (Ref. 37) and piperidine (Ref. 35) in the absence of a condensing reagent. Using the latter reaction the good yields of nucleoside-5'-phosphopiperidides were obtained and a clean procedure for the preparation of nucleosidedi- and triphosphates was proposed avoiding the use of strongly allergenic DCC (Ref. 38).

In contrast with all above mentioned nucleophiles alcohols react with III with a moderate rate measurable by ^{31}P NMR technique. Phosphodiester is expected to be the primary product of the reaction. However it reacts readily with second molecule of III giving pyrophosphate triester IV

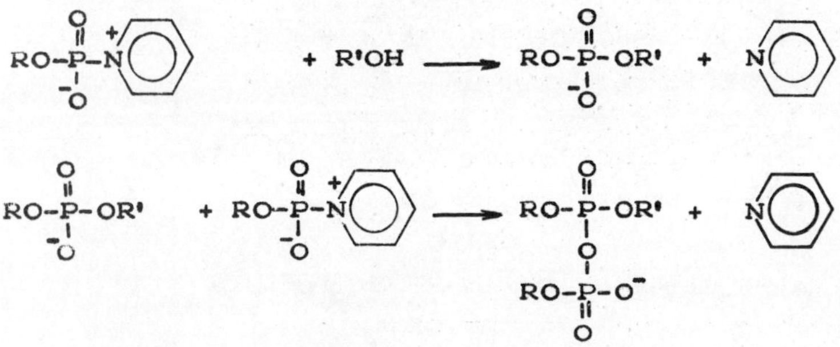

$$\text{RO-P(=O)(-O}^-\text{)-N}^+\text{C}_5\text{H}_5 + \text{R'OH} \longrightarrow \text{RO-P(=O)(-O}^-\text{)-OR'} + \text{NC}_5\text{H}_5$$

$$\text{RO-P(=O)(-O}^-\text{)-OR'} + \text{RO-P(=O)(-O}^-\text{)-N}^+\text{C}_5\text{H}_5 \longrightarrow \text{RO-P(=O)(-OR')-O-P(=O)(-O}^-\text{)-OR} + \text{NC}_5\text{H}_5$$

(IV)

Therefore, IV accumulates in the reaction mixture as a first observable by ^{31}P NMR technique product of phosphorylation of alcohols by III (Ref. 32).

Due to this process some complications arise when dinucleotides (or longer oligonucleotides) are treated with TPS. The reaction of the activated phosphomonoester group with phosphodiester groups of the same or other molecules leads to a set of cyclic pyrophosphates. E.g. due to intramolecular reaction dinucleotide is in the significant extent converted to V (Refs. 39 & 40).

$$^-\text{O-P(=O)(-O}^-\text{)-O-dT-O-P(=O)(-O}^-\text{)-O-dT(Ac)} \xrightarrow{\text{TPS}} \text{cyclic} \;\;^-\text{O-P(=O)-O-dT-O-P(=O)-O-dT(Ac) (bridged by O)}$$

(V)

The severalfold excess of TPS is necessary to convert phosphomonoester group to phosphorylpyridinium one. The latter may appear only after phosphodiester group is protected by transformation to pyrophosphate tetraester, the final compound being

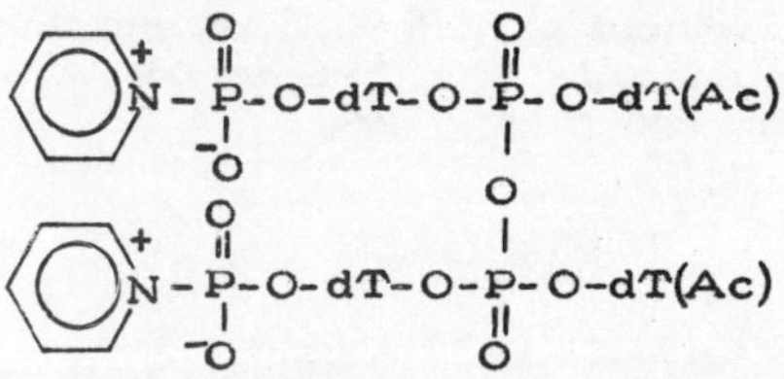

(VI)

Compound of the type VI may be also obtained by treatment of dinucleotides with cross-linked polystyrenesulfonyl chloride and separated from the excess of the condensing reagents. Using this method some simple oligonucleotide synthesis were performed. It was found that the reaction mixtures are significantly less coloured than with the use of TPS and that subsequent separation of the product is essentially simplified (Ref. 41).

THE KINETIC PROOF OF THE ROLE OF PHOSPHORYLPYRIDINIUM DERIVATIVES AS THE REACTIVE PHOSPHORYLATING INTERMEDIATES

The accumulation of phosphorylpyridinium derivatives in the reaction of phosphomonoesters with arenesulfonyl chlorides (Ref. 32) as well as with the mixture of triphenylphosphine and dipyridyldisulfide (Ref. 34) was found to be a stepwise process proceeding via intermediate accumulation and consumption of the symmetric pyrophosphate diesters and tripolyphosphate triesters. The extremely simplified scheme of the process with arenesulfonyl chloride as a condensing reagent may be represented as follows

$$^-\text{O-P(=O)(OR)-O}^- + \text{ArSO}_2\text{Cl} \longrightarrow \;^-\text{O-P(=O)(OR)-OSO}_2\text{Ar} + \text{Cl}^- \qquad (4)$$

(VII)

$$^-O-\underset{\underset{OR}{|}}{\overset{\overset{O}{\|}}{P}}-OSO_2Ar + N\langle\bigcirc\rangle \longrightarrow {}^-O-\underset{\underset{OR}{|}}{\overset{\overset{O}{\|}}{P}}-\overset{+}{N}\langle\bigcirc\rangle + ArSO_3^- \qquad (5)$$

$$^-O-\underset{\underset{OR}{|}}{\overset{\overset{O}{\|}}{P}}-O^- + {}^-O-\underset{\underset{OR}{|}}{\overset{\overset{O}{\|}}{P}}-\overset{+}{N}\langle\bigcirc\rangle \longrightarrow {}^-O-\underset{\underset{OR}{|}}{\overset{\overset{O}{\|}}{P}}-O-\underset{\underset{OR}{|}}{\overset{\overset{O}{\|}}{P}}-O^- + N\langle\bigcirc\rangle \qquad (6)$$

$$^-O-\underset{\underset{OR}{|}}{\overset{\overset{O}{\|}}{P}}-O-\underset{\underset{OR}{|}}{\overset{\overset{O}{\|}}{P}}-O^- + ArSO_2Cl \longrightarrow {}^-O-\underset{\underset{OR}{|}}{\overset{\overset{O}{\|}}{P}}-O-\underset{\underset{OR}{|}}{\overset{\overset{O}{\|}}{P}}-OSO_2Ar + Cl^- \qquad (7)$$

(VIII)

$$^-O-\underset{\underset{OR}{|}}{\overset{\overset{O}{\|}}{P}}-O-\underset{\underset{OR}{|}}{\overset{\overset{O}{\|}}{P}}-OSO_2Ar + 2N\langle\bigcirc\rangle \longrightarrow 2\,{}^-O-\underset{\underset{OR}{|}}{\overset{\overset{O}{\|}}{P}}-\overset{+}{N}\langle\bigcirc\rangle + ArSO_3^- \qquad (8)$$

(stepwise reaction)

$$^-O-\underset{\underset{OR}{|}}{\overset{\overset{O}{\|}}{P}}-O-\underset{\underset{OR}{|}}{\overset{\overset{O}{\|}}{P}}-O^- + {}^-O-\underset{\underset{OR}{|}}{\overset{\overset{O}{\|}}{P}}-\overset{+}{N}\langle\bigcirc\rangle \longrightarrow {}^-O-\underset{\underset{OR}{|}}{\overset{\overset{O}{\|}}{P}}-O-\underset{\underset{OR}{|}}{\overset{\overset{O}{\|}}{P}}-O-\underset{\underset{OR}{|}}{\overset{\overset{O}{\|}}{P}}-O^- + N\langle\bigcirc\rangle \qquad (9)$$

$$^-O-\underset{\underset{OR}{|}}{\overset{\overset{O}{\|}}{P}}-O-\underset{\underset{OR}{|}}{\overset{\overset{O}{\|}}{P}}-O-\underset{\underset{OR}{|}}{\overset{\overset{O}{\|}}{P}}-O^- + ArSO_2Cl \longrightarrow {}^-O-\underset{\underset{OR}{|}}{\overset{\overset{O}{\|}}{P}}-O-\underset{\underset{OR}{|}}{\overset{\overset{O}{\|}}{P}}-O-\underset{\underset{OR}{|}}{\overset{\overset{O}{\|}}{P}}-OSO_2Ar + Cl^- \qquad (10)$$

(IX)

$$^-O-\underset{\underset{OR}{|}}{\overset{\overset{O}{\|}}{P}}-O-\underset{\underset{OR}{|}}{\overset{\overset{O}{\|}}{P}}-O-\underset{\underset{OR}{|}}{\overset{\overset{O}{\|}}{P}}-OSO_2Ar + 3N\langle\bigcirc\rangle \longrightarrow 3\,{}^-O-\underset{\underset{OR}{|}}{\overset{\overset{O}{\|}}{P}}-\overset{+}{N}\langle\bigcirc\rangle + ArSO_3^- \qquad (11)$$

(stepwise reaction)

The first step of the reaction may proceed without participation of pyridine. This may be demonstrated by treatment of mononucleotide with TPS in DMFA solution. The kinetic curves of the reaction are presented in Fig. 2A

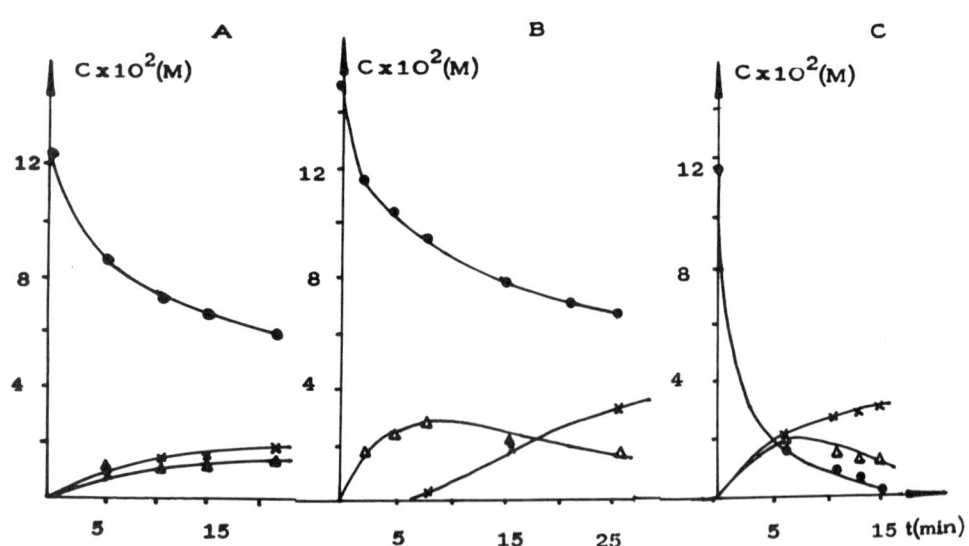

Fig. 2. The kinetic curves of the consumption of pdT(Ac) (•) and of the accumulation of pyrophosphate diester (△) and tripolyphosphate triester (×) in the reaction of pdT(Ac) (Et_3NH^+ salt) with TPS in dimethylformamide at $30°$. A) 0.066M TPS; B) 0.148M TPS; C) 0.139M TPS + 0.78M pyridine

It is seen that with small amounts of TPS pyrophosphate accumulates as a first observable product followed by the appearance of tripolyphosphate triester. This means that phosphomonoester may be phosphorylated directly with VII thus giving pyrophosphate diester. The latter may react with VII giving tripolyphosphate triester in parallel with the reactions (6) and (9). However as is seen in Fig. 2B and 2C the consumption of the starting phosphomonoester is significantly accelerated in the presence of 7% of pyridine. Thus the step (4) is subjected to pyridine catalysis most probably due to formation of the arenesulfonylpyridinium cation

$$ArSO_2Cl + N\bigcirc \longrightarrow ArSO_2-\overset{+}{N}\bigcirc \cdot Cl^-$$

$$ArSO_2-\overset{+}{N}\bigcirc + {}^-O-\underset{OR}{\overset{\overset{O}{\|}}{P}}-O^- \longrightarrow {}^-O-\underset{OR}{\overset{\overset{O}{\|}}{P}}-OSO_2Ar + N\bigcirc$$

Thus different routes of the formation of the intermediate mixed anhydrides as well as of the formation of substituted pyro- and tripolyphosphate exist in the system under consideration. Their relative role in the overall process may be established by special kinetic measurements.

Anyhow a set of transient intermediates (VII-IX) appear in parallel with phosphorylpyridinium derivative. Therefore, the question arises, whether in the ternary system alcohol reacts preferentially with these transient intermediates or it is phosphorylated mainly with III. In order to answer this question we have measured the rate constant of the reaction of IIIa with (Tr)dT

$$\overset{+}{\bigcirc}N-\underset{O^-}{\overset{\overset{O}{\|}}{P}}-O-dT(Ac) + (Tr)dT \longrightarrow (Tr)dT-O-\underset{O^-}{\overset{\overset{O}{\|}}{P}}-O-dT(Ac) + N\bigcirc \qquad (12)$$

in the pyridine solution at $2°$ (Ref. 42). As it was already mentioned pyrophosphate triester X accumulates in the course of this reaction

$$(Tr)dT-O-\underset{\underset{\underset{O}{\overset{\|}{P}}-O\ dT(Ac)}{\overset{|}{O}}}{\overset{\overset{O}{\|}}{P}}-O-dT(Ac)$$

$$(X)$$

At the same temperature the kinetics of the accumulation of IIIa and X was studied simultaneously for ternary system pdT(Ac), (Tr)dT and condensing reagent the latter being either TPS or the mixture of triphenylphosphine and dipyridyldisulfide. The data are presented in Fig. 3.

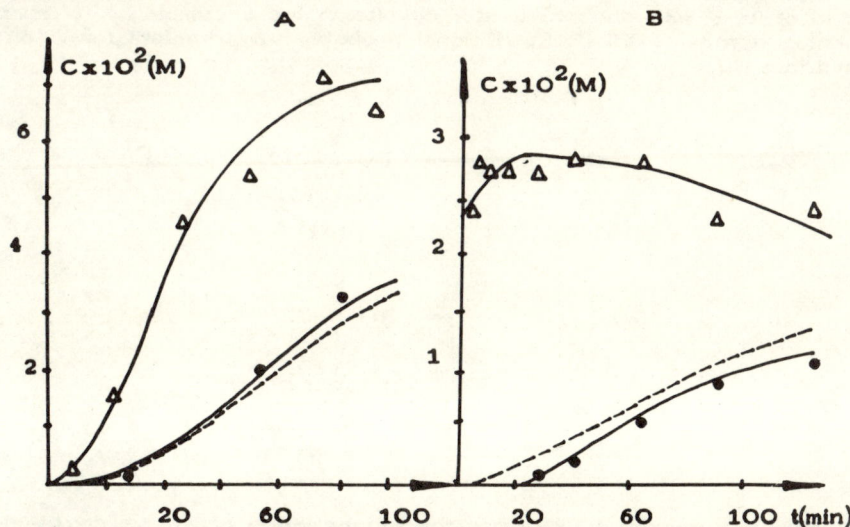

Fig. 3. The kinetic curves of the accumulation of IIIa (△) and X (•) in the reaction of (Tr)dT with pdT(Ac) in pyridine in the presence of TPS (A) and the mixture of triphenyl phosphine and dipyridyldisulfide (B) at $0°$. Dotted line - kinetic curve of the accumulation of X calculated using (13). (The other components are not shown).

Using the kinetic curves for IIIa we may easily calculate the accumulation of X due to reaction between IIIa and (Tr)dT as

$$[X] = \int_0^t k\,[\text{IIIa}]\,[(\text{Tr})\text{dT}]\,dT \tag{13}$$

where (Tr)dT is easily estimated as the difference between the initial concentration and the amount of X already formed. The calculated curves for both systems are presented in the dotted lines. It is seen that experimental and calculated curves coincide within the limits of accuracy of the experiments. This means that the reaction of IIIa with (Tr)dT is the main route of the phosphodiester bond formation. Therefore, phosphorylpyridinium derivatives are the main phosphorylating intermediates in the course of the formation of phosphodiesters in the ternary system for both condensation with arenesulfonyl chlorides and oxidative-reductive condensation with the mixture of triphenylphosphine and dipyridyldisulfide.

Phosphorylation of hydroxygroups may proceed in the absence of any measurable amount of phosphorylpyridinium derivative in the presence of pyrophosphate triester. The rate constant of the reaction of X with (Tr)dT was measured and found to be $\sim 10^2$ lower than that of IIIa (Ref.12). However this does not mean necessarily that the direct reaction of X with OH-group takes place. The same result may be due to the presence of $\sim 1\%$ of IIIa in the reaction mixture as a result of the reversibility of the reaction (12). In favour of this concept pyrophosphate triesters were found to be very low reactive towards water in the absence of pyridine. E.g. the rate constants of the hydrolysis of cyclic pyrophosphate triester V in the aqueous and pyridine solution at 30^0 were found to be 6.9×10^{-7} and $2.1 \times 10^{-2} M^{-1} s^{-1}$ respectively thus differing more than 10^4 times.

PHOSPHODIESTER AZOLIDES AS PHOSPHORYLATING INTERMEDIATES IN THE PHOSPHOTRIESTER SYNTHESIS

The interest to phosphotriesters in the field of the oligonucleotide synthesis has raised significantly in the last few years. Those scientists who elaborated and used phosphodiester methodology of the oligonucleotide synthesis have realised that phosphodiester groups of both nucleoside and nucleotide components behave as a sort of traps for reactive phosphorylating intermediates thus enhancing significantly the necessary excess of the nucleotide component and of the condensing reagent. Esterification of internucleotide phosphates was proposed to overcome the resulting complications (Refs. 43-47). In the same time the phosphotriester methodology of the oligonucleotide synthesis was essentially ameliorated due to introduction of new condensing reagents - arenesulfonyltriazolides (Ref.48) and especially tetrazolides (Ref.49). The formers were found to give better yields and cleaner reaction mixtures as compared with previously used arenesulfonyl chlorides. The latters were found to accelerate the overall process respective to chlorides.

The reaction of TPS with phosphodiester results in the accumulation of respective pyrophosphate tetraester XII (Ref.50) most probably via transient formation of the mixed anhydrides XI

$$\text{RO-P(=O)(O}^-\text{)-OR'} + \text{ArSO}_2\text{Cl} \longrightarrow \text{RO-P(=O)(OSO}_2\text{Ar)-OR'} + \text{Cl}^- \tag{14}$$

(XI)

$$\text{XI} + \text{RO-P(=O)(O}^-\text{)-OR'} \longrightarrow \text{RO-P(=O)(OR')-O-P(=O)(OR')-OR} \tag{15}$$

(XII)

^{31}P NMR technique permits to observe the accumulation of XII as compounds which are presented in the spectrum by two close singlets shifted 12-14 ppm upfield as compared with the starting phosphodiester (Ref.51) corresponding to two diastereoisomers (Ref.52) (assuming $R \neq R'$ and either R or R' or both are nucleosides with assymetric carbon atoms). The formation of XII proceeds rather fast the rate constant being only few times lower than that for respective phosphomonoester (Ref.12). In the presence of alcohol the formation of XII is followed by significantly slower phosphorylation

$$XII + R''OH \longrightarrow RO-\underset{\underset{OR''}{|}}{\overset{\overset{O}{||}}{P}}-OR' + RO-\underset{\underset{O^-}{|}}{\overset{\overset{O}{||}}{P}}-OR' \qquad (16)$$

the rate constant being of the same order of magnitude as that for pyrophosphate triesters (Ref. 12). In the same time the excess of TPS should be present during this step in order to convert liberating phosphodiester back to XII. This leads to an anavoidable long contact between R"OH and $ArSO_2Cl$ resulting in the significant extent of sulfonylation of alcohol. Sulfonylation of the nucleoside component was proposed to be one of the reasons of the moderate and some times even low yields of phosphotriesters with TPS as condensing reagent (Ref. 48).
When alcohol may be taken in some excess the yield of phosphotriester may be sufficiently high. Thus we have elaborated a procedure for preparation of dicyanoethyl esters of nucleotides with TPS as condensing reagent using the excess of 2-cyanoethanol (Ref. 53). However traditionally in the oligonucleotide synthesis the growing oligonucleotide chain plays the role of the nucleoside component and, therefore, the latter can not be taken in the excess.
To elucidate the reason of the advantages of arenesulfonyl triazolide as compared with respective chlorides we have studied the kinetics of the process of the protection of internucleotide phosphate in (Tr)dTpdT(Ac) with 2-cyanoethyl or with 2-phenylthioethyl residues using mesithylenesulfonyl triazolide (MST) as condensing reagent (Ref. 54). The kinetic curves are presented in Fig. 4.

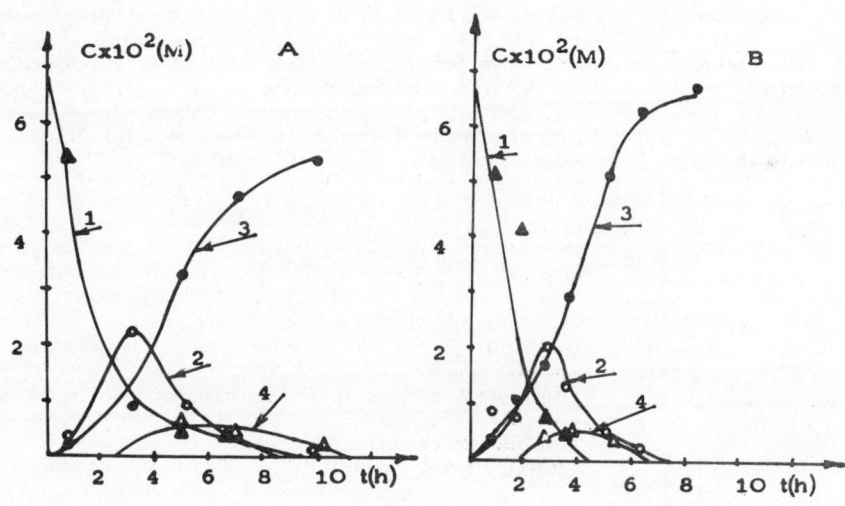

Fig. 4. The kinetic curves of the reaction of (Tr)dTpdT(Ac) with 0.5 M 2-cyanoethanol (A) and 0.5 M 2-phenylthioethanol (B) in the pyridine solution at 50° with 0.35 M mesithylenesulfonyl triazolide as a condensing reagent. 1. (Tr)dTpdT(Ac); 2. pyrophosphate tetraester; 3. phosphotriester; 4. triazolide of (Tr)dTpdT(Ac).

It is seen that the accumulation of XII with MST proceeds significantly slower than that with TPS (maximum is reached in several hours at 50° instead of several minutes at room temperature). However the following phosphorylation step is at least partially the same with both condensing reagent and therefore proceeds with nearly the same rate. The single difference is the formation with MST of the compound with $\delta = 10.5$ ppm in ^{31}P NMR spectrum. The same signal appears when the solution of pyrophosphate derivative of (Tr)dTpdT(Ac) is treated with triazole. Therefore this signal may be reasonable ascribed to triazolide of (Tr)dTpdT(Ac). The kinetic curves of this compound are typical of the intermediates and therefore it may be concluded that phosphorylation of alcohols proceeds partially via phosphodiester triazolide. To the same conclusion have come the authors of (Ref. 55) who have found that the signal of triazolide of $(DMTr)dTp(C_6H_4Cl)$ appears in the ^{31}P NMR spectrum when phosphodiester is treated with p-nitrobenzenesulfonyl triazolide in the pyridine solution. The phosphorylating ability of phosphodiester triazolides is in agreement with the previously found phosphorylating ability of phosphodiester imidazolides. Thus

dibenzylphosphorylimidazolide was found to react readily with n-propanol (Ref.56). Recently imidazolide of (Tr)dTpdT(Ac) was demonstrated to phosphorylate easily 2-cyanoethanol (Ref.47).
Therefore, the main change in the overall process of the phosphotriester formation with arenesulfonyl triazolide as a condensing reagent in comparison with arenesulfonyl chloride is the change of the step (14) for the significantly slower reaction

$$\begin{array}{c} \text{RO-P-OR}' \\ \parallel \\ \text{O}^- \end{array} + \text{ArSO}_2\text{-N}\diagup\diagdown\text{N} \xrightarrow{(H^+)} \begin{array}{c} \text{RO-P-OR}' \\ \parallel \\ \text{OSO}_2\text{Ar} \end{array} + \text{HN}\diagup\diagdown\text{N}$$

This process is followed by reactions (15) and (16). XII partially converts to phosphodiester triazolide

$$\text{XII} + \text{HN}\diagup\diagdown\text{N} \longrightarrow \begin{array}{c} \text{O} \\ \parallel \\ \text{RO-P-OR}' \\ | \\ \text{N}\diagdown \\ \text{N}\diagup\text{N} \end{array} + \begin{array}{c} \text{O} \\ \parallel \\ \text{RO-P-OR}' \\ | \\ \text{O}^- \end{array} \qquad (17)$$

The latter participates in the phosphorylation of alcohol in parallel with XII. Consequently only the first fast step of the condensation is slowered and the duration of the overall process does not increase significantly when arenesulfonyl chloride is changed for respective triazolide. In the same time the latter was demonstrated to be rather poor sulfonylating reagent. Essential decrease of the sulfonylation rate with unsignificant raise of the whole time of condensation results in the enhancement of the yield of desired phosphotriester.
To elucidate the reason of the powerful condensing action of arenesulfonyl tetrazolides in the phosphotriester formation we have first of all compared the kinetics of the activation of $(ClC_6H_4)pdT(Ac)$ with TPS and with respective tetrazolide (TPS tetrazolide). The reaction results in the formation of the derivative represented in ^{31}P NMR spectrum by two close signals with the center at δ = 19.2ppm shifted 14.1 ppm upfield as compared with starting phosphodiester. These signals may be reasonably assigned to pyrophosphate tetraester. As is seen in Fig. 5 the reaction rate with TPS tetrazolide only slightly exeeds that with chloride

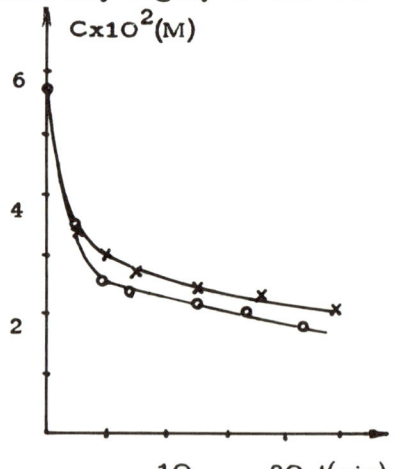

Fig. 5. The kinetic curves of the consumption of $(ClC_6H_4)pdT(Ac)$ in the reaction with 0.05 M TPS (x) and 0.05 M TPS tetrazolide (o) in the pyridine solution at 2°.

Therefore, the main difference in the condensing properties of these two reagents is not connected with the formation of pyrophosphate tetraester.
More essential difference between these reagents concerning the activation step was found when the influence of the addition of acids and bases on the pyrophosphate tetraester formation was studied. The data are presented in Fig. 6. It is seen that the reaction of TPS is slightly accelerated by the addition of either p-toluenesulfonic acid or triethylamine. In the same time the rate of the consumption of $(ClC_6H_4)pdT(Ac)$ in the presence of TPS tetrazolide is significantly enhanced in the presence of p-toluenesulfonic acid and falls nearly to zero in the presence of triethylamine. This result is not unexpected as the protonation of the tetrazole moiety of arenesulfonyltetrazolide should strongly enhance the elctrophilicity of the arenesulfonyl residue.

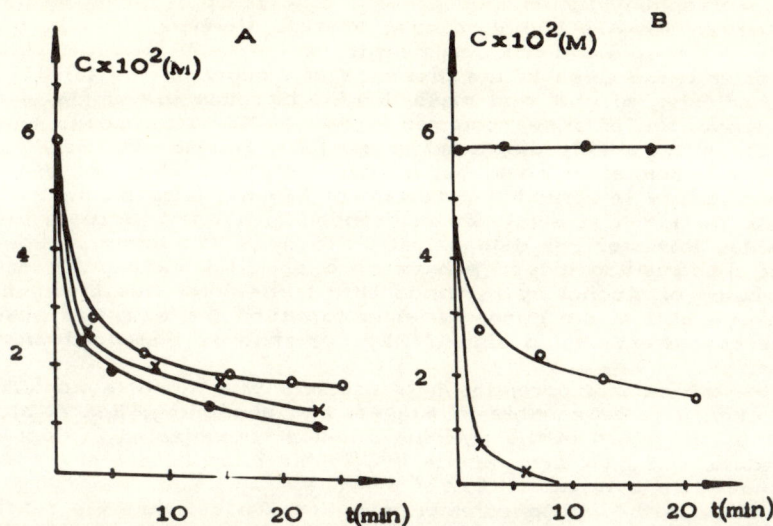

Fig. 6. The kinetic curves of the consumption of $(ClC_6H_4)pdT(Ac)$ in the reaction with 0.06 M TPS (A) and with 0.06 M TPS tetrazolide (B) in the pyridine solution at $2°$: (o) - without any additional component; (●) - with 0.2M triethylamine; (x) - with 0.15M p-toluene sulfonic acid

As the use of arenesulfonyl tetrazolides instead of chlorides accelerates essentially the overall process of the phosphotriester formation it should in the first line enhance the rate of the slower stage namely that of phosphorylation. It seems reasonable to suggest that tetrazole liberated in the first stage

$$2\ RO\text{-}\overset{O}{\underset{O^-}{P}}\text{-}OR' + ArSO_2\text{-}N\underset{N=N}{\overset{\diagup=N}{\diagdown}} \xrightarrow{(H^+)} XII + ArSO_3^- + HN\underset{N=N}{\overset{\diagup=N}{\diagdown}}$$

reacts with XII similar to reaction (17) with triazole giving phosphodiester tetrazolide.

$$XII + HN\underset{N=N}{\overset{\diagup=N}{\diagdown}} \rightleftharpoons RO\text{-}\overset{O}{\underset{O^-}{P}}\text{-}OR' + RO\text{-}\overset{O}{\underset{\underset{N-\!\!-N}{\overset{|}{\underset{\diagdown\diagup}{N}}}}{P}}\text{-}OR' \tag{18}$$

$$(XIV)$$

The latter should be powerful phosphorylating reagent due to strong electronegativity of the tetrazole moiety. Indeed, we have found that the transfer of the arenesulfonyl residue to phosphodiester proceeds with arenesulfonyl tetrazolide even slightly faster than with the corresponding chloride. It seems reasonable to expect by analogy that the phosphorylating activity of XIV should be close to that of phosphodiester chloridates.

To elucidate whether tetrazole really catalyses the reaction of pyrophosphate tetraesters with alcohols we have studied the reaction of tetraphenylpyrophosphate with $(Tr)dT$. To escape to the maximal extent the complications connected with the unavoidable traces of water especially dangerous in the absence of the condensing reagent we have prepared tetraphenylpyrophosphate by the reaction of $(PhO)_2POCl$ with diphenylphosphoric acid in pyridine using the slight excess of the former. The addition of 9 eqv. of $(Tr)dT$ both in the absence and in the presence of 6 eqv. of tetrazole results in the consumption of tetraphenylpyrophosphate and in the accumulation of the nearly stoichiometric amounts of diphenylphosphoric acid and $(Tr)dT(OPh)_2$. However, in the absence of tetrazole half time of the disappearance of tetraphenylpyrophosphate is about 200 min whereas in the presence of tetrazole over 70% of tetraphenylpyrophosphate is consumed within 5 min. Hence, tetrazole strongly catalyses the phosphotriester phormation.

To prove directly that phosphodiester tetrazolides are powerful phosphorylating entities we have prepared diphenylphosphotetrazolide by treatment of $(PhO)_2POCl$ in pyridine with the excess tetrazole. The signal of the starting chloridate immediately disappears from the ^{31}P NMR spectrum and is changed for the singlet signal at δ = 21.4 ppm ascribed to diphenylphosphotetrazolide. The subsequent addition of p-nitrophenol results in the fast (within few minutes) conversion of tetrazolide to phosphotriester (p-nitrophenyldiphenylphosphate, δ = 18.6 ppm). This means that phosphodiester tetrazolides are highly reactive towards alcohols. This is confirmed by the fact that in no case we could demonstrate by ^{31}P NMR technique the presence of phosphodiester tetrazolides in the presence of alcohol. In the ternary system containing phosphodiester, alcohol and arylsulfonyl tetrazolide the single signal relating to reactive derivatives of phosphodiester is that of XII. This means also that the equilibrium (18) is essentially displaced to the left. Due to extremely fast reaction of phosphodiester tetrazolide we could not up till now perform the necessary kinetic measurements to elucidate whether the reaction of XII with tetrazole is the main or even single cause for the high efficiency of arenesulfonyl tetrazolides in the phosphotriester formation. However our data permit us to state that tetrazole liberated in the first stage of condensation indeed behave as a powerful nucleophilic catalyst in the phosphorylation of alcohol by pyrophosphate tetraesters. Due to significant decrease of the overall time of the phosphotriester formation the extent of sulfonylation of the nucleoside component should significantly decrease as compared with arenesulfonyl chlorides.

By analogy with the reaction of pyrophosphate triesters with alcohols considered in the preceding section it is reasonable to suggest that phosphorylation of alcohols with pyrophosphate tetraesters in the pyridine solution is subjected to nucleophilic catalysis by pyridine. Our data demonstrate that in the phosphotriester formation catalytic action of tetrazole overpasses that of pyridine.

This is not the case in the phosphodiester formation. To compare the reactivities of pyridinium derivatives of phosphomonoesters III with that phosphomonoester tetrazolide we have studied the effect of tetrazole on the formation of (Tr)dTpdT(Ac). IIIa was obtained by treatment of the pyridine solution of pdT(Ac) with cross-linked polystyrenesulfonyl chloride. The separated solution of IIIa was treated by tetrazole and (Tr)dT and accumulation of internucleotide phosphates was followed by ^{31}P NMR technique. The main P-containing components of the reaction mixture were IIIa, tetrazolide of pdT(Ac) and X. The kinetic curve of the formation of X due to reaction of IIIa with (Tr)dT may be easily calculated using (13). The experimental data and the results of calculation are presented in Fig. 7.

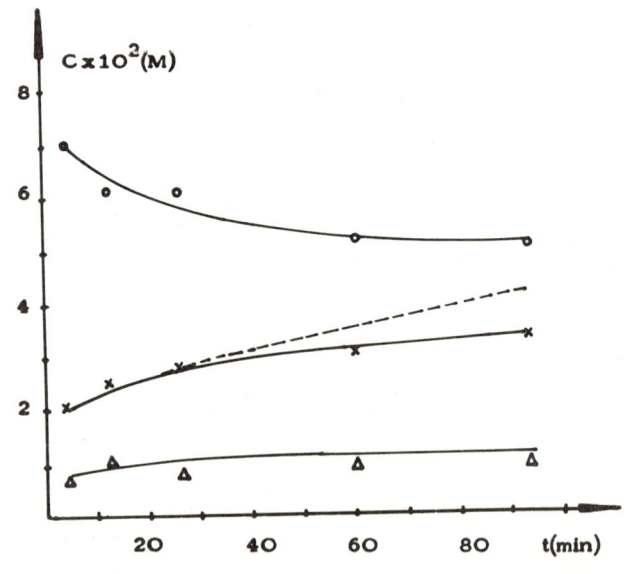

Fig. 7. The kinetic curves of the formation of internucleotide phosphates (x) in the reaction of IIIa ($\triangle$) with (Tr)dT in the presence of 0.2 M of tetrazole at 30° in the pyridine solution. (o) - concentration of pdT(Ac) tetrazolide. The dotted line is calculated by (13).

It is seen that both curves coincide with in the limits of accuracy. This means that the reaction of IIIa with (Tr)dT is the main route of the phosphodiester formation pdT(Ac) tetrazolide being essentially less reactive.

Therefore, there is no reason to expect that phosphodiester tetrazolides are significantly more reactive than phosphodiester pyridinium derivative. The possible explanation of the high catalytic power of tetrazole in the phosphotriester formation may be the significantly greater concentration of tetrazolide XIV as compared with the phosphodiester pyridinium derivative. It should be emphasized that while the phosphorylpyridinium derivatives of phosphomonoesters III are neutral particles those of phosphodiesters are positively charged. In the pyridine solution with rather small dielectric constant phosphodiester pyridinium cation exists mainly as an ion pair with counterpart anion. E.g. the reaction of XII with pyridine should result in the formation of the ion pair XVI

$$XII + N\text{-Py} \rightleftharpoons RO-\overset{O}{\underset{\underset{N^+\text{-Py}}{|}}{P}}-OR' \cdot RO-\overset{O}{\underset{\underset{O^-}{|}}{P}}-OR'$$

(XVI)

This is equivalent to the great raise of local concentration of nucleophilic phosphodiester anion and should result in the strong displacement of the equilibrium to the left. Therefore, the concentration of phosphodiester pyridinium cation may be extremely small. This is not the case with the phosphodiester tetrazolide molecule which is electrically neutral and may exist as a separate particle.

In this connection it is worth mention that even phosphodiester chloridate does not react with pyridine in the measurable extent. $(PhO)_2POCl$ remains unchanged in the pyridine solution as revealed by ^{31}P NMR spectrum. This means that equilibrium

$$(RhO)_2POCl + N\text{-Py} \rightleftharpoons PhO-\overset{O}{\underset{\underset{N^+\text{-Py}}{|}}{P}}-OPh \cdot Cl^-$$

is nearly completely displaced to the left. In the same time we have already mentioned that equilibrium

$$RO-\overset{O}{\underset{\underset{O^-}{|}}{P}}-Cl + N\text{-Py} \rightleftharpoons RO-\overset{O}{\underset{\underset{O^-}{|}}{P}}-N^+\text{-Py} + Cl^-$$

(III)

is displaced to the right as III exists as the single observable in the ^{31}P NMR spectrum compound in the presence of the significant amount of Cl^-.

DIMETHYLAMINOPYRIDINIUM DERIVATIVES OF PHOSPHODIESTERS AS PHOSPHORYLATING REAGENTS

Having found out that the reaction of pyrophosphate tetraesters with alcohols is subjected to the strong nucleophilic catalysis by tetrazole we have tried to estimate the catalytic effect on the same reaction of the other well known nucleophilic catalyst namely -N,N-dimethylaminopyridine (DMAP) (Ref. 57). The treatment of $(PhO)_2POCl$ with DMAP in the methylene chloride solution results in the immediate conversion of $(PhO)_2POCl$ to the compound represented in ^{31}P NMR spectrum by singlet signal at = 13.0 ppm (Ref. 58). With the stoichiometric amount of DMAP the reaction at room temperature is incomplete. At $-40°$ the signal of $(PhO)_2POCl$ entirely disappears. ^1H NMR spectrum of the reaction mixture at $-40°$ is presented in Fig. 8. This spectrum completely agrees with that expected for dimethylaminopyridinium derivative of diphenylphosphate which should be formed due to reaction

$$PhO-\overset{O}{\underset{\underset{Cl}{|}}{P}}-OPh + N\text{-Py}-N(CH_3)_2 \rightleftharpoons PhO-\overset{O}{\underset{\underset{N^+\text{-Py-}N(CH_3)_2}{|}}{P}}-OPh \cdot Cl^-$$

(XVII)

The signals of - and -protons of the pyridine ring as well as the protons of CH_3-groups are shifted downfield as compared with DMAP as it should be due to

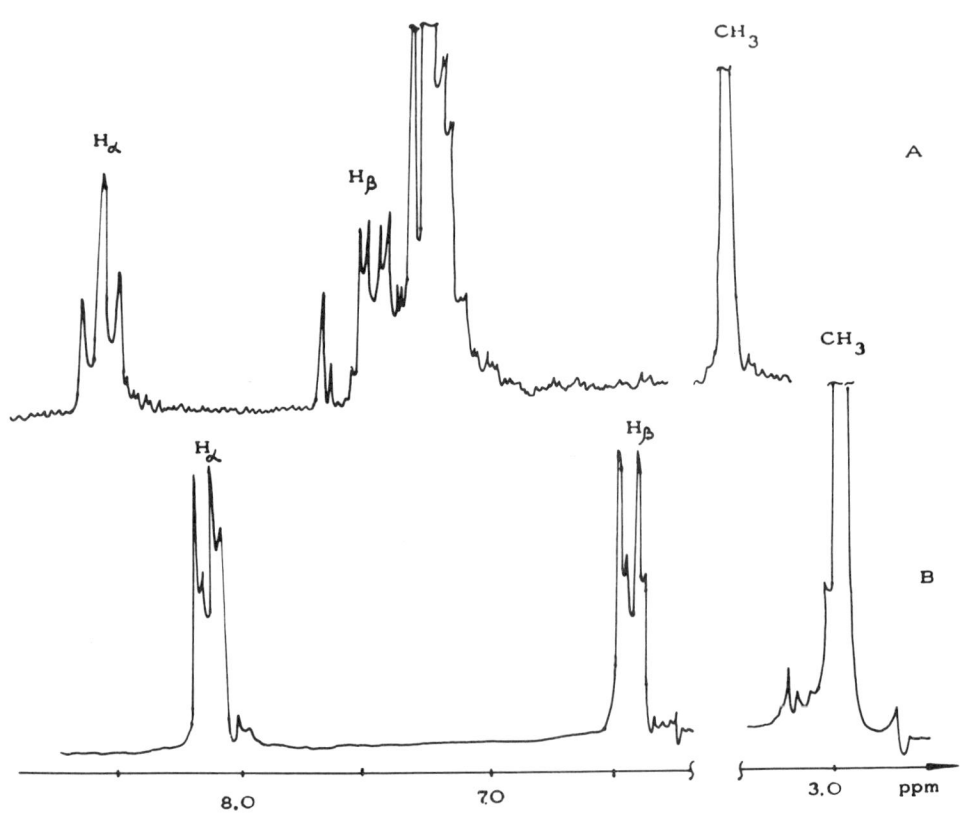

Fig. 8. ^{1}H NMR spectrum of the DMAP derivative of diphenylphosphoric acid (A) and DMAP (B) in CH_2Cl_2 at $-40°$.

the appearance of the positive charge at the DMAP moiety. Besides mutual splitting of α- and β-protons additional splitting of both signals with J = 7.5 Hz and 2.5 Hz, respectively is seen as it should be due to spin-spin coupling $^1H_\alpha - ^{31}P$ and $^1H_\beta - ^{31}P$. The signal of ^{31}P resonance is partially splitted to triplet at two vicinal protons. At room temperature the spin-spin coupling $^1H - ^{31}P$ is not seen presumably due to the rapid exchange between the DMAP moiety of XVII and unreacted DMAP.

Hence, in contrast with the case of pyridine the DMAP derivative of phosphodiester may be accumulated in significant amounts in spite of the presence of Cl^- as the counterion. This may be easily understood as due to significantly higher nucleophilicity of DMAP in comparison with pyridine XVII is essentially poorer electrophile than the respective pyridinium cation. The same compound XVII is formed when diphenylphosphoric acid is treated with TPS and DMAP in the pyridine solution. The addition of 2-cyanoethanol to pyridine solution of XVII results in the immediate conversion to diphenyl-2-cyanoethyl phosphate thus indicating that XVII is a powerful phosphorylating reagent. Therefore we may expect that DMAP should behave as a strong catalyst in the phosphotriester formation.

The kinetics of the accumulation of diphenyl-2-cyanoethanol in the reaction mixture containing diphenylphosphoric acid, 2-cyanoethanol and TPS with and without DMAP is presented in Fig. 9. As DMAP should react in the unprotonated form tri-n-butylamine is added to neutralize arylsulfonic acid and hydrogen chloride liberated in the activation step. It is seen that DMAP strongly accelerates the phosphotriester formation.

The similar results were obtained in our laboratory when the influence of DMAP on the internucleotide bond formation was studied. In these experiments dianilidate of pdC^{An} (Ref. 59) was used as a nucleoside and p-chlorophenyl ester of pdT(Ac) as a nucleotide component. The kinetic curves of the accumulation of dinucleotide in the reaction mixture containing $(PhNH)_2pdC^{An}$, $(ClC_6H_4)pdT(Ac)$ and TPS in the pyridine solution in the absence and in the presence of DMAP are presented in Fig. 10.

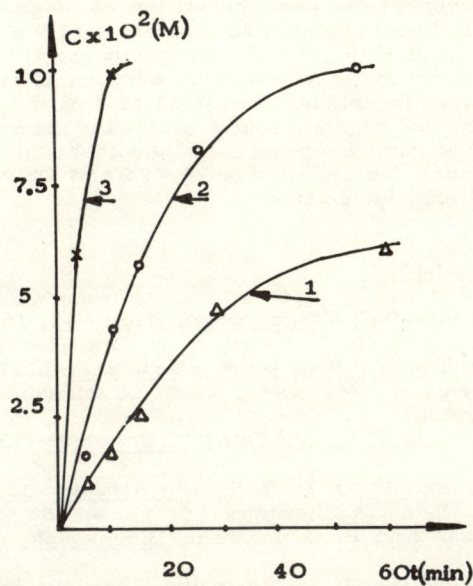

Fig. 9. The kinetic curves of the accumulation of 2-cyanoethyl-diphenylphosphate in the reaction of diphenylphosphate (0.15 mmol) with 2-cyanoethanol (0.3 mmol) in the presence of 0.45 mmol TPS and 1.05 mmol of tri-n-butylamine in 1.5 ml of pyridine at 30°
1) without DMAP; 2) in the presence of 0.015 mmol DMAP;
3) in the presence of 0.15 mmol DMAP.

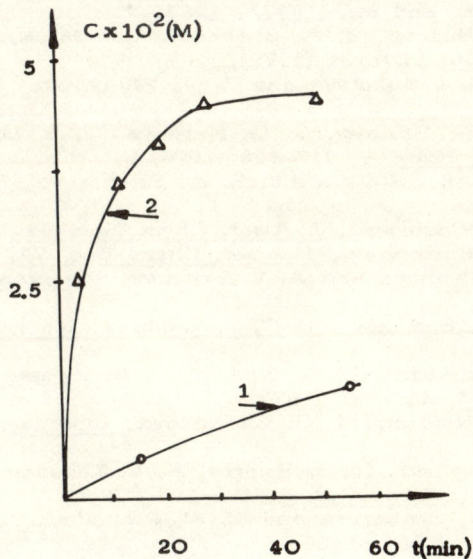

Fig. 10. The kinetic curves of the accumulation of $(PhNH)_2pdC^{An}(ClC_6H_4)pdT(Ac)$ in the reaction of the respective monomers with TPS in pyridine at 30° in the absence (1) and in the presence of DMAP (2). Concentrations used: 1) $(PhNH)_2pdC^{An}$ 0.09 M, $(ClC_6H_4)pdT(Ac)$ 0.07 M, TPS 0.25 M; 2) $(PhNH)_2pdC^{An}$ 0.05 M, $(ClC_6H_4)pdT(Ac)$ 0.06 M, TPS 0.21 M, DMAP 0.49 M.

It is seen that in the presence of DMAP TPS perform the condensation within the time not exceeding that commonly used with arenesulfonyl tetrazolides as condensing reagents. The protected dinucleotide was isolated in 65% yield from the reaction mixture. In the recently appeared paper of Dobrynin et al. (Ref. 60) it was demonstrated that DMAP catalyses the reaction (1) of phosphorylation of 5'-group of nucleosides with phosphodiester triazolides. The authors also indicate that the reaction between $(DMTr)dC^{bz}p(C_6H_4Cl)$ and $dTp(C_6H_4Cl)(C_2H_4CN)$ with triisopropyl benzenesulfonyl triazolide as a condensing reagent is accelerated by addition of DMAP and the yield of the condensation product is enhanced from 33 to 46%.
Therefore the use of DMAP in the oligonucleotide synthesis seems to be rather promising. Certainly the accompanying by-processes should be investigated in details prior to give final recommendation for the mixture of TPS and DMAP as a condensing reagent for the preparation of oligonucleotides.

REFERENCES

1. N. I. Grineva and T. S. Lomakina, Zhurn. Obsh. Khim. 42, 1630-1634 (1972).
2. S. K. Vassilenko, N.I. Grineva, G. G. Karpova, A. Ya. Kozorovitsky, M. Ya. Saarma and M. G. Tiunov, Dokl. Akad. Nauk SSSR 212, 1227-1230 (1973).
3. V. G. Budker, N. I. Grineva, N. D. Kobets, T. S. Lomakina and S. P. Mayev, Biokhimiya 43, 761-763 (1978).
4. P. S. Miller, J. C. Barrett and P. O. P. Ts'o, Biochemistry 13, 4887-4896 (1974).
5. J. C. Barrett, P. S. Miller and P. O. P. Ts'o, Biochemistry 13, 4897-4906 (1974).
6. R. C. Pless and P. O. P. Ts'o, Biochemistry 16, 1239-1250 (1977).
7. P. S. Miller, L.T. Braiterman and P. O. P. Ts'o, Biochemistry 16, 1988-1996 (1977)
8. H. G. Khorana, Some recent Developments in the Chemistry of Phosphate Esters of Biological Interest, Wiley, New York (1961).
9. A. M. Michelson, The Chemistry of Nucleosides and Nucleotides, Acad. Press. London and New York (1966).
10. Z. A. Shabarova and A. A. Bogdanov, The Chemistry of Nucleic Acids and their Components, Khimiya, Moscow (1978).
11. D. G. Knorre and V. F. Zarytova, Nucl. Acids. Res. 3, 2709-2729 (1976).
12. D. G. Knorre, V. F. Zarytova, A. V. Lebedev, L. M. Khalimskaya and E. A. Sheshegova, Nucl. Acids Res. 5, 1253-1272 (1978).
13. K. L. Ararwal and F. Rifina, Nucl. Acids Res. 5, 2809-2823 (1978).
14. A. M. Michelson, Chem. and Ind. (1960), 1267.
15. N. I. Sokolova, V. V. Nosova, Z. A. Shabarova and M. A. Prokofiev, Dokl. Akad. Nauk. SSSR 206, 129-131 (1972).
16. V. V. Shumyantzeva, N. I. Sokolova and Z. A. Shabarova, Nucl. Acids Res. 3, 903-916 (1976).
17. O. I. Gimautdinova, N. I. Grineva, G. G. Karpova, T. S. Lomakina and E. L. Shelpakova, Bioorgan. Khim. 4, 917-926 (1978).
18. M. G. Ivanovskaya, N. I. Sokolova and Z. A. Shabarova, Bioorgan. Khim. 5, 35-40 (1979).
19. T. Mukaiyama and M. Hashimoto, J. Amer. Chem. Soc. 94, 8528-8532 (1972).
20. J. G. Moffatt and H. G. Khorana, J. Amer. Chem. Soc. 79, 3741-3746 (1957).
21. V. F. Zarytova, E. M. Ivanova and A. V. Lebedev, Bioorgan. Khim. 2, 1196--1204 (1976).
22. G. F. Mishenina, V. V. Samukov and T. N. Shubina, Bioorgan. Khim. 5, 886--894 (1979).
23. V. G. Budker, V. F. Zarytova, D. G. Knorre, N. D. Kobets and O. I. Ryazankina, Bioorgan. Khim. 3, 618-625 (1977)
24. V. F. Zarytova, V. C. Ryte and T. S. Chernikova, Bioorgan. Khim. 3, 1626-1631 (1977).
25. V. L. Drutsa, V. F. Zarytova, D. G. Knorre, A. V. Lebedev, N. I. Sokolova and Z. A. Shabarova, Nucl. Acids Res. 5, 185-193 (1978).
26. V. G. Shestakov, Z. A. Shabarova and M. A. Prokofiev, Vestnik Mosk. Universiteta, ser.khim. (1964), n. 4, 81-84
27. T. Glonek, R. A. Kleps and T. C. Myers, Science 185, 352-355 (1974)
28. V. F. Zarytova, D. G. Knorre, V. A. Kurbatov, A. V. Lebedev, V. V. Samukov and G. V. Shishkin, Biborgan. Khim. 1, 793-798 (1975)
29. G. T. Babkina, V. F. Zarytova and D. G. Knorre, Bioorgan. Khim. 1, 611-615 (1975).
30. A. Todd, Proc. Nat. Acad. Sci. USA, 45, 1389-1397 (1959).
31. V. F. Zarytova, D. G. Knorre, A. V. Lebedev, A. S. Levina and A. I. Rezvukhin, Dokl. Akad. Nauk SSSR 212 , 630-633 (1973).
32. D. G. Knorre, A. V. Lebedev, A. S. Levina, A. I. Rezvukhin and V. F. Zarytova, Tetrahedron 30, 3073-3079 (1974).

33. V. F. Zarytova, D. M. Graifer, E. M. Ivanova, D. G. Knorre, A. V. Lebedev and A. I. Rezvukhin, Nucl. Acids Res. Sp. Publ. N 4, s209-s212 (1978).
34. D. G. Knorre, G. F. Mishenina, V. V. Samukov and T. N. Shubina, Dokl. Akad. Nauk SSSR 236, 613-616 (1977).
35. S. N. Zagrebelny, V. F. Zarytova, A. S. Levina, L. N. Semenova, E. V. Yarmolinskaya and S. M. Yasnetskaya, Bioorgan. Khim. 4, 729-734 (1978)
36. V. F. Zarytova, D. G. Knorre, V. K. Potapov, A. I. Rezvukhin, S. I. Turkin and Z. A. Shabarova, Izv. Sib. Otd. Akad. Nauk. SSSR, ser. khim. nauk, (1974) iss. 4, 152-155
37. S. I. Turkin, V. K. Potapov, Z. A. Shabarova, V. F. Zarytova and D. G. Knorre, Bioorgan. Khim. 1, 1430-1433 (1975).
38. S. N. Zagrebelny, V. F. Zarytova, A. S. Levina, E. G. Lubenets, S. A. Pozdnyakovich, A. G. Khmelnitsky and S. M. Yasnetskaya, Izv. Sib. Otd. Akad. Nauk SSSR, ser.khim. nauk (1978) iss. 1, 143-147
39. D. G. Knorre, A. V. Lebedev and V. F. Zarytova, Nucl. Acids Res. 3, 1401--1418 (1976)
40. V. F. Zarytova and A. V. Lebedev, Bioorgan. Khim. 3, 1211-1218 (1977)
41. S. N. Zagrebelny, S. M. Yasnetskaya, V. F. Zarytova, E. G. Lubenets and A. G. Khmelnitsky, Bioorgan. Khim. 5, 1133-1139 (1979)
42. V. F. Zarytova, D. G. Knorre, V. P. Starostin and L. M. Khalimskaya, Izv. Sib. Otd. Akad. Nauk SSSR, ser. khim. nauk (1979), iss. 1, 88-93
43. J. Smrt, Tetrahedron Letters, (1972), 3437-3438
44. W. S. Zielinski and J. Smrt, Collect. Czech. Chem. Communs. 39, 2483-2490 (1974).
45. D. G. Knorre, G. F. Mishenina and T. N. Shubina, Bioorgan. Khim. 2, 1189--1195 (1976).
46. N. S. Bystrov, V. N. Dobrynin, M. N. Kolossov and B. K. Chernov, Bioorgan. Khim. 2, 1271-1272 (1976).
47. N. F. Sergeyeva, Z. A. Shabarova and M. A. Prokofiev, Dokl. Akad. Nauk SSSR 234, 607-609 (1977).
48. N. Katagiri, K. Itakura and S. A. Narang, J. Amer. Chem. Soc. 97, 7332-7337 (1975).
49. J. Stawinski, T. Hozunn, S. A. Narang, C. P. Bahi and H. Wu, Nucl. Acids Res. 4, 353-371 (1977).
50. F. Eckstein and I. Rizk, Chem. Ber. 102, 2362-2377 (1969).
51. V. F. Zarytova, D. G. Knorre, A. V. Lebedev, A. S. Levina and A. I. Rezvukhin, Izv. Sib. Otd. Akad. Nauk SSSR, ser. khim. nauk (1974), iss. 3, 126-131
52. A. V. Lebedev and A. I. Rezvukhin, Izv. Sib. Otd. Akad. Nauk SSSR, ser. khim. nauk (1974), iss. 2, 149-154
53. V. F. Zarytova, L. M. Kuznetsova, T. S. Lomakina and V. P. Starostin, Izv. Sib. Otd. Akad. Nauk SSSR, ser. khim. nauk (1979), iss. 1, 93-100
54. V. F. Zarytova and E. A. Sheshegova, Bioorgan. Khim. 4, 901-909 (1978).
55. V. N. Dobrynin, E. F. Boldyreva, N. S. Bystrov, I. V. Severtseva, B. K. Chernov and M. N. Kolossov, Bioorgan. Khim. 4, 523-534 (1978).
56. R. Blakeley, F. Kerst and F. H. Westheimer, J. Amer. Chem. Soc. 88, 112-119 (1966).
57. G. Höfle, W. Steglich and H. Vorbrüggen, Angew. Chem. 90, 602-615 (1978)
58. V. F. Zarytova, E. M. Ivanova, D. G. Knorre, A. V. Lebedev, A. I. Rezvukhin and E. V. Yarmolinskaya, Dokl. Akad. Nauk SSSR 248, n. 5 (1979)
59. G. F. Mishenina, V. V. Samukov, L. N. Semenova and T. N. Shubina, Bioorgan Khim. 4, 735-739 (1978).
60. V. N. Dobrynin, N. S. Bystrov, V. K. Chernov, I. V. Severtseva and M. N. Kolossov, Bioorgan. Khim. 5, 1254-1256 (1979).

SYNTHESIS OF tRNA$_f^{Met}$ FROM E. COLI

M. Ikehara, E. Ohtsuka, S. Tanaka, S. Nishikawa, T. Tanaka,
T. Miyake, E. Nakagawa, T. Wakabayashi, R. Fukumoto,
H. Uemura, Y. Taniyama, T. Doi, A. F. Markham and
J. Antkowiak

Faculty of Pharmaceutical Sciences, Osaka University, Suita Osaka, 565 Japan

ABSTRACT

A synthetic approach to the formylmethionine-tRNA from E. coli is described. The methods employed include 1) phosphodiester synthesis using block condensation or stepwise addition methods, 2) a mixed method combining di- and triester approaches, 3) the phosphotriester method using o-nitrobenzyl for 2'-OH, p-chlorophenyl and anilidate for phosphate and acyl groups for base protection, and 4) joining of the synthetic fragments with RNA ligase.

INTRODUCTION

It is of great interest to elucidate the structure-function relationship of tRNAs, which act as carrier molecules in protein biosynthetic systems.

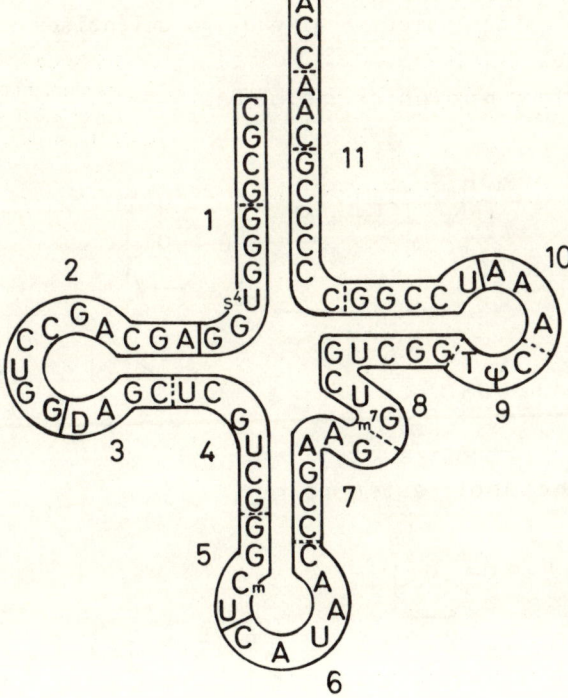

Fig. 1. Structure of formylmethionine tRNA. Boxes indicates fragments synthesized by joining chemical oligonucleotides at the places of dotted line.

In 1965 Holley and coworkers (1) elucidated the structure of yeast alanine tRNA. This was the first nucleic acid with defined sequence. Since then a number of tRNA structures have been determined and studies on their structure-function relationship have been performed. One approach to such studies is to modify the tRNA structure either directly with chemical reagents or by mutation. However, information obtained by this approach has been rather limited. In this regard we attempted to synthesize parts of this molecule by using the aromatic phosphroamidate method to obtain suitably protected ribonucleotide blocks which could be condensed to give a 3'-end nonanucleotide and two 5'-end hexamers (2).

We then turned our attention to formylmethionine tRNA from E. coli (3). This tRNA acts as the initiator in prokaryotic protein biosynthesis and has to be recognized by a number of enzymes and proteins to play this role.

CHEMICAL SYNTHESIS OF RIBOOLIGONUCLEOTIDES

For synthesizing the desired tRNA molecule, we undertook 1) chemical synthesis of fragments and 2) their joining with RNA ligase to afford longer oligonucleotides and hopefully the total molecule. The synthesis of ribo-oligonucleotides is more difficult than that of deoxyribo counterparts, because ribonucleotides have cis-diol systems, which are not present in deoxyribonucleotides. For the chemical synthesis of ribooligonucleotides we employed two approaches 1) the so-called phosphodiester method and 2) the phosphotriester method. The first method involves activation of phosphomonoesters to give phosphodiesters directly. In the triester method, di-esterified phosphate is activated to give oligonucleotides with protected internucleotidic bonds (Fig. 2).

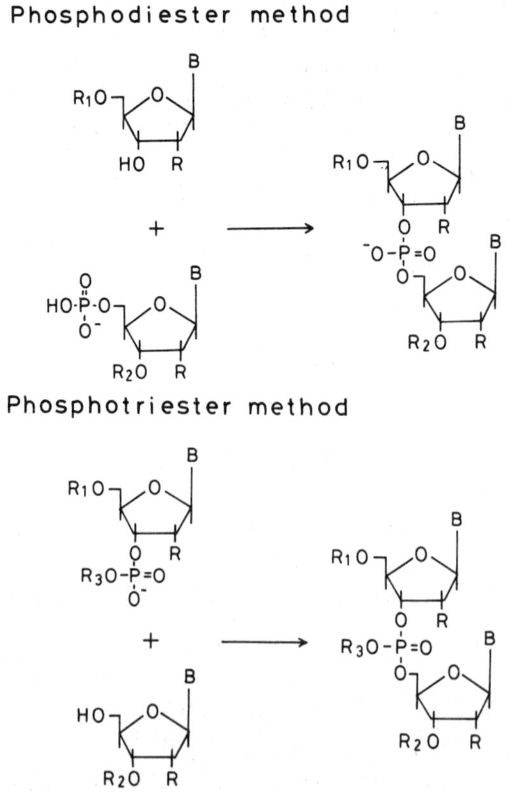

Fig. 2. Phosphodiester and phosphotriester approaches.

1) Synthesis by the diester method

We have previously reported the synthesis of fragments corresponding to bases 1-4 (4), 5-10 (5), 58-60 (6), 61-71 (7) and 72-77 (8). In these cases the phosphodiester approach was employed and various improvements in isolation procedures of the desired oligonculeotides were developed. The stepwise addition of mononucleotide units to the 3'-end nucleoside afforded the desired product relatively easily, e. g. a hexanucleotide corresponding to bases 41-46 of the tRNA (Fig. 3). In order to investigate limitations of

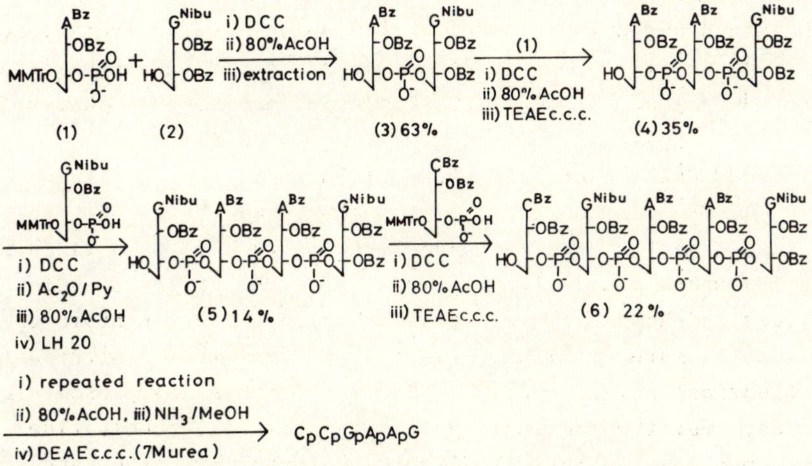

Fig. 3. Synthesis of fragment 7 by the diester method.

the approach an octanucleotide of bases 47-54, GUCGUCGG, was also synthesized by this method. In the latter case a rapid separation of the products was performed by using TEAE-cellulose column chromatography.

This method was also applied to the case of a trinucleotide containing Ψ-uridine. This nucleoside is known to be isomerized to alpha- and pyranösyl derivatives by acid treatment and special care has to be taken. As shown in Fig. 4, when the N-1 position of pseudouridine was protected by a benzoyl group, condensation successively with C and ribo-T gave pure TΨC, whose structure was confirmed by enzymatic digestion. Thus, the diester method is suitable for obtaining oligonucleotides containing modified bases.

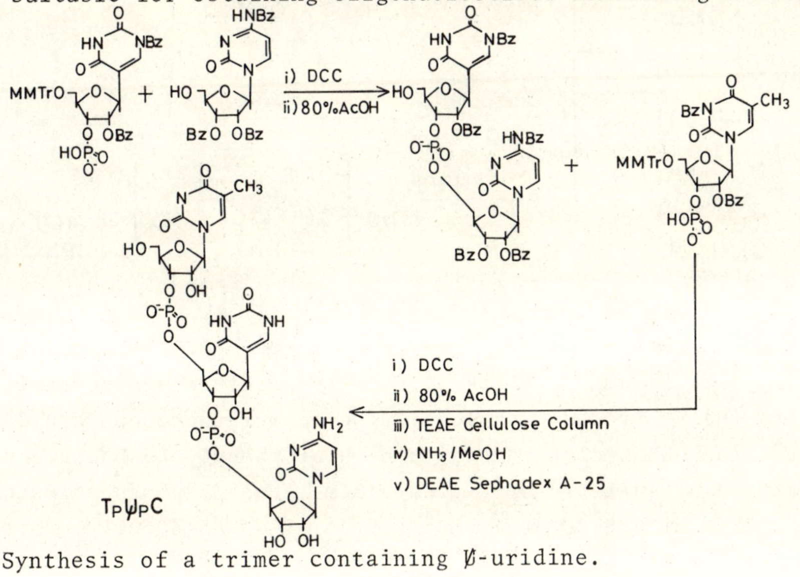

Fig. 4. Synthesis of a trimer containing Ψ-uridine.

2) Synthesis by the triester method

The triester method is widely utilized in deoxyribonucleotide synthesis. We first intended to apply this method to ribooligonucleotides by developing a new protective group for 2'-OH. The o-nitrobenzyl group, which is removable by photoactivation seems to be suitable because the deprotection procedure does not interfere with other deprotection methods. Using o-nitrobenzyl bromide with sodium hydride or dibutyltin oxide we were able to introduce o-nitrobenzyl groups to A, C (8), U and N^2-isobutyryl-G (9) in yields of 25-40% after appropriate separations. (Fig. 5).

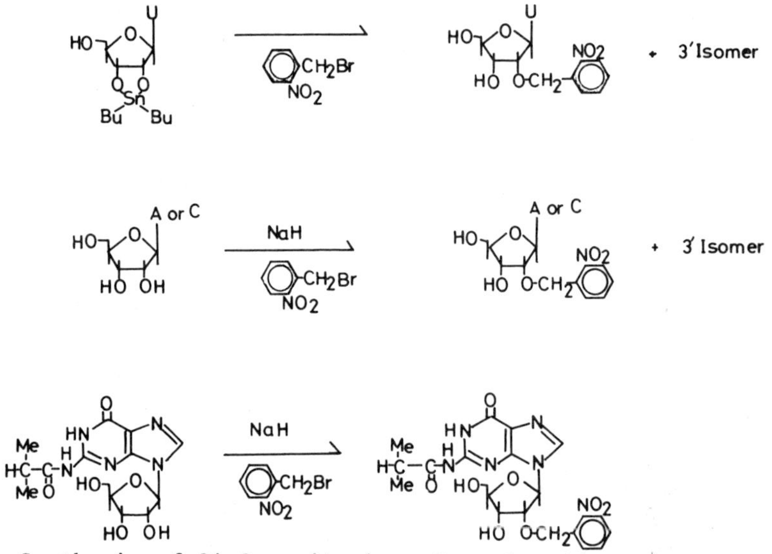

Fig. 5. Synthesis of 2'-O-o-nitrobenzyl nucleosides.

We then attempted to develop a new phosphorylating reagent, which is useful for 3'-phosphorylation of 2'-o-nitrobenzyl nucleosides. Thus, p-chlorophenylphosphoroanilido chloride was synthesized and used for the phosphorylation of 2'-,5'-(and base) protected nucleosides (10) (Fig. 6).

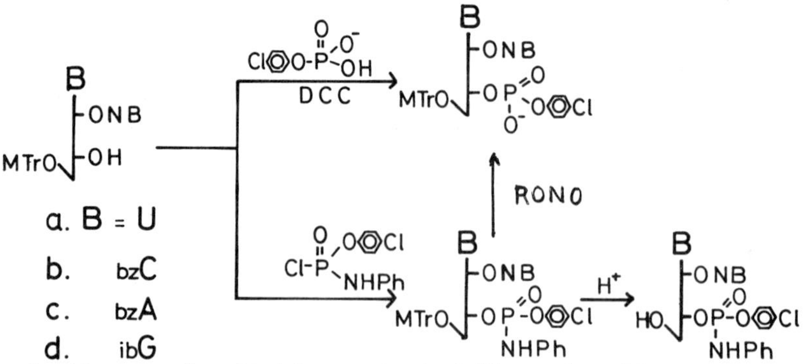

Fig. 6. Synthesis of suitably protected 3'-nucleotides.

The resulting nucleotides were easily transformed to 3'-phenyl phosphates by treatment with isoamyl nitrite at room temperature and neutrality and with acid to the 5'-free nucleotides. The former compounds could also be synthesized from oNB-nucleosides with p-chlorophenyl phosphate and DCC.

For obtaining suitably protected dimers, the 3'-phosphodiesters were condensed either with G, U or C having free 5'-OHs and fully protected 3'-

phosphates (Fig. 7.). Using slight excesses of 5'-units and mesitylene-
sulfonyl triazolide, the dinucleotides were obtained in high yields between
75-85%. These dinucleotides were analyzed by elemental analysis and enzymatic digestion to give mononucleotides in correct ratios.

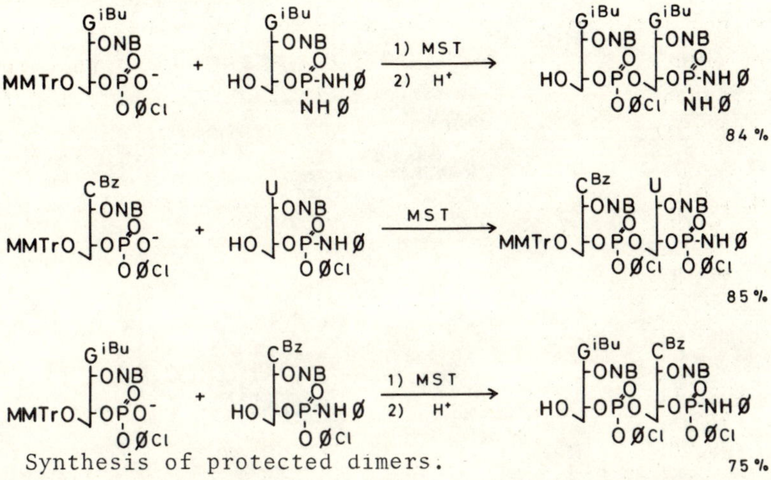

Fig. 7. Synthesis of protected dimers.

These dinucleotides were utilyzed to synthesize a decanucleotide corresponding to bases 11-20 from the 5'-end as shown in Fig. 8. Thus the

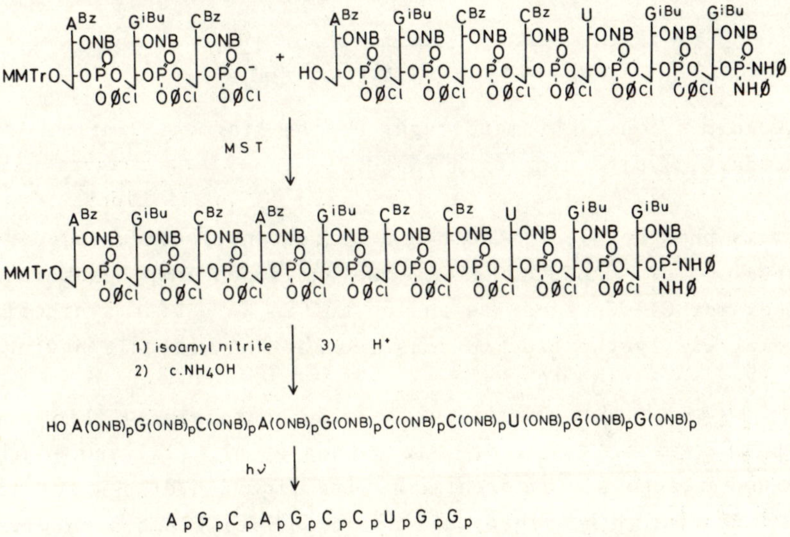

Fig. 8. Synthesis of fragment 2 spanning bases 11-20.

CpUp unit was condensed with the GpGp unit by using MST to give the tetramer
CUGGp in a yield of 60%. The repeating trinucleotide units AGCp was synthesized from GCp and Ap units using MST as the condensing agent in a yield
of 63%. After isoamyl nitrite treatment this trimer was condensed with the
CUGGp unit previously obtained to give a heptamer in a yield of 38%. Repeated condensation of AGCp to the hexamer gave rise to a decanucleotide
AGCAGCCUGGp in a yield of 27% after total removal of the protecting groups
and DEAE-cellulose column chromatography with 7M urea. Thus the present
method has proved suitable for obtaining longer oligonucleotides in reasonable amounts. The product was proved to be structurally correct by RNase

digestion and base sequence analysis using the post-labeling method as shown in Fig. 9.

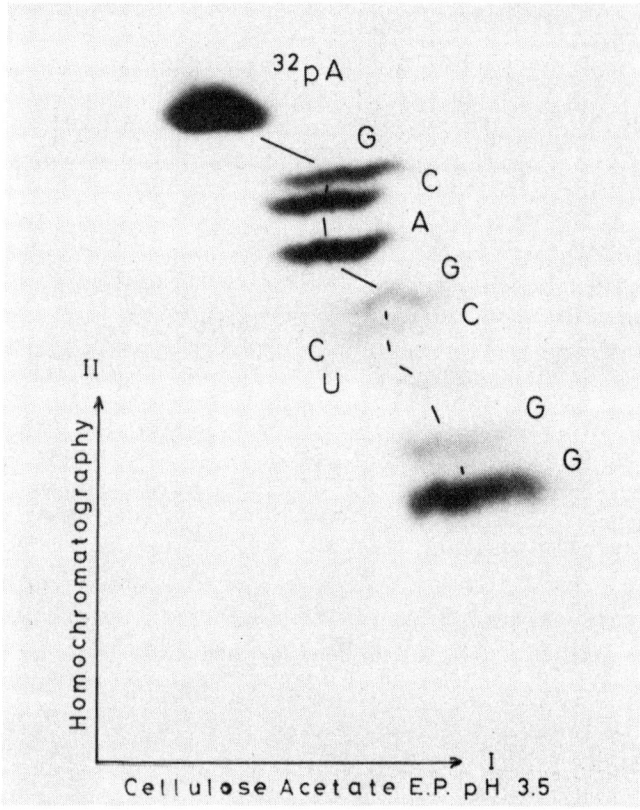

Fig. 9. Two dimensional chromatography of the fragment 2 after labeling at 5'-OH and partial digestion with RNase P1.

Using this phosphotriester method a tetrdecanucleotide adjacent to the former decanucleotide UAGCUCGUCGGGCU, the anticodon loop hexamer CAUAAC, the 3'-end hexamer CAACCA and its analog CGACCA were also synthesized. The latter oligomer was synthesized to construct a tRNA molecule having a "matched" end as shown in Fig. 10.

In order to synthesize an oligonucleotide containing 4-thioU, a pentanucleotide pGGCGG was synthesized. Since Ueda et al. (11) showed that cytidine was converted to s^4U by treatment with liquid hydrogen sulfide, the pentamer was heated with hydrogen sulfide at 30° for 140 hrs to give pGGs^{4}U-GG in a yield of 40% as shown in Fig. 11.

Two pentanucleotides having sequences of 5'-end of the tRNA, UGGCGGp and CGCGGp, which also cound be utilized in ligation to synthesize "matched end" tRNA (see Fig. 10) and a heptanucleotide AUCGAAp, which has the sequence of eukaryotic loop IV were also synthesized (12).

3) Joining of the chemical fragments using RNA ligase

RNA ligase which was first isolated by Hurwitz and coworkers (13) from T4-infected E. coli, seems to be extremely suitable for joining the ribo-oligonucleotide fragments synthesized above.

We tested this enzyme for joining shorter oligonucleotides such as trimers and found that the enzyme is able to join them with some base pre-

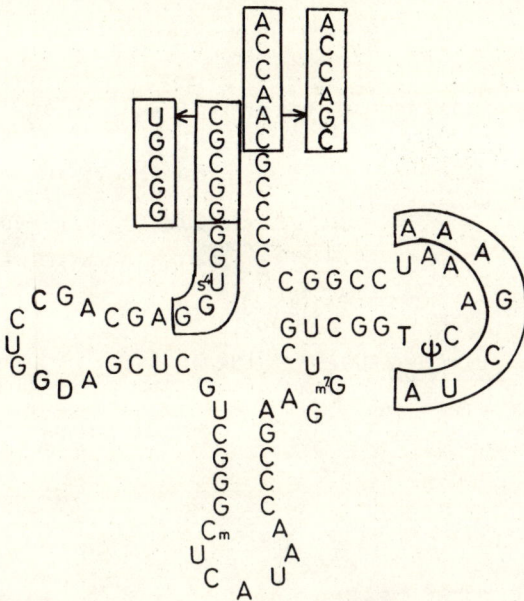

Fig. 10. Illustration of base changes at the 5'- and 3'-end of tRNA$_f^{Met}$. The eukaryotic loop IV sequence and the s^4U containing pentamer were shown.

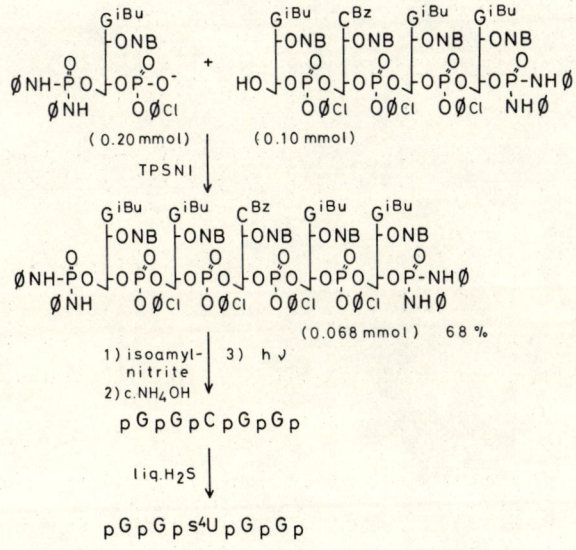

Fig. 11. Synthesis of s^4U containing pentamer.

ferences (14). Furthermore, we observed that an intermediate adenylated 5'-oligonucleotide was accumulated in poor yield reactions (15).

This enzyme was exploited to synthesize the 5'-end eicosanucleotide as shown in Fig. 12. Thus the chemically synthesized fragment 1-b was phosphorylated at 5'-OH by using [γ-^{32}P]-ATP and polynucleotide kinase and joined to 1-a in the conditions as shown in Fig. 12. The ligation proceeded in a yield of 36%. The synthetic fragment 2 was again phosphorylated with ^{32}P and allowed to react with the decamer 1 to yield an eicosamer, CGC-GGGGUGGAGCAGCCUGGp in a yield of 23%. The final product was isolated by gel electrophoresis either on a slab gel or preparative disks. The chain length of this product was confirmed by a 20% polyacrylamide gel electrophoresis as shown in Fig. 13.

Synthesis of an eicosanucleotide from E.coli tRNA$_f^{Met}$ 5'-end with RNA ligase

CGCG + GGGUGG
1-a   1-b
    ↓
  *pGGGUGG

CGCGGGGUGG   AGCAGCCUGGp
1   36%   2
                ↓
             *pAGCAGCCUGGp
              ↓
    CGCGGGGUGGAGCAGCCUGGp
            23%

reaction condition

acceptor	100-200 μM
donor	100 μM
HEPES-NaOH (pH8.3)	50 mM
ATP	200 μM
$MgCl_2$	10 mM
DTT	10 mM
BSA	10 μg/ml
RNA ligase	140 u/ml

25°C, 0.5-2 hrs.

Fig. 12.

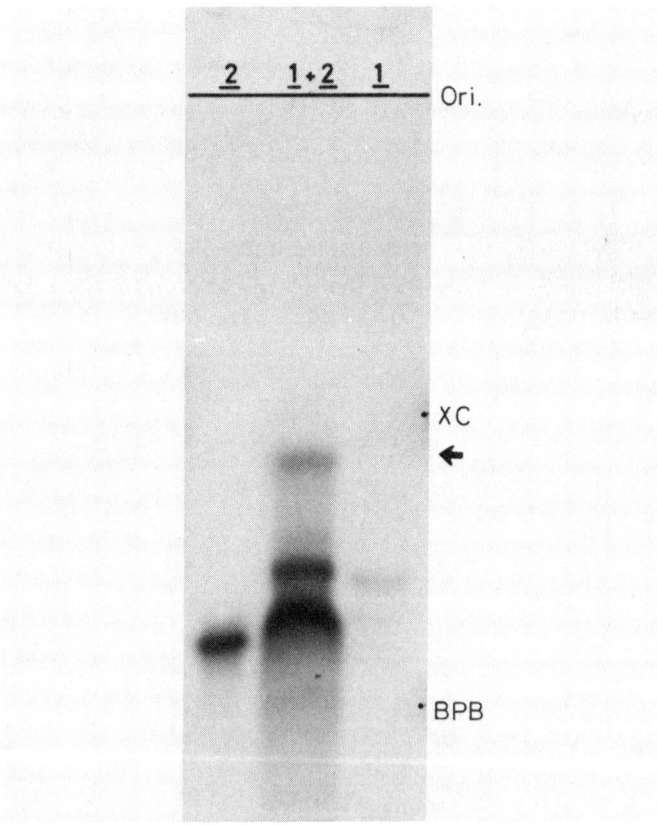

Fig. 13. Acrylamide gel electrophoresis of synthesis of the eicosamer corresponding to bases 1-20, which is indicated by an arrow.

The structure of this eicosamer was further supported by reconatitution experiments. As shown in Fig. 14, when the eicosamer was incubated with purified methionyl-tRNA synthetase in the presence of a 3/4 molecule obtained from the intact tRNA by partial digestion with RNase T1 (16), amino acid accepting activity of 84% with respect to the intact tRNA$_f^{Met}$ was observed. Thus the eicosanucleotide was proved to completely replace natural 1/4 molecule. The fact that even 5'-end decanucleotide showed 11%

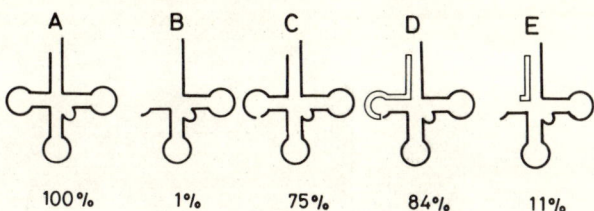

Fig. 14. Schematic representation of the reconstitution experiments. Heavy lines indicated natural fragments and boxes indicated synthetic fragments. Amino acid acceptor activity was presented as % activity relative to the intact tRNA.

may be interesting.

A longer fragment corresponding to bases 35-60 (26-mer) was further synthesized by essentially the same method. As shown in Fig. 15, a dodecanucleotide was first synthesized by joining CAUAAC and ^{32}pCCGAAG in a yield of 39%. GUC and ^{32}pGUCGG, as well as UUC and ^{32}pAAAp, were joined in yields of 40 and 56%, respectively. UUCpAAAp, thus obtained, was kinated and joined with GUCpGUCGG in 36% yield. Finally the former dodecamer was joined to the tetradecamer to give a 26-mer in a yield of 52%. Thus we were able to obtain an oligonucleotide spanning from the anticodon loop to the TѰC loop. A tetradecamer spanning bases 21-34 was also synthesized from fragments 3, 4 and 5 (Fig. 3) using RNA ligase. In this case, However, the joining reaction proceeded in rather low yield.

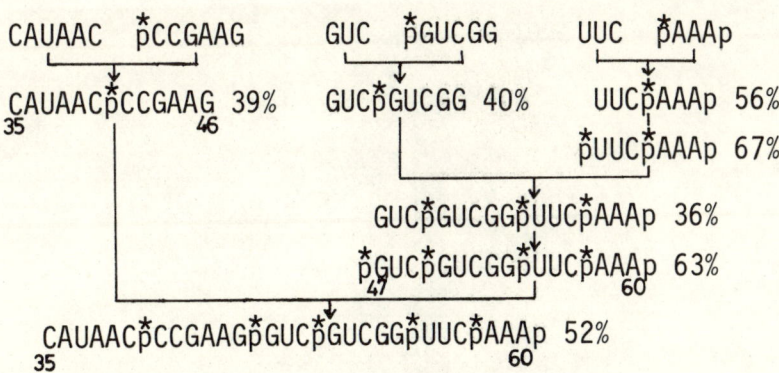

Fig. 15. Synthesis of a hexaeicosanucleotide (bases 35-60).

The joining of the 5'-end eicosamer and the adjacent tetradecamer (Fig. 16, 1 and 2) proceeded rather smoothly and gave rise to the 5'-half molecule in a yield of 39%. By the slab gel electrophoresis we could detect the product at correct position as shown in Fig. 17.

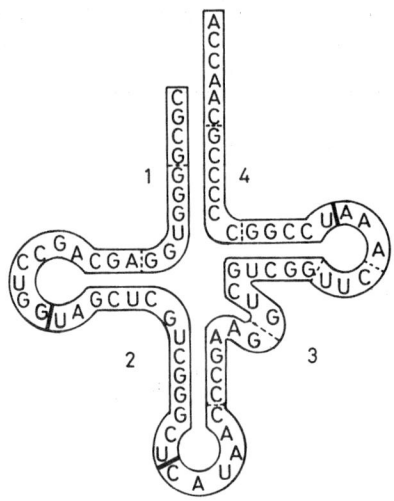

Fig. 16. A nascent strand of tRNA$_f^{Met}$. Four large fragments were joined at the position of solid lines.

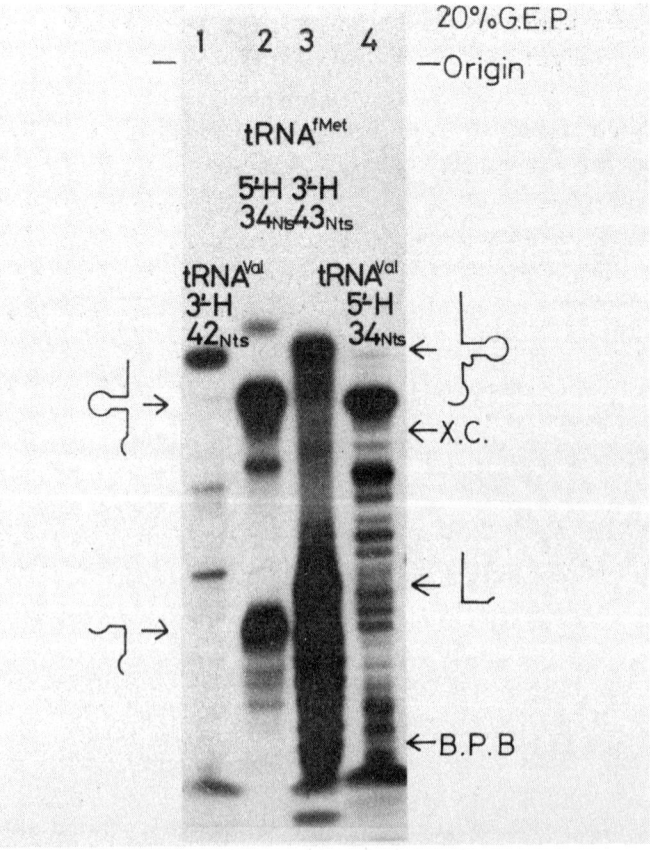

Fig. 17. Radioautogram of slab gel electrophoresis of synthesis of the 5'-half and 3'-half molecules. The half molecules obtained by limited digestion of tRNA^{Val} were used as references.

The 3'-half molecule was also synthesized by the joining of the hexaeicosanucleotide (3) and the heptadecanucleotide (4), though the yield was low. The product was detected on gelelectrophoresis as shown in Fig. 17. When migration distance of these half molecules were plotted against logarithm of the molecular weight, a straight line was observed, which proves that the chain length of these half molecules might be correct (Fig. 18).

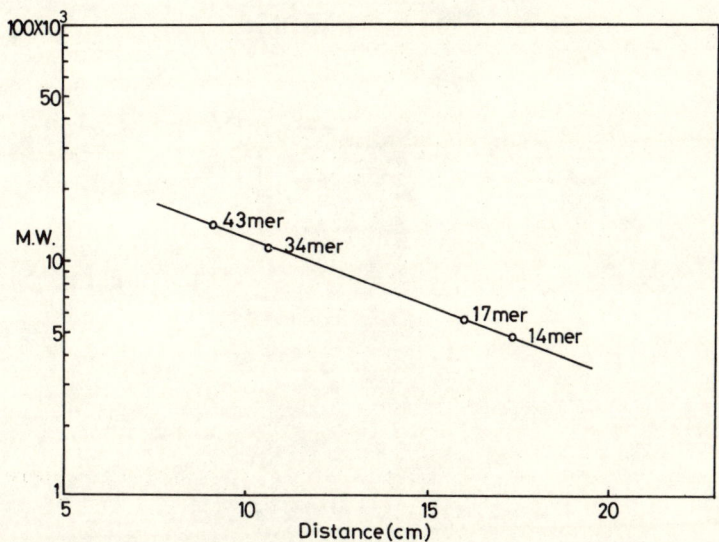

Fig. 18. A plot of migration distance of oligonucleotides against logarithm of molecular weight.

Finally we incubated two half molecules obtained above to synthesize the total molecule of the tRNA. As shown in the 20% polyacrylamide gel electrophoretogram (Fig. 19), the joined product was detected at the position of same migration distance with the intact $tRNA_f^{Met}$. Although the yield was very low, we could thus obtain the total molecule of a nascent strand of the $tRNA_f^{Met}$ from E. coli. The density scanning pattern of this electrophoretogram is shown in Fig. 20 (lower half), which indicated the product is clearly resolved. When we plot the molecular weight of the product together with the starting materials against the migration distance, a straight line was obtained. Thus the chain length of the total molecule was confirmed.

We are currently conducting sequence analysis of this tRNA and testing the amino acid acceptor activity of this specimen. Incorporation of various altered oligonucleotide strands into proper positions to investigate the structure-function relationship is now in progress.

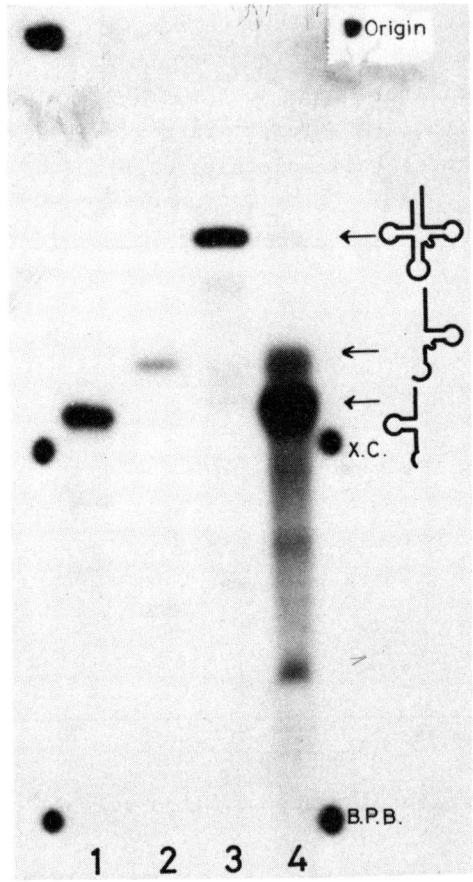

Fig. 19. Electrophoretogram of synthesis of the total tRNA molecule.

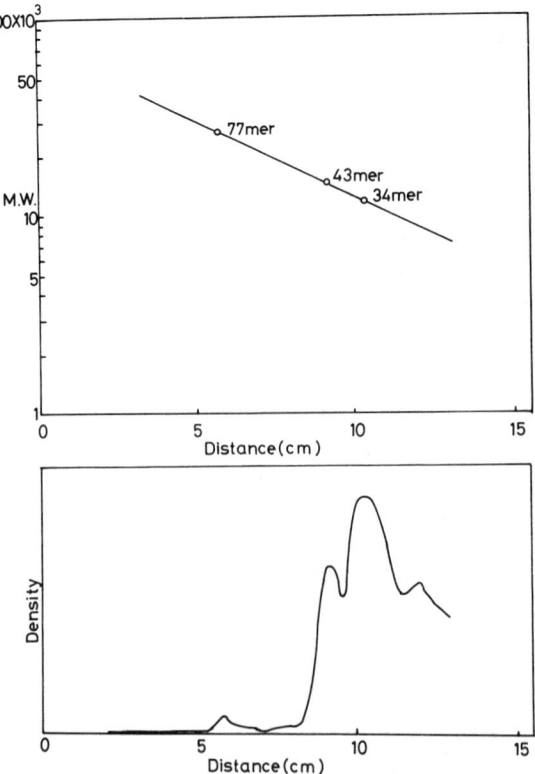

Fig. 20. Molecular weight-migration distance plot and density scanning.

REFERENCES

1. R. W. Holley, J. Apgar, G. A. Everett, J. T. Madison, M. Marquisee, S. H. Merrill, J. R. Penswick and A. Zamir, Science 147, 1462-1465 (1965).
2. E. Ohtsuka, M. Ubasawa, S. Morioka and M. Ikehara, J. Am. Chem. Soc. 95, 4725-4733 (1973).
3. S. Corey, K. A. Marcker, S. K. Dube and B. F. C. Clark, Nature 220, 1039-1040 (1968).
4. E. Ohtsuka, T. Miyake and M. Ikehara, Chem. Pharm. Bull. 27, 341-345 (1979).
5. E. Ohtsuka, E. Nakagawa, T. Tanaka, A. F. Markham and M. Ikehara, Chem. Pharm. Bull. 26, 2998-3006 (1978).
6. S. Uesugi, S. Tanaka, E. Ohtsuka and M. Ikehara, Chem. Pharm. Bull., 26, 2396-2406 (1978).
7. A. F. Markham, T. Miyake, E. Ohtsuka and M. Ikehara, Heterocycles 8, 229-235 (1977).
8. E. Ohtsuka, S. Tanaka, T. Tanaka and M. Ikehara, Chem. Pharm. Bull. 25, 949-959 (1977).
9. E. Ohtsuka, S. Tanaka and M. Ikehara, Nucleic Acids. Res. 1, 1351-1357 (1974); idem., Synthesis 453-454 (1977).
10. E. Ohtsuka, T. Tanaka, T. Wakabayashi, Y. Taniyama and M. Ikehara, J. Chem. Soc. Chem. Comm. 824-825 (1978).
11. T. Ueda, M. Imazawa and K. Miura, Tetrahed. Lett. 2507-2511 (1971).
12. E. Ohtsuka, S. Tanaka and M. Ikehara, J. Am. Chem. Soc. 100, 8210-8213 (1978).
13. R. Silver, V. G. Malathi and J. Hurwitz, Proc. Nat. Acad. Sci. U. S. A. 69, 3009-3013 (1972).
14. E. Ohtsuka, S. Nishikawa, R. Fukumoto, S. Tanaka, A. F. Markham, M. Ikehara and M. Sugiura, Eur. J. Biochem. 81, 285-291 (1976).
15. E. Ohtsuka, S. Nishikawa, M. Sugiura and M. Ikehara, Nucleic Acids Res. 3, 1613-1623 (1976).
16. T. Seno, M. Kobayashi and S. Nishimura, Biochim. Biophys. Acta 190, 285-303 (1969).
17. E. Ohtsuka, S. Nishikawa, A. F. Markham, S. Tanaka, T. Miyake, T. Wakabayashi, M. Ikehara and M. Sugiura, Biochemsitry 17, 3894-4899 (1978).

THE SYNTHESIS OF A POTENTIALLY METABOLICALLY-LABILE DERIVATIVE OF 2'-DEOXY-5-FLUOROURIDINE 5'-PHOSPHATE

R. N. Hunston, A. S. Jones and R. T. Walker

Department of Chemistry, University of Birmingham, Birmingham B15 2TT, U.K.

Abstract - 2'-Deoxy-5-fluoro-5'-O-(tetrahydro-2H-1",3",2"-oxaphosphorine 2"-oxide-2"-yl)uridine (11) has been synthesised by two routes from 2'-deoxy-5-fluorouridine. The product, after metabolic hydroxylation, is a potential source of 2'-deoxy-5-fluorouridine 5'-phosphate, the active metabolite of the anticancer drug 2'-deoxy-5-fluorouridine and has been tested in vivo against Sarcoma S-180 in mice.

Tumour cells can be regarded as mutants of normal cells which can coexist in the same host, but in general no biochemical differences can be shown between these cells. Effective cancer chemotherapy can be defined as a selective cytotoxicity of anticancer drugs for the tumour cells as compared to vital normal cells in the same host such that the tumour cell population is reduced to less than the number of cells necessary to establish the fatal disease. This number is not known, but it can be shown that by implantation of lymphatic leukaemias L1210 or P388 in mice, the life span of the animals is directly related to the number of leukaemia cells implanted and this relationship holds from $10^7 \rightarrow 1$ cell (Ref. 1).

Thus one requires a basis for selective anticancer activity and in the nucleoside analogue field where one is necessarily concerned with antimetabolites which will interfere with anabolic reactions necessary for cell multiplication in all cells, one requires some method of targetting or selection for tumour cells. Among the factors which can be considered are: (i) preferential transport into the neoplastic cell, (ii) preferential degradation of the drug by normal tissues, (iii) preferential activation by the neoplastic cell, (iv) preferential interaction with a biochemical target in the neoplastic cell, (v) preferential activation of the drug by the conditions of lower pH in some cancer cells, and (vi) selectivity by controlling the drug metabolism by taking advantage of the hypoxic cells often found in solid tumours (Ref. 2).

One of the most widely used antineoplastic drugs is 5-fluorouracil (I) and although this is not a nucleoside, its major effect is exerted because of its conversion to 2'-deoxy-5-fluorouridine 5'-phosphate (Ref. 2). It is a primary drug for the treatment of solid tumours, particularly breast, ovarian and gastrointestinal carcinomas but is ineffective in the treatment of leukaemia and lymphomas. The deoxynucleoside (Ref. 3) has never achieved wide clinical application as it has no superior toxicological or therapeutic properties when compared with the free base (Ref. 3). 2'-Deoxy-5-fluorouridine 5'-phosphate (FdUMP) is

1. R = H
2. R = 2'-deoxyribosyl 5'-phosphate

3. $R_1 = R_2 = H$
12. $R_1 = H, R_2 = $ Trityl

4.

ineffective as it does not cross the cell membrane but the nucleoside analogue Ftorafur (Ref. 4) (1-[tetrahydro-2-furanyl]-5-fluorouracil) (4) does show some interesting pharmacological properties and is in clinical use particularly in the Soviet Union, where it was first synthesised, and in Japan.

The metabolism of 5-fluorouracil proceeds in two primary directions to produce 2'-deoxy-5-fluorouridine 5'-phosphate and 2'-deoxy-5-fluorouridine (FUDR). The former compound is a potent inhibitor of thymidylate synthetase, the enzyme responsible for catalysing the conversion of 2'-deoxyuridine 5'-phosphate (dUMP) to 2'-(deoxy)thymidine 5'-phosphate and hence the synthesis of DNA ceases (Ref. 5) although this may not account for all the tumour cell inhibition caused by 5-fluorouracil. The reaction catalysed by this enzyme involves the transfer and subsequent reduction of a one-carbon unit from the co-factor N^5-, N^{10}-methylenetetrahydrofolate to the 5-position of dUMP. The drug methotrexate inhibits this enzyme indirectly by blocking dihydrofolate reductase and hence limiting the availability of the cofactor but FdUMP and N^5, N^{10}-methylenetetrahydrofolate form a ternary complex with the thymidylate synthetase. Normally a hydrogen atom is abstracted from C-5 of the uracil moiety of dUMP but the marked stability of the C-F bond present in FdUMP prevents the reaction (Ref. 6).

The first chemical synthesis of 5-fluorouracil was performed by Heidelberger in 1957 by constructing the pyrimidine ring using ethyl fluoroacetate (Ref. 7). The deoxynucleoside was originally prepared enzymatically (Ref. 8) but the patent for its commercial production from the base and suitably protected sugar derivative was granted in 1960 (Ref. 9). In 1971, 5-fluorouracil and FUDR were synthesised by Robins and Naik directly from uracil and 2'-deoxyuridine respectively by the action of trifluoromethylhypofluorite (Ref. 10). More recently, 5-fluorouracil has been prepared by the action of elemental fluorine on uracil in hydrogen fluoride in the presence of an aliphatic carboxyiic acid as a nucleophile (Ref. 11).

There are several potential ways in which the use of 5-fluorouracil could be extended and improved. One would be to produce a derivative which enables the concentration of the drug in the bloodstream to remain at a significant level for a reasonable length of time. After intravenous administration, the plasma clearance of 5-fluorouracil is very rapid (half-life 11 - 20 min) (Ref. 5). Following systemic administration, as much as 80% of the drug is eliminated through hepatic metabolic degradation and urinary excretion during the first few hours. Ftorafur however is removed more slowly, having an observed half-life of 5 hours and appreciable levels of 5-fluorouracil can be detected even 24 hours after a single dose of Ftorafur has been administered (Ref. 5).

Another method of improving the efficacy of 5-fluorouracil would be to try to counteract the inevitable appearance of resistant tumours. It has been shown that 5-fluorouracil-resistant tumours can have one of several altered biochemical characteristics, including decreased activities of uridine phosphorylase or uridine kinase or pyrimidine phosphoribosyl transferase. Other resistant cell lines have been found which have a less sensitive thymidylate synthetase (Ref. 12) and require much higher levels of FdUMP before they are inhibited. Many of these resistant lines could be attacked if only the active species, FdUMP, itself could be used but this cannot cross the cell membrane.

Attempts to synthesise analogues of FUDR which overcome some of these problems have met with little success although one of the first analogues to be synthesised, 2'-deoxy-5-trifluoromethyluridine (5) is a potent inhibitor of thymidylate synthetase (Ref. 13) and also possesses useful antiviral properties (Ref. 14). Its mode of action has still to be fully resolved (Ref. 12). More typical is the case of 3-deaza-2'-deoxy-5-fluorouridine (Ref. 15) (6) which does not inhibit thymidylate synthetase at all. Hydrolysis of the glycosyl moiety has been overcome by replacing 2-deoxyribosyl with arabino- (Ref. 16), lyxo- (Ref. 17) and 2'-deoxy-2'-fluoro-ribofuranosyl (Ref. 18) derivatives but they possess no appreciable therapeutic activity.

5. R = 2'-deoxyribosyl

6. R = 2'-deoxyribosyl

Other derivatives such as 2',3'-dideoxy-2',3'-dehydro-5-fluorouridine (Ref. 7) and 2',3'-dideoxy-5-fluorouridine (Ref. 19) (8) demonstrated appreciable activity but it is now thought that these compounds are cleaved non-enzymatically to 5-fluorouracil and therefore are serving as rather expensive depots of this latter compound (Ref. 20). This is thought to be the case with Ftorafur (4) but this has the advantage of being cheap to synthesise, is more lipid soluble than 5-fluorouracil and thus may cross the blood-brain barrier more easily and also release 5-fluorouracil over a period of time (Refs. 21 & 22).

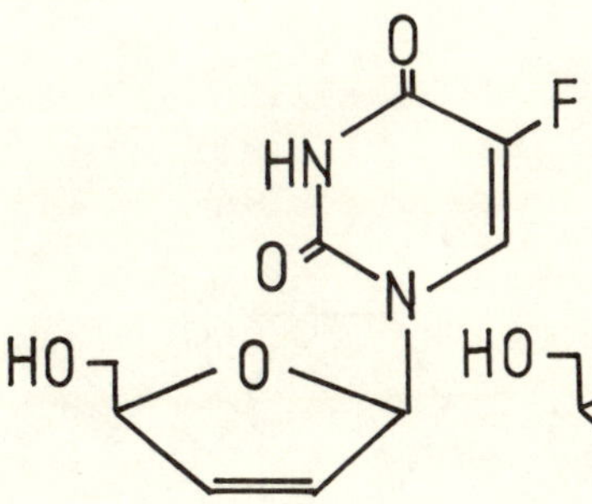

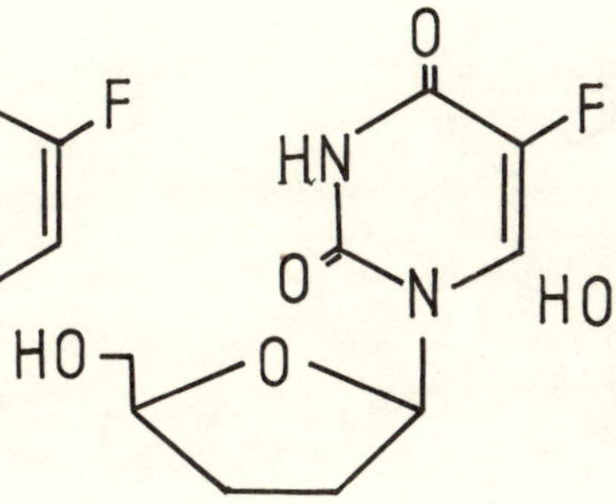

Thus it was the intention of the present work to synthesise chemically a derivative of FdUMP which might cross the cell membrane and then act as a depot of the active species, FdUMP, such that not only would many 5-fluorouracil-resistant cell lines be susceptible but also by encouraging a steady production of FdUMP from the depot analogue over a period of time, the drug would also be more effective.

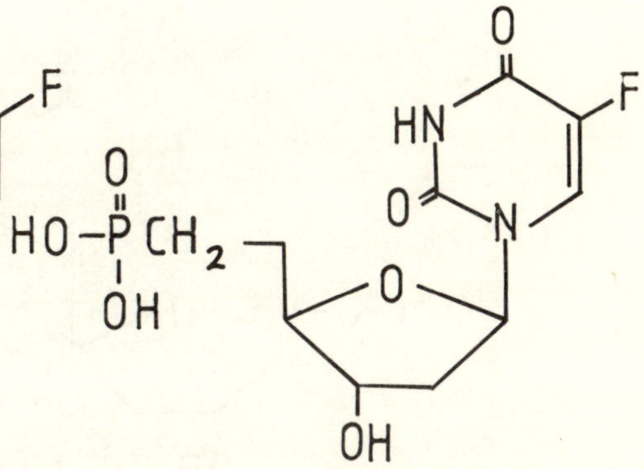

Many derivatives of FdUMP have been synthesised in such an attempt to deliver the active antimetabolite to the required site but none of a large group of phospho di- and tri-esters, amides and basic exters (Ref. 23) were more active than FdUMP itself (Ref. 24). One compound recently synthesised however which does seem to have some activity is the phosphonate analogue 1-(2,5,6-trideoxy-β-$\underline{D}$-ribohexofuranosyl)-5-fluorouracil 6'-phosphoric acid (Ref. 25) (9).

Cyclophosphamide (10) is a well known antineoplastic agent which provides therapeutic activity against a relatively broad spectrum of human cancers (Ref. 26). In vitro the compound is inactive and it requires an initial activation step of hydroxylation by the liver microsomal oxidase at the C-4 position (Ref. 27). This metabolite is active in vivo and in vitro. It is proposed that the cytotoxic selectivity of the drug towards cancerous as opposed to normal cells may result from rate inequalities between the competing enzymatic detoxofication and non-enzymatic toxification pathways, the latter leading to acrolein and the formation of the alkylating agent nitrogen mustard phosphoric acid diamide as shown below (Ref. 28).

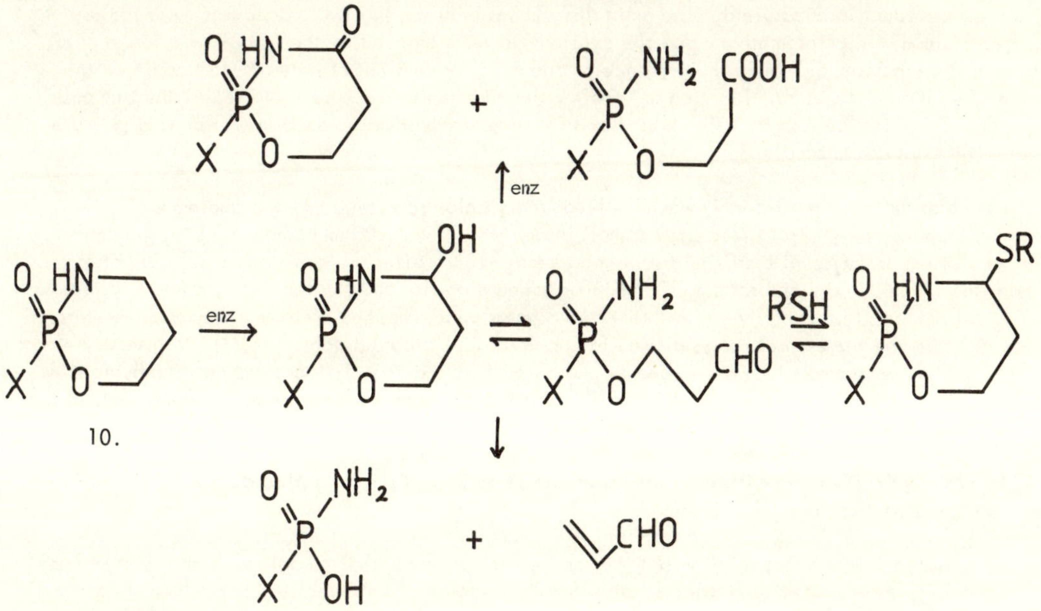

$X = N(CH_2CH_2Cl)_2$

Thus it was our intention to use the oxaphosphorine ring system to act as a transport system for the charged FdUMP which would enable the neutral analogue to cross the cell membrane and then to be enzymatically hydroxylated to the labile 4-hydroxy derivative which would then liberate the active antimetabolite FdUMP. The compound whose synthesis is required is thus 2'-deoxy-5-fluoro-5'-O-(tetrahydro 2H, 1", 3", 2"-oxaphosphorine-2"-oxide-2"-yl)uridine (11) (CPFUDR).

11. $R_1 = H$
15. $R_1 = CH_3CO-$

16.

The compound (11) was synthesised by two routes. 2'-Deoxy-5-fluorouridine (3) was tritylated to give 12 which was then acetylated (13) and detritylated to give 3'-O-acetyl-2'-deoxy-5-fluorouridine (14) in an overall yield of 40%. This was then reacted with phosphoryl chloride and N-methylmorpholine in 1,2-dimethoxyethane at 0° and the reaction allowed to warm up to room temperature. The reaction was monitored by TLC and when all the starting material had disappeared, the mixture was recooled to 0° and a solution of 3-aminopropan−1-ol in 1,2-dimethoxyethane, added. The solution was allowed to stand for 12h at room temperature, filtered, the solvent removed under reduced pressure and the residue separated on a silica column to give what was assumed to be an inseparable mixture of diasteriomers (15) in 13% yield. The 3'-O-acetyl group could be removed with methanolic ammonia to give the required product. The overall yield was only 5.5% and was not suitable for the production of sufficient material for screening purposes.

The 2-stage construction of the oxaphosphorine ring was achieved in only 20% yield and it was decided to preform this ring and to then transfer it to the 5'-position of the nucleoside. The compound required (16) whose preparation had been described by Iwamoto et al. (Ref. 29) was synthesised although many attempts following the published details failed. An alternative unpublished procedure (Ref. 30) was however found to give a pure product in good yield. The main differences between the two procedures were the use of chloroform instead of dichloromethane and the cyclization was achieved by the initial reaction of the hydroxyl of 3-aminopropan-1-ol in the presence of one molar equivalent of base at -15° and then the subsequent reaction of the amino function at 0° after the addition of a further molar equivalent of base. The product could then be recrystallised and stored for several months at room temperature and pressure without significant decomposition.

Phosphoryl chloride (76.7 g, 0.5 mol) was dissolved in dry chloroform (400 ml) and cooled to -15°. A solution of 3-aminopropan-1-ol (37.5 g, 0.5 mol) in dry chloroform (100 ml) containing triethylamine (70 ml) was added dropwise with stirring to the mixture at -10°. After the addition, a solution of triethylamine (70 ml) in dry chloroform (100 ml) was added dropwise and the temperature was kept below 0°. The solution was left at 0° overnight, taken to dryness under reduced pressure, the residue extracted with dry acetone and the extract recrystallised from chloroform carbontetrachloride, 1:3 to give a yield of 38 g (49%). This reagent could now be used to phosphorylate 2'-deoxy-5-fluorouridine in anhydrous pyridine. Phosphorylation only occurred at the 5'-position and a product identical in all respects to that previously described was isolated in 36% yield.

The biological results obtained with this compound compared with those for 5-fluorouracil (FU) and FUDR under the same conditions are given below.

These results are somewhat disappointing but it is as well to remember that no animal tumour system has been objectively demonstrated to predict reliably for the response of any specific human tumour to drug treatment (Ref. 1) and further work to produce more material for testing and the synthesis of other analogues is in progress.

TUMOUR	CPFUDR mmol/Kg	act.*	FU mmol/Kg	act.	FUDR mmol/Kg	act.
S-180	2.74	2.0	0.77	2.3	2.03	9.6
			1.54	2.9	6.10	13.2
			2.31	4.4	8.14	4.5

* Tumour weight of control/tumour weight of treated mice

REFERENCES

1. F.M. Schabel, Nucleoside Analogues: Chemistry, Biology and Medical Applications, R.T. Walker, E. De Clercq and F. Eckstein eds., Plenum, New York (1979).
2. W.H. Prusoff and P.H. Fischer, in ref. 1.
3. A. Rossi, in ref. 1.
4. S.A. Hiller, R.A. Zhuk and M. Lidak, Amer. Chem. Soc. Abs. 73, 109793 (1970).
5. B.A. Chabner, C.E. Meyers and V.T. Oliverio, Semin. Oncol. 4, 165 (1977).
6. P.V. Danenberg, Biochim. Biophys. Acta 473, 73 (1977).
7. R. Duschinsky, E. Pleven and C. Heidelberger, J. Amer. Chem. Soc. 79, 4559 (1957).
8. C. Heidelberger, Progress in Nucleic Acid Res. and Molec. Biol. 4, 1 (1965).
9. M. Hoffer, U.S. Patent 2, 949, 451 (August 16, 1960).
10. M.J. Robins and S.R. Naik, J. Amer. Chem. Soc. 93, 5277 (1971).
11. S. Misak, S. Ishii and T. Takohara, Amer. Chem. Soc. Abs. 89, 129535 (1978).
12. C. Heidelberger, Antineoplastic and Immunosuppressive Agents, Springer, Berlin, 2, 193 (1975).
13. C. Heidelberger, D. Parsons and D.C. Remy, J. Amer. Chem. Soc. 84, 3597 (1962).
14. H.E. Kaufman and C. Heidelberger, Science 145, 585 (1964).
15. S. Nesnow and C. Heidelberger, J. Heterocyclic Chem. 10, 779 (1973).
16. N.C. Yung, J.H. Burchenal, R. Fecher, R. Duschinsky and J.J. Fox, J. Amer. Chem. Soc. 83, 4060 (1961).
17. J.J. Fox and N.C. Miller, J. Org. Chem. 28, 936 (1963).
18. J.F. Codington, I.L. Doerr and J.J. Fox, J. Org. Chem. 29, 558 (1964).
19. T.A. Khwaja and C. Heidelberger, J. Med. Chem. 10, 1066 (1967).
20. R.J. Kent and C. Heidelberger, Biochem. Pharmacol. 19, 1095 (1970).
21. A.M. Cohen, Drug Metab. Disposit 3, 103 (1975).
22. C.E. Myers, R. Diasio, H.M. Eliot and B.A. Chabner, Cancer Treat. Rev. 3, 175 (1976).
23. D.C. Remy, A.V. Sunthankar and C. Heidelberger, J. Org. Chem. 27, 2491 (1962).
24. K.L. Mukherjee and C. Heidelberger, Cancer Res. 22, 815 (1962).
25. J.A. Montgomery, H.J. Thomas, R.L. Kisliuk and Y. Gaumont, J. Med. Chem. 22, 109 (1979).
26. H. Arnold and F. Bourseaux, Angew. Chem. 70, 539 (1958).
27. M. Colvin, C.A. Padgett and C. Feuselan, Cancer Res. 33, 915 (1973).
28. P.J. Cox, B.J. Philips and P. Thomas, Cancer Treat. Rep. 60, 321 (1976).
29. R.H. Iwamoto, E.M. Acton, L. Goodman and B.R. Baker, J. Org. Chem. 26, 4743 (1961).
30. E. Barber, personal communication.

SYNTHESIS AND BIOLOGICAL ACTIVITY OF SOME ANALOGUES OF NUCLEIC ACIDS COMPONENTS

A. Holý

Institute of Organic Chemistry and Biochemistry, Czechoslovak Academy of Sciences, 166 10 Praha 6, Czechoslovakia

<u>Abstract</u> - Three novel types of nucleotide analogues have been synthetized and investigated in vitro, namely, (a) inosine 2´(3´)-phosphate derivatives containing 5´-O-linking group for the binding to Sepharose supports, (b) 5´-O-phosphorylmethyl derivatives of ribonucleotides which are effective competitive inhibitors of some 5´-nucleotidases and (c) phosphomonoester and phosphodiester derivatives of four 9-(2,3,4-trihydroxybutyl)adenines; ribonuclease T2 cleaves 2´,3´-cyclic phosphodiesters with absolute configuration 2S.

Nucleotide analogues modified both in the alcoholic part of the molecule and in the phosphorus moiety are useful tools for investigation and isolation of enzymes connected with nucleic acids metabolism. The present contribution describes three novel types of such analogues and their possible application for the above purposes.

1. INOSINE 2´(3´)-THIOPHOSPHATE DERIVATIVES FOR AFFINITY CHROMATOGRAPHY

Affinity chromatography is regarded as one of the most efficient methods for the purification of enzymes. Its principle consists in binding of a compound with an increased affinity towards the particular enzyme, to an insoluble support. On the basis of an intermolecular interaction, the protein is tightly bound or retarded on the support column and, after the removal of contaminants, is eluted under the conditions which dissociate the complexes involved (increased ionic strength, pH etc.). For nucleases, satisfactory results have been obtained with thiophosphate analogues as active ligands (1,2). The binding of an analogue to a polymer support requires the presence of an additional grouping of a sufficient length mediating the covalent link; the reaction leading to such a linkage must not affect the inhibitory activity of the ligand, or the properties of the support either.
In our particular case we have been dealing with ribonuclease Streptomyces aureofaciens which is known to split specifically the ester bonds of 3´-guanylic and 3´-inosinic acid (3,4). The synthesis of possible inhibitors was therefore undertaken in the series of inosine derivatives only. It was possible to predict that the 2´(3´)-thiophosphates of this series might exhibit inhibitory activity, suitable for affinity chromatography. Keeping the hypoxanthine moiety intact (preliminary studies have shown that any change of this part results in loss of affinity) (4), the only other position left for the link to the support is the 5´-position. We have chosen 5´-O-carboxymethylinosine derivatives as compounds with a stable (ether) bond to the nucleoside, which offers various methods for binding to the support. Two different types of binding have been employed: (a) binding of an activated carboxymethyl group to the 6-aminohexyl-Sepharose 4A and (b) binding of an aminoalkylamide of the former nucleoside derivative to the cyanogen-bromide-activated Sepharose 4A (Scheme 1). Both types do not practically differ in the character of the link, but they do differ in the length of the spacer group. In the both series, the ligands tested contained both 2´(3´)-phosphate groups and/or thiophosphate groups.

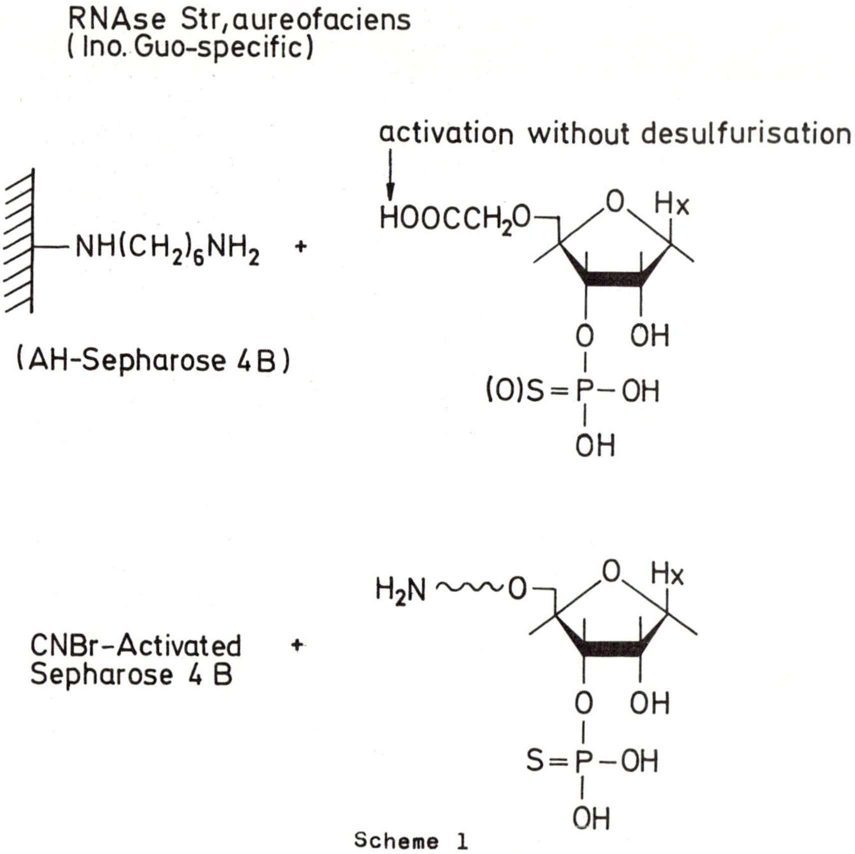

RNAse Str. aureofaciens
(Ino. Guo-specific)

Scheme 1

For comparatory purposes, the derivatives of inosine 5′-phosphate and 5′-thiophosphate linked via the levulinate spacer group (5) were also prepared by the reaction of the corresponding 5′-nucleotides with ethyl orthoformate and ethyl levulinate followed by alkaline hydrolysis (Scheme 2).

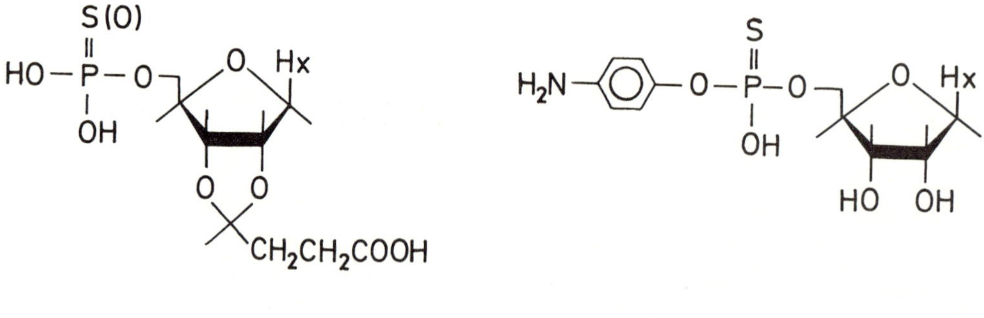

Scheme 2

5′-O-Carboxymethylinosine was prepared by the reaction of 2′,3′-protected inosine with sodium chloroacetate in the presence of sodium hydride (cf.6). The optimum conditions involve the use of two molar equivalents of sodium hydride and dimethyl sulfoxide as a solvent. After removal of the protecting group, the material obtained was purified by anion exchange chromatography. The formation of the N(1)-substituted derivative and/or the 5′,N(1)-disubstituted derivative was minimal under the conditions employed. The 5′-O-carboxymethyl derivative was transformed to a 2′(3′)-phosphite by acid catalyzed transesterification with triethyl phosphite (7) and then to the 2′(3′)-thiophosphate by the elegant method of Hata (8), i.e. reaction with trimethylsilyl chloride and sulfur (Scheme 3).

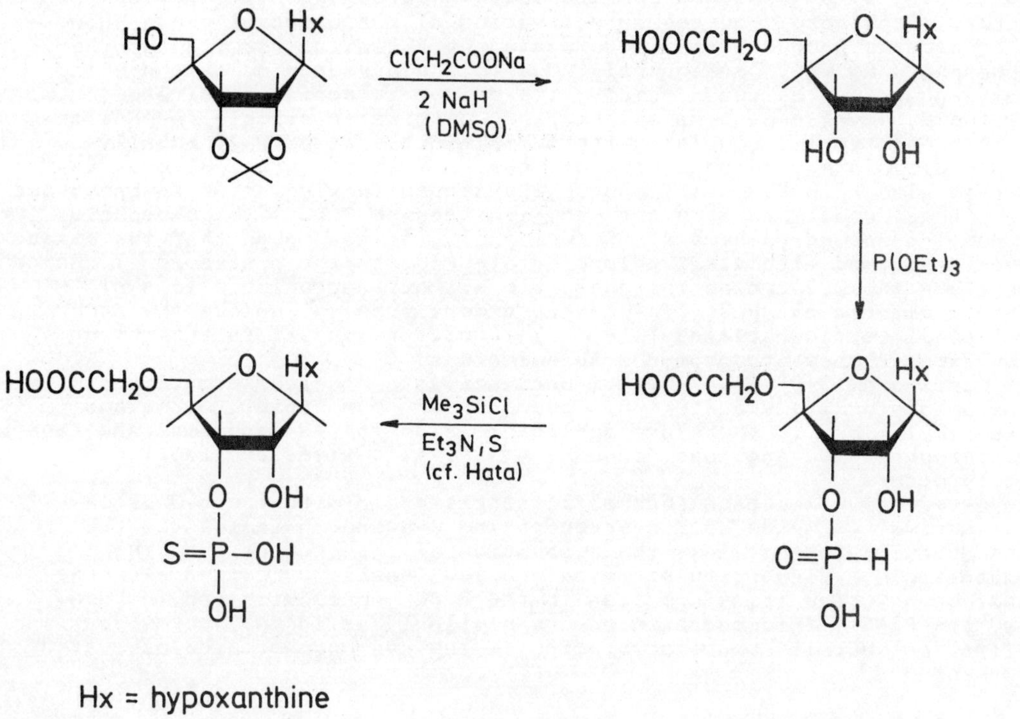

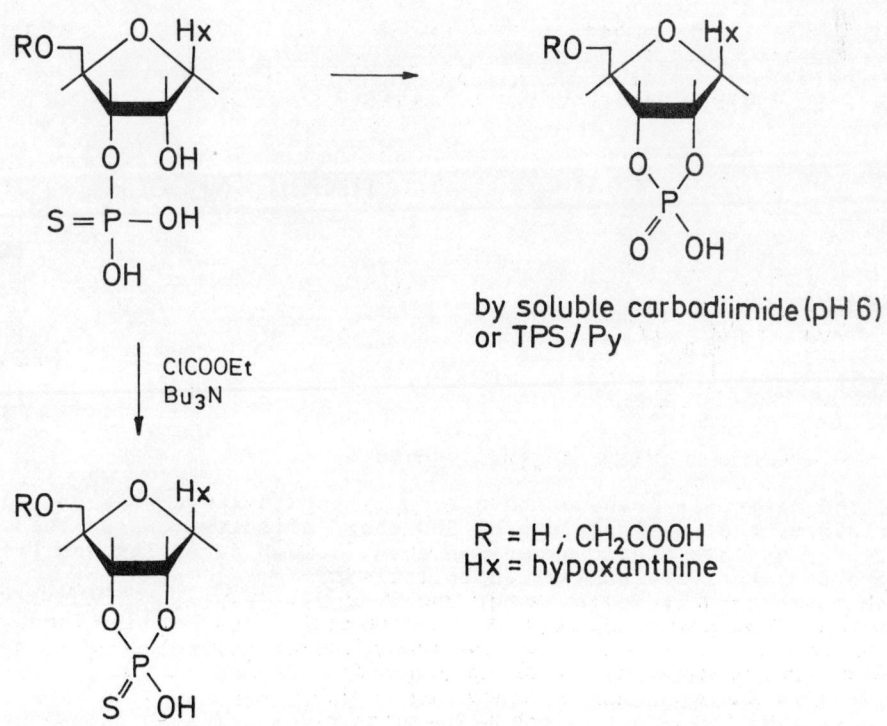

Hx = hypoxanthine

Scheme 3

R = H; CH$_2$COOH
Hx = hypoxanthine

Scheme 4

The binding of an analogue bearing the 5′-O-carboxymethyl function towards aminohexyl-Sepharose requires an activation of the carboxyl group. However, this process is unequivocally accompanied by a similar reaction at the thiophosphate moiety. Consequently, due to the presence of neighbouring hydroxyl group at the sugar moiety, the sulfur is easily eliminated under the 2′,3′-cyclic phosphate formation. With water-soluble carbodiimides, this reaction is complete within few minutes at pH 6-7, equally with N,N-dicyclohexylcarbodiimide or, contrary to the previous observation (9), also with 2,4,6-triisopropylbenzenesulfonylchloride in anhydrous medium. These reactions were checked with inosine 2′(3′)-thiophosphate as a model compound (Scheme 4). Surprizingly, it was found that the mixed-anhydride method with ethyl chloroformate (10) is the procedure of choice which gives the 2′,3′-cyclic thiophosphate without appreciable loss of sulfur (monitored by HPLC); the compound mentioned is, unlike the other guanyl-specific ribonuclease T1 (cf.11), quite resistant to the action of the ribonuclease Streptomyces aureofaciens.

The binding of 5′-O-carboxymethyl derivatives of nucleosides to proteins by the chloroformate procedure has been already described in previous papers (12,13) and is therefore applicable also for AH-Sepharose and the above thiophosphate analogue, though, at the same time, cyclisation probably occurs.

The alternative approach (Scheme 5) starts again with the 5′-O-carboxymethyl derivative which is converted to the 2-aminoethylamide via the p-nitrophenyl ester (cf.14); the separation of the product from the accompanying N,N-disubstituted amide proceeds easily by cation exchange chromatography. The transformation to the 2′(3′)-phosphite and, subsequently, the 2′(3′)-thiophosphate occurs similarly as in the Scheme 4. The final product is ready for binding to the cyanogen-bromide-activated Sepharose.

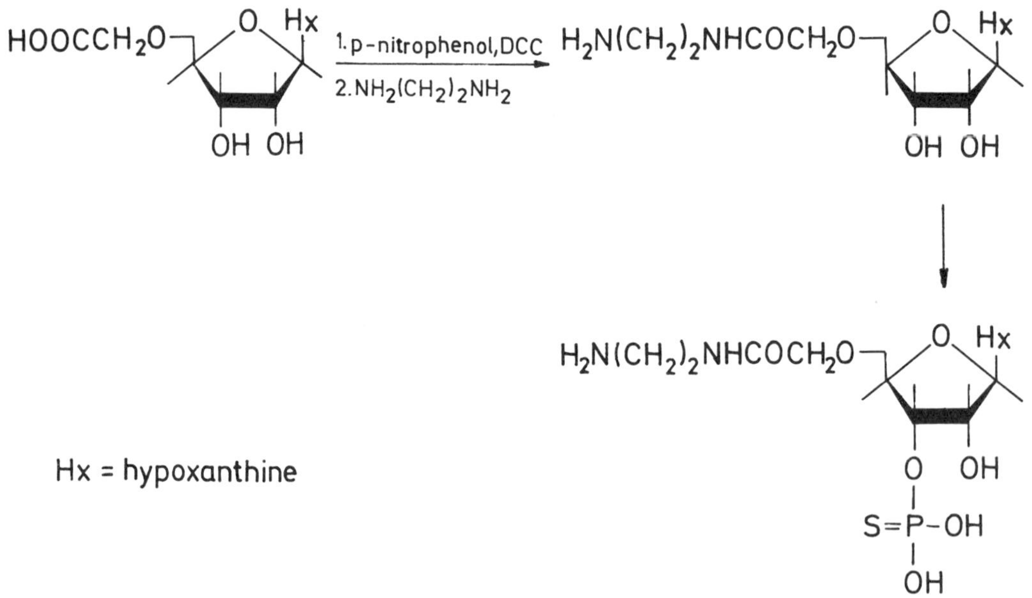

Scheme 5

All the materials prepared have been linked to the corresponding support materials by the usual procedures and their affinity towards the enzyme was tested by both static experiments and column runs. The results obtained are qualitatively summarized in Table 1.

From these data it follows that the ligands containing 2′(3′)-thiophosphate are more effective than 5′-thiophosphate derivatives, both of them being more active than the phosphate containing molecules. It seems that the most active material is the AH-Sepharose loaded by 5′-O-carboxymethylinosine 2′(3′)-thiophosphate (or 2′,3′-cyclothiophosphate). At the same time, the differences mentioned confirm that the affinity cannot be due to the ion exchanger character of the modified Sepharose. However, since the additional factors (retardation of inactive proteins, recovery of enzyme and/or release of contaminating pigments) are of importance in the practical evaluation, the final choice of the best material will await further investigation.

Table 1. Affinity binding of Str.aureofaciens RNAse to the modified Sepharose 4B (pH 6.5-8.0)

Ligand	Sepharose 4B	Binding
5'-CM-Ino-p_s	AH	strong
5'-CM-Ino-p	AH	medium
5'-AECM-Ino-p_s	BrCN	strong
2',3'-Lev-IMP	AH	medium
2',3'-Lev-IMP$_s$	AH	medium
IMP$_s$-(p-aminophenyl)	BrCN	weak

Abbreviations: p_s... 2'(3')-thiophosphate, p... 2'(3')-phosphate, CM... carboxymethyl, AECM... 2-aminoethylaminocarbonylmethyl, Lev... levulinyl, IMP$_s$... inosine 5'-thiophosphate

2. 5'-O-PHOSPHONYLMETHYL DERIVATIVES OF RIBONUCLEOTIDES- A NOVEL TYPE OF 5'-NUCLEOTIDE ANALOGUES

The idea of nucleotide analogues resistant towards the action of dephosphorylating enzymes was followed for many years in various laboratories. The compounds investigated include nucleoside phosphites (7,15-17), alkanephosphonates (18-21), hydroxy-(22) and aminoalkanephosphonates (23) 5'-deoxynucleoside phosphonates (24-29) and the already mentioned thiophosphate analogues (9,11,30-34). Many of these compounds also exhibit significant inhibitory activity towards various enzymes. In the course of our work we have been stimulated by the finding (35) that the response of phosphomonoesterases can be to a large extent affected by the presence or absence of an oxygen atom in the vicinity of phosphorus thus causing a decrease in inhibitory activity between the alkyl(aryl)phosphates and phosphonates. Therefore, we undertook the synthesis of novel nucleotide analogues derived from hydroxymethanephosphonic acid containing the nucleoside moiety bound by a (comparatively) stable ether linkage to the hydroxymethyl group. The reaction scheme for the synthesis (Scheme 6) of the 5'-isomers consists in treatment of ribonucleoside 2',3'-isopropylidene derivatives with disodium salt of chloromethanephosphonic acid. The reaction is best performed with sodium hydride in dimethyl sulfoxide; the compounds containing acid NH-linkage (uracil, hypoxanthine) require excess of sodium hydride to generate the nucleophilic anion of the 5'-hydroxyl group. After removal of the protecting group at pH 2, the products are isolated by the usual ion exchange techniques. The substitution at 5' can be achieved also with cytidine and adenosine derivatives with or without protection of the exo-amino function; however, due to the simultaneous substitution at N(3) or N(1), respectively, some losses of the required products are at hand.

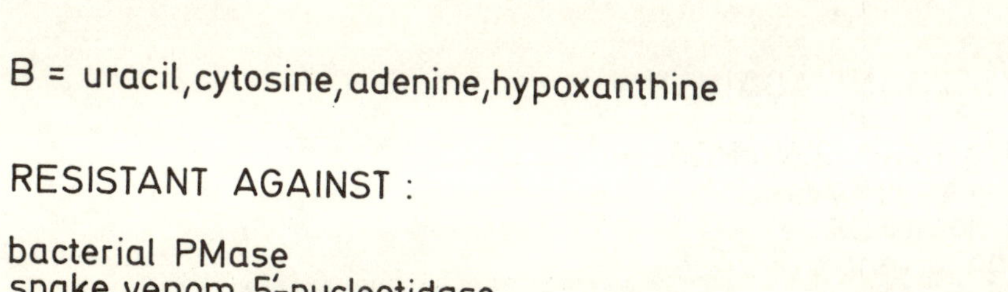

B = uracil, cytosine, adenine, hypoxanthine

RESISTANT AGAINST:

bacterial PMase
snake venom 5'-nucleotidase
bull semen 5'-nucleotidase

Scheme 6

The 5'-ribonucleotide analogues thus prepared have many features in common with their natural counterparts: they possess two dissociable groups important for enzyme recognition and do not differ sterically too much from nucleotides. They are reasonably stable in acidic and alkaline solutions but decompose in boiling acetic acid. The expected stability towards dephosphorylating enzymes (bacterial alkaline phosphatase, intestinal alkaline phosphatase, seminal and snake venom 5'-nucleotidases) was fully confirmed. On the other hand, they behave like very potent competitive inhibitors of the two latter enzymes. The data compiled in Table 2 demonstrate the inhibitory action of three such analogues on seminal and snake venom 5'-nucleotidases (the both enzymes display also significant substrate inhibition; the corresponding data computed from Dixon-plots reflect relative inhibitory effect referred to the concentration of substrate and inhibitor).

In practical terms, an equimolar amount of the analogue fully inhibits the dephosphorylation of UMP under the assay conditions. The experiments on the binding of such nucleotide analogues on affinity support materials are now in progress.

Table 2. Inhibition of UMP dephosphorylation by nucleotide analogues.

5'-O-Phosphorylmethyl derivative of	$(\frac{i}{s})_{v/2}$	
	$s=5{,}65\times 10^{-3}$M	$s=16{,}95\times 10^{-3}$M
Snake venom 5'-nucleotidase		
Uridine	0,723	0,525
Cytidine	2,685	2,491
Adenosine	0,338	0,496
Bull semen 5'-nucleotidase		
Uridine	0,584	0,394
Cytidine	0,763	0,541
Adenosine	0,087	0,089

$(\frac{i}{s})_{v/2}$ ... inhibitor to substrate ratio for $v=v_0/2$, v_0 ... initial rate of splitting in the absence of inhibitor.

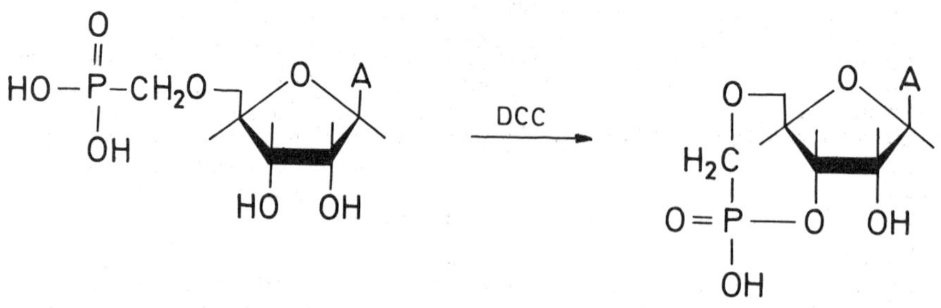

RESISTANT AGAINST:

dog heart 3',5'-cPDAse
spleen cPDAse
pollen cPDAse
snake venom PDAse
T 2 RNAse

Scheme 7

The above nucleotide analogues, (5'-deoxyribonucleoside-5'-yl)oxymethane-phosphonic acids, can be used for further syntheses: the reaction of the adenosine derivative with N,N'-dicyclohexylcarbodiimide in dilute pyridine solution results in cAMP analogue (Scheme 7) containing a seven-membered cyclic phosphonate ring. This compound is quite resistant towards the cAMP-phosphodiesterase from dog heart as well as towards other phosphodiesterases. Its further implications are now under investigation.

Finally, a synthesis of a UpU-analogue was performed via classical phosphodiester approach (Scheme 8): 2,5-di-O-benzoyluridine on condensation with uridine analogue in the presence of TPS, followed by methanolysis, affords the desired compound in a fair yield. Necessarily, the analogue is quite stable not only against ribonucleases (pancreatic RNAse, T2-RNAse), but also against 3'- (calf spleen) or 5'- (snake venom) phosphodiesterases.

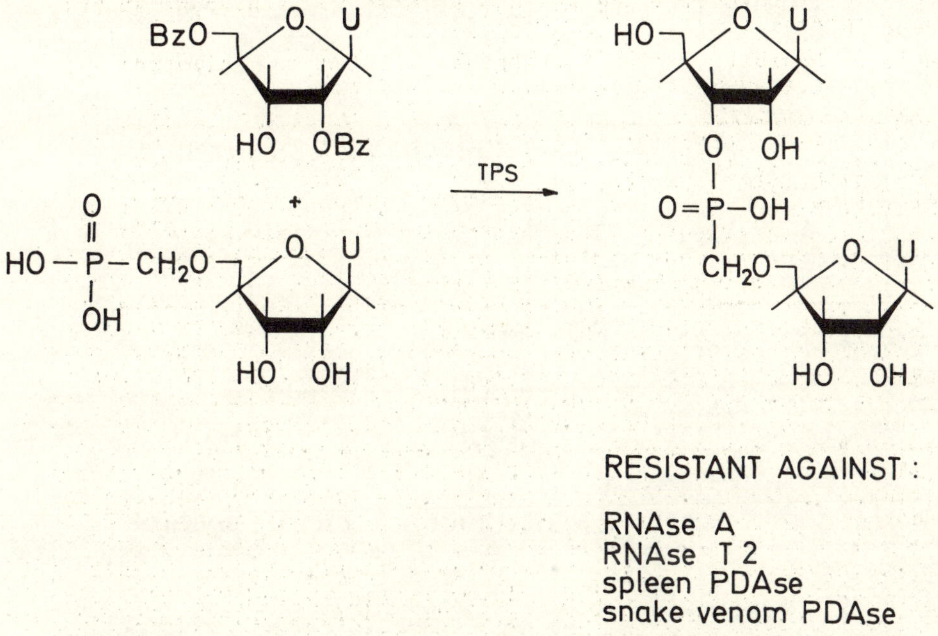

RESISTANT AGAINST:

RNAse A
RNAse T 2
spleen PDAse
snake venom PDAse

Scheme 8

The chemistry and biochemical implications of this type of nucleotide analogues is encouraging: the linkages are stable both chemically and enzymatically and the oligonucleotide analogues eventually obtained would almost surely exhibit different physico-chemical and biological properties dependent upon the steric changes involved. These and other problems including analogues of 3'-nucleotides, 2'-deoxyribonucleotides or 5'-diphosphates, are now under exhaustive investigation.

3. OPEN-CHAIN ANALOGUES OF NUCLEOTIDES

The chemistry of open-chain analogues of nucleic acids components (derivatives of nucleic acids bases bearing hydroxyl-containing aliphatic chain) has been investigated in this Laboratory in the recent years in connection with the hypothesis of minimum necessary similarity (36). The leading idea of this work is the assumption that such analogues might sufficiently differ from the natural metabolites to be stable under the in vivo conditions and, on the other hand, might fulfill the substrate requirements of other enzymes involved in the nucleic acids metabolism to display inhibitory or substrate activities. This assumption was confirmed by preliminary studies on some enzymes (37) and, mainly, by the recent finding of a substantial antiviral activity of two compounds: 9-(2-hydroxyethyloxymethyl)guanine ("acycloguanosine") (38) and 9-(S)-(2,3-dihydroxypropyl)adenine ("S-DHPA")(39,40). Also eritadenine (a natural hypocholesterolemic agent)(41) and 9-erythro--(2-hydroxy-3-nonyl)adenine ("EHNA", a powerful adenosine deaminase inhibitor)(42) belong to this category.

In our previous paper (37) we discovered the homochiral affinity of the cyclic phosphodiester derivatives of (S)-(2,3-dihydroxypropyl) uracil, -thymine or -adenine to some ribonucleases. This finding stimulated a detailed investigation of the homologous 2,3,4-trihydroxybutyl series of compounds with regard to their behaviour in enzymatic systems. Bearing two centres of chirality, the mentioned compounds can exist in two enantiomeric pairs (Scheme 9). The synthesis of parent "nucleoside analogues" has been already described elsewhere (43).

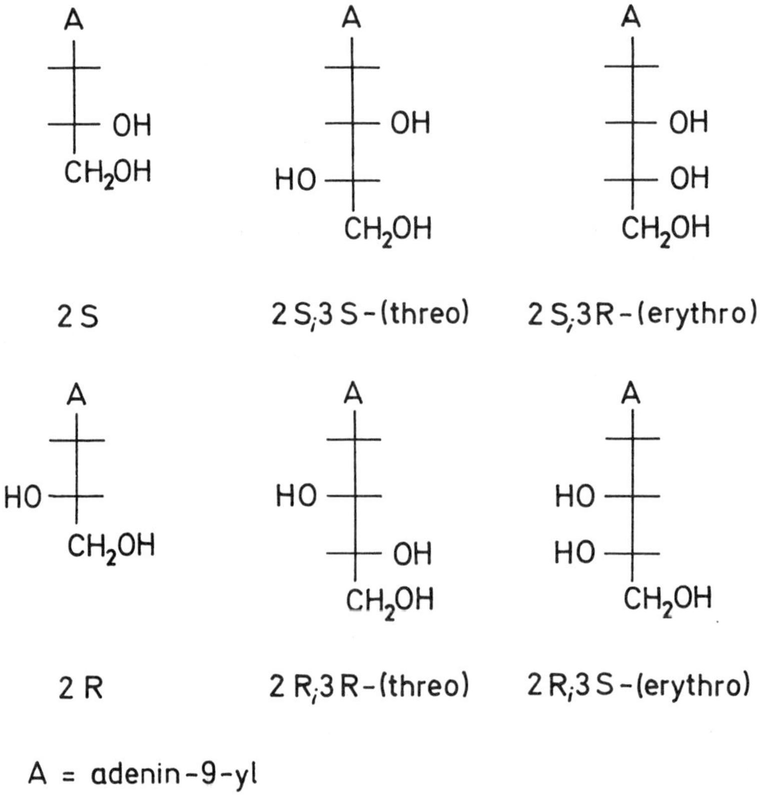

2 S   2 S,3 S-(threo)   2 S,3 R-(erythro)

2 R   2 R,3 R-(threo)   2 R,3 S-(erythro)

A = adenin-9-yl

Scheme 9

The assymetric syntheses (Scheme 10) start either from sugar derivatives by the degradative procedures, or, alternatively, from accesible optically active carboxylic acid derivatives by build-up pathways.

The phosphorylation of these compounds by phosphoryl chloride/triethyl phosphate procedure (44) affords mainly the 4'-phosphate accompanied by the 2'- and 3'-isomers. These isomers are easily separated by paper chromatography in borate-containing systems, or by HPLC technique. The preparative separation was achieved by anion-exchange chromatography on Dowex 1 resin. The phosphomonoesters thus obtained were easily converted to the cyclic phosphodiesters by the action of water-soluble carbodiimides. In all cases, only the five-membered phosphodiesters were detected; the separation of 2',3'- and 3',4'-isomers was finally performed by preparative paper chromatography and the constitution proved by the weak acidic cleavage to the known isomeric phosphomonoesters mentioned (Scheme 11).

These procedures were performed in two series of compounds, derived from uracil and adenine. The preliminary experiments revealed that the response to the ribonuclease T2 is markedly higher in the latter series and the further examinations were henceforth carried out with the adenine derivatives only. In this instance, the substrate activity of the 2',3'- and 3',4'-cyclic phosphates belonging to the two (erythro and threo) enantiomeric series was examined with non-specific ribonuclease T2 (Asp.oryzae). (The enzyme used was isolated according to the previous paper (45) and finally purified by isoelectric focusing; the preparation contained only traces of T1 RNAse and phosphomonoesterase activity.) As marker compounds,

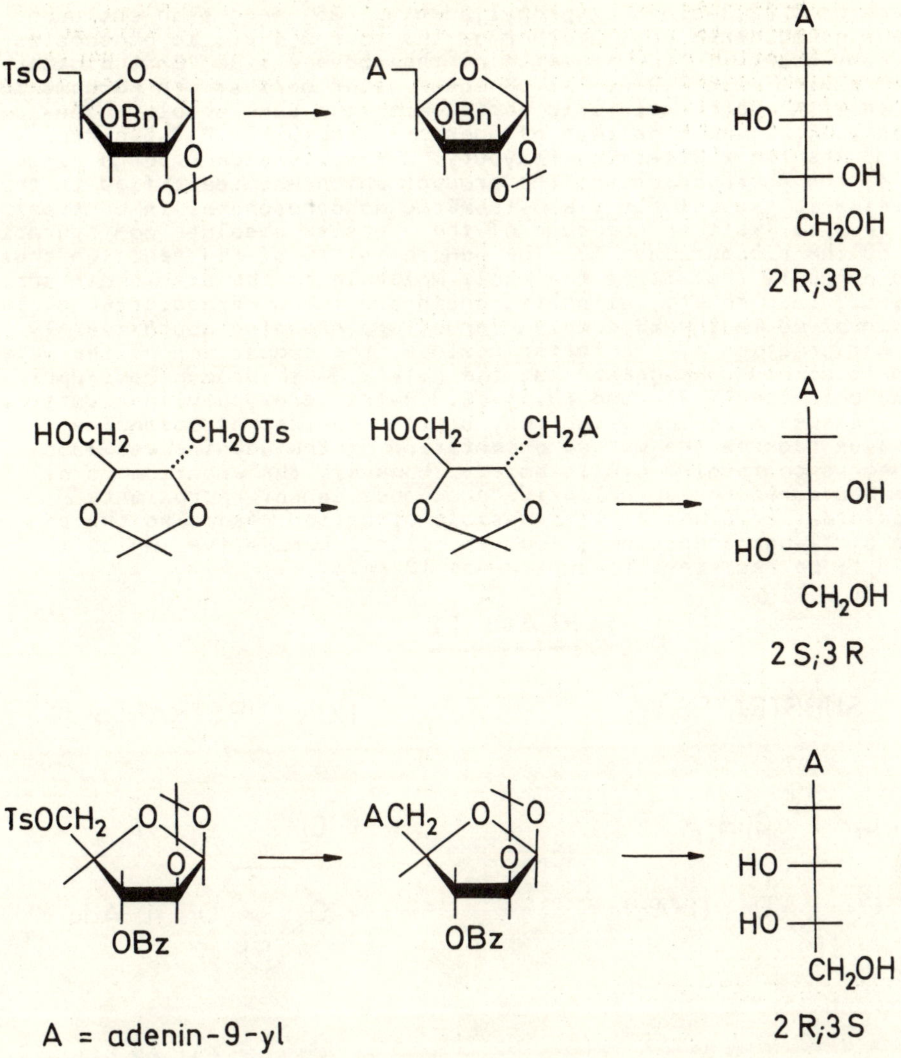

A = adenin-9-yl

Scheme 10

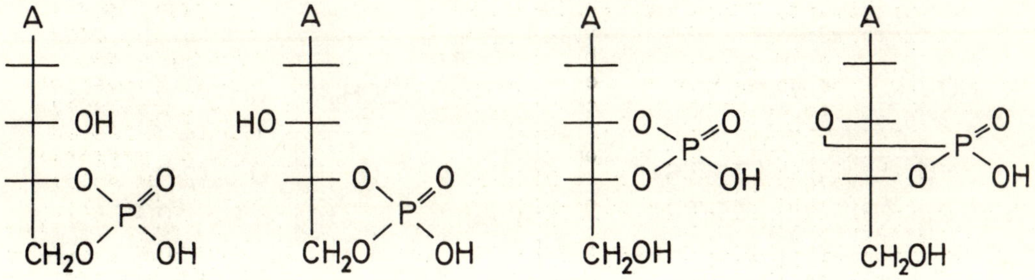

3,4-cyclic phosphates

2,3-cyclic phosphates

Scheme 11

the 2′,3′-cyclic phosphates of adenosine, L-adenosine (46) and of the two enantiomers of 9-(2,3-dihydroxypropyl)adenine (36) were also included. It has been established that neither of the four 3′,4′-cyclic phosphates underwent any change in the presence of the above enzyme. On the other hand, the results with 2′,3′-cyclic phosphodiester derivatives (Scheme 12) indicate that in the enantiomeric series with the same absolute configuration at C(2′)-atom as that of adenosine, both the 2,3-dihydroxypropyl- and the two 2,3,4-trihydroxybutyl derivatives do undergo ring-opening of the phosphodiester. The product which was identified in the latter series is exclusively the 3-isomeric monophosphate. In contrast to these results, neither compound of the opposite absolute configuration is split by the ribonuclease T2. The homochirality of the reaction thus confirmed suggests that it is the whole molecule of the open-chain analogue, i.e. its base moiety, aliphatic chain and the phosphodiester cycle, which is involved in the ES-complex formation. Assuming approximately the same conformation of the latter residue, the comparison of the molecular models allows to suggest that the (S)-(2,3-dihydroxypropyl)derivative and both the (S,R)- and (S,S)-(2,3,4-trihydroxybutyl)derivatives can easily assume a conformation very similar to that of adenosine derivative as regards the mutual orientation of the adenine base and the phosphorus containing cyclic moiety. However, the enantiomers of the above mentioned three nucleotide analogues cannot approximate such a conformation (the nearest possible situation resembles the conformation of "1-homoadenosine", i.e. an allitol derivative, which is also known to be resistant towards RNAse T2 (47)).

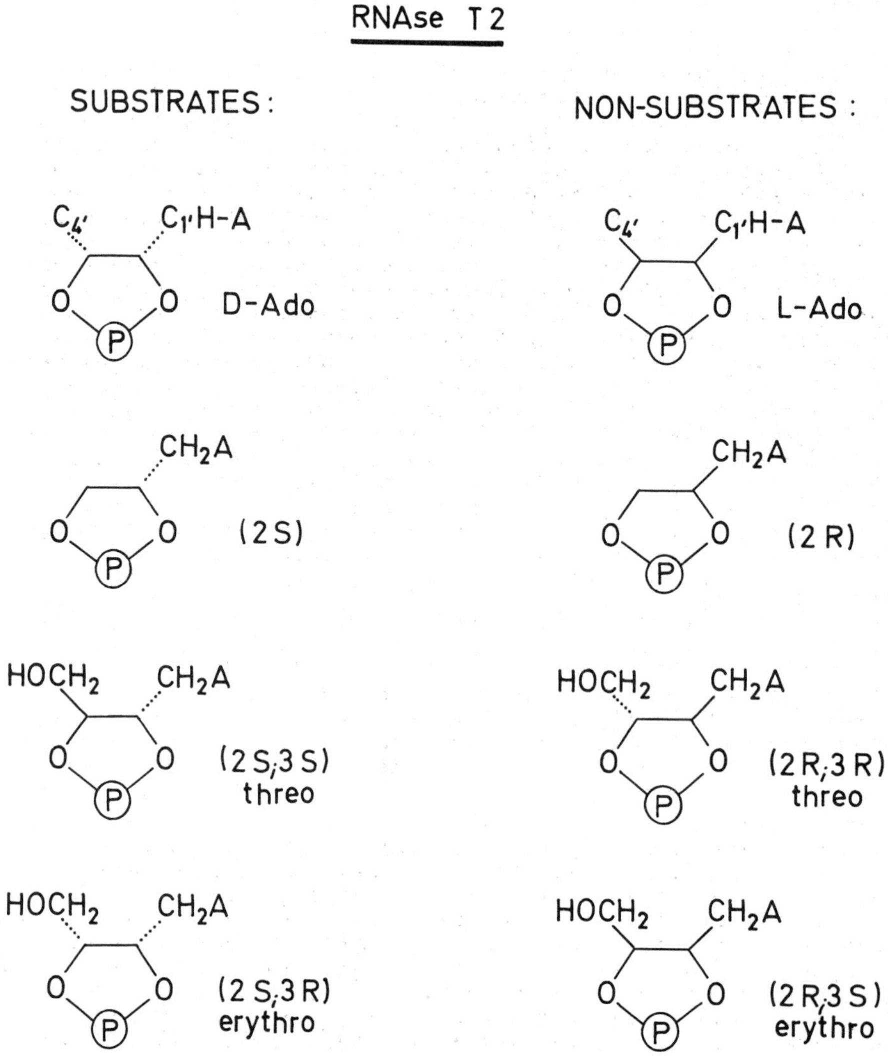

RNAse T2

SUBSTRATES: NON-SUBSTRATES:

D-Ado   L-Ado

(2S)   (2R)

(2S,3S) threo   (2R,3R) threo

(2S,3R) erythro   (2R,3S) erythro

Scheme 12

The kinetic measurements performed so far with the four substrates are summarized in Table 3. The K_m constants calculated from the linear Lineweaver-Burke plots are very similar; in other words, as far as the K_m value can be regarded as a measure of the substrate affinity, the compounds tested exhibit affinity similar to the "natural" adenosine derivative. However, the maximum rate constants V_{max} differ dramatically within four orders of magnitude from this substrate. Whereas the three-carbon-chain derivative still exhibits a medium rate of splitting, the homologous 2,3,4-trihydroxybutyl derivative of the threo (2S,3S)-configuration is split very slowly; the erythro (2S,3R)-derivative is cleaved still more slowly and its kinetic parameters so far obtained are very inaccurate.

Table 3. Kinetic parameters of RNAse T2 splitting

2',3'-Cyclic phosphate	$K_m \cdot 10^3$ M	$V_{max} \cdot 10^6$ (mol.min^{-1})
Adenosine	3.561	63720.0
9-(S)-(2,3-Dihydroxypropyl)adenine	3.105	67.0
9-(2S,3S)-(2,3,4-Trihydroxybutyl)adenine	4.873	5.8
9-(2S,3R)-(2,3,4-Trihydroxybutyl)adenine	+	+

Anyway, it is perhaps possible to conclude that the presence and relative orientation of the substituent (CH_2OH) group at the 3'-carbon atom of the aliphatic chain brings an additional negative factor affecting the reaction at the active site of the enzyme. This seems to be understandable in the light of the results obtained with 9-(o-L-lyxofuranosyl)adenine derivative in which other hydroxylic function is introduced into the vicinity of the phosphorus atom and which is hydrolyzed more slowly than the adenosine derivative (48). Thus, the structure of the nucleoside derivative is optimal, since the energy required for its conformational change in the ES-complex is smaller than that required for the open-chain analogues with the freedom of rotation around the sp3-hybridized C(1')-atom. However, as proven by the above data, even the latter analogues are acceptable as substrates for the RNAse T2 and form the same type of product by the enzymatically catalyzed reaction.

The situation might be similar also with other enzymes and the open-chain analogues; it is however by no means of general validity: e.g., the above 2'-, 3'- or even 4'-monophosphates are not substrates for 5'-nucleotidases either from snake venom or from bull semen. Owing to the general interest in the open-chain analogues, other enzymes connected with nucleic acids metabolism should be closely investigated. On the other hand, the above examples demonstrate that such analogues are suitable for studies leading to better understanding the processes connected with the substrate requirements and mode of action of enzymes in general.

Acknowledgement.- My sincere thanks are due to Dr. I.Rosenberg from this Institute who provided the data connected with the enzyme assays and kinetical measurements, and to Mr. P.Koiš (Department of Biochemistry, Komenský University, Bratislava) who performed some experiments included in the first part of this paper.

REFERENCES

1. A.M.Frischauf and F.Eckstein, Eur.J.Biochem. 32, 479-485 (1973).
2. A.Holý and I.Rosenberg, Collect.Czech.Chem.Commun.44, 957-975 (1979).
3. M.Bačová, E.Zelinková and J.Zelinka, Biochim.Biophys.Acta. 235, 335-342 (1971).
4. G.S.Ivanova, A.Holý, E.Zelinková, D.I.Bezborodova, N.M.Abrosimova- -Amelyanchik and R.I.Tatarskaya, Collect.Czech.Chem.Commun. 39, 2986-2997 (1974).
5. H.Rosemeyer and F.Seela, Carboh.Res. 62, 155-163 (1978).
6. M.H.Halford and A.S.Jones, J.Chem.Soc. 1968, 2667-2670.
7. A.Holý and F.Šorm, Collect.Czech.Chem.Commun. 31, 1562-1568 (1966).
8. T.Hata and M.Sekine, Tetrahedron Lett. 1974, 3943-3946.
9. F.Eckstein, J.Am.Chem.Soc. 92, 4718-4723 (1970).
10. A.M.Michelson, Biochim.Biophys.Acta 114, 460-468 (1966).
11. F.Eckstein, H.H.Schulz, H.Rüterjans, W.Haar, W.Mauer, Biochemistry 11, 3507-3512 (1972).
12. H.Pischel, A.Holý and G.Wagner, Collect.Czech.Chem.Commun. 39, 3773-3781 (1974).
13. G.A.Williams and M.W.Chase in Methods in Immunology and Immunochemistry, Vol.I., p.149ff, Academic Press, New York/London (1967).
14. H.Pischel, A.Holý and G.Wagner, Collect.Czech.Chem.Commun. 44, 1634-1641 (1979).
15. J.A.Schofield and A.R.Todd, J.Chem.Soc. 1961, 2316-2320.
16. A.Holý and F.Šorm, Collect.Czech.Chem.Commun. 31, 1544-1561 (1966).
17. A.Holý, Collect.Czech.Chem.Commun. 32, 3064-3067 (1967).
18. N.Anand and A.R.Todd, J.Chem.Soc. 1951, 1867-1872.
19. T.C.Myers, K.Nakamura and A.B.Danielzadeh, J.Org.Chem. 30, 1517-1520 (1965).
20. T.C.Myers and L.N.Simon, J.Org.Chem. 30, 443-446 (1965).
21. A.Holý, Collect.Czech.Chem.Commun. 32, 3713-3718 (1967).
22. A.Holý and Ng.D.Hong, Collect.Czech.Chem.Commun. 36, 316-317 (1971).
23. N.N.Gulyaev and A.Holý, FEBS Lett. 22, 294-296 (1972).
24. A.Holý, Tetrahedron Lett. 1967, 881-884.
25. D.H.Rammler and L.Yengoyan, Biochemistry 6, 3629-3638 (1966).
26. D.H.Rammler, A.Bagdasarian and F.Morris, Biochemistry 11, 9-12 (1972).
27. A.Hampton, T.Sasaki and B.Paul, J.Am.Chem.Soc. 95, 4404-4414 (1973).
28. H.P.Albrecht, G.H.Jones and J.G.Moffatt, J.Am.Chem.Soc. 92, 5511-5513 (1970).
29. G.H.Jones, H.P.Albrecht, N.P.Damodaran and J.G.Moffatt, J.Am.Chem.Soc. 92, 5510-5511 (1970).
30. P.M.J.Burgers and F.Eckstein, Proc.Nat.Acad.Sci.U.S.A. 75, 4798-4800, (1978).
31. F.Eckstein, Accounts Chem.Res. 12, 204-210 (1979).
32. A.Hampton, L.W.Brox and M.Bayer, Biochemistry 8, 2303-2308 (1969).
33. D.Dunaway-Marino, Tetrahedron 32, 2991-2996 (1976).
34. A.W.Murray and M.R.Atkinson, Biochemistry 7, 4023-4029 (1968).
35. A.Holý in Nucleosides, Nucleotides and Their Biological Applications (Montpellier 1978) in press.
36. M.Landt, S.C.Boltz and L.G.Butler, Biochemistry 17, 915-919 (1978).
37. A.Holý and G.S.Ivanova, Nucleic Acids Res. 1, 19-34 (1974).
38. G.B.Elion, P.A.Furman, J.A.Fyfe, P.deMiranda, L.Beauchamp and H.J.Schaeffer, Proc.Nat.Acad.Sci.U.S.A. 74, 5716-5722 (1977).
39. E.DeClercq, J.Descamps, P.DeSomer and A.Holý, Science 200, 563-565 (1978).
40. E.DeClercq and A.Holý, J.Med.Chem. 22, 510-513 (1979).
41. I.Chibata, K.Okumura, S.Takeyama and K.Kodera, Experientia 25, 1237-1239 (1969).
42. H.J.Schaeffer and Ch.F.Schwender, J.Med.Chem. 17, 6-8 (1974).
43. A.Holý, Collect.Czech.Chem.Commun. 44, 593-612 (1979).
44. M.Yoshikawa, T.Kato and T.Takenishi, Tetrahedron Lett. 1967, 5065-5068.
45. M.Naoi-Tada, K.Sato-Asano and F.Egami, J.Biochem.(Tokyo) 46, 757-765 (1969).
46. A.Holý and F.Šorm, Collect.Czech.Chem.Commun. 34, 3383-3401 (1969).
47. A.Holý, Collect.Czech.Chem.Commun. 35, 81-88 (1970).
48. A.Holý and F.Šorm, Collect.Czech.Chem.Commun. 34, 3523-3532 (1969).

THE SYNTHESIS OF BIOLOGICALLY RELEVANT MOLECULES VIA PHOSPHORUS YLIDS

H. J. Bestmann

Institut für Organische Chemie der Universität Erlangen-Nürnberg, D-8520 Erlangen, Henkestr. 42, Federal Republic of Germany

Abstract: The synthesis of molecules in the vitamin-A series, carotinoids, pheromones, chinic and shikimic acids using phosphorus ylids is reported.

The discovery of the Wittig-reaction 25 years ago made possible the synthesis of many biologically relevant natural products. The mechanism of this reaction has recently been reinterpreted. The stereochemistry of the resulting double bond can now be predicted with some confidence.[1][2][3][4]

Ylids, 1, react with aldehydes, 2, to oxaphosphetanes, 3, which undergo pseudorotation to give structure 4. The original ylid C-P bond is then cleaved to give the betaine, 5. The electronic nature of the substituent R^2 determines the stereochemistry of the product olefins. Electron donor groups R^2 cause a fast elimination of the phosphine oxide, giving Z-olefins, 8. Electron withdrawing groups R^2 enhance the stability of 5 and allow conversion to the thermodynamically more stable form, 6, which now eliminates phosphine oxide to give the E-olefin, 7. These observations lead to the following rule: Phosphorus ylids with R^2 = Alkyl yield stereoselectively Z-olefins in the Wittig-reaction. If, however, R^2 is COR or CN E-olefins are formed. This rule applies only to ylid solutions which are free of lithium salts; e.g. those prepared using the sodium amide[5] or sodium silazide[6] method.

The intermediacy of the betaine, 5, allows deuterium or tritium to be incorporated on the double bond of the product olefin during the Wittig-reaction[2,7].

Retinoids have recently assumed great importance as prophylactics in oncology and dermatology.[8] Compounds of the vitamin-A series are now prepared on an industrial scale by means of the Wittig-reaction [9].

The tertiary alcohol, 9, is reacted with triphenylphosphine, 10, and an acid to give the phosphonium salt, 11, whose corresponding ylid reacts with aldehyde, 13, to give the vitamin-A acetate, 14. If the ester-aldehyde, 15, is used in place of 13 vitamin-A acid-ester is produced. This can then be saponified.

The therapeutically important retinoid, 17, can be synthesised from the ylid 16 and the aldehyde, 15.[8]

When ylids of type 18 are oxidised, olefins, 19, are produced, whereas phosphoranes of structure 20 (R^1, $R^2 \neq H$) are converted to ketones, 21. Oxygen,[10] periodate[11] and the adduct of ozone and triphenylphosphite[12] have been used as oxidants.

A phosphonium salt, 23, can be prepared from vitamin-A, 22, triphenylphosphine, 10, and an acid. This salt can be converted to the ylid 24 by base. Oxidation of the ylid leads to β-carotin, 25, (with D replaced by H).[14] H_2O_2 has proved to be the best, and simplest, oxidant in this case.[15] If the oxidation is performed at -70°C with the ozone-triphenylphosphite adduct and C_2H_5OD is added the mono deuterated β-carotin, 25, is obtained with very high incorporation of deuterium.

The intramolecular Wittig-reaction has recently been employed in the synthesis of a new type of β-lactam antibiotic. The β-lactam, 26, is converted to the ylid, 27, which after saponification and oxidation yields the aldehyde, 28. Thienamycin derivative, 29, is formed from 28 via an intramolecular Wittig-reaction.[16] Analogously the descysteaminylthienamycin derivative, 31, is prepared from the β-lactamic-ylidic thioformic acid ester, 30.[17]

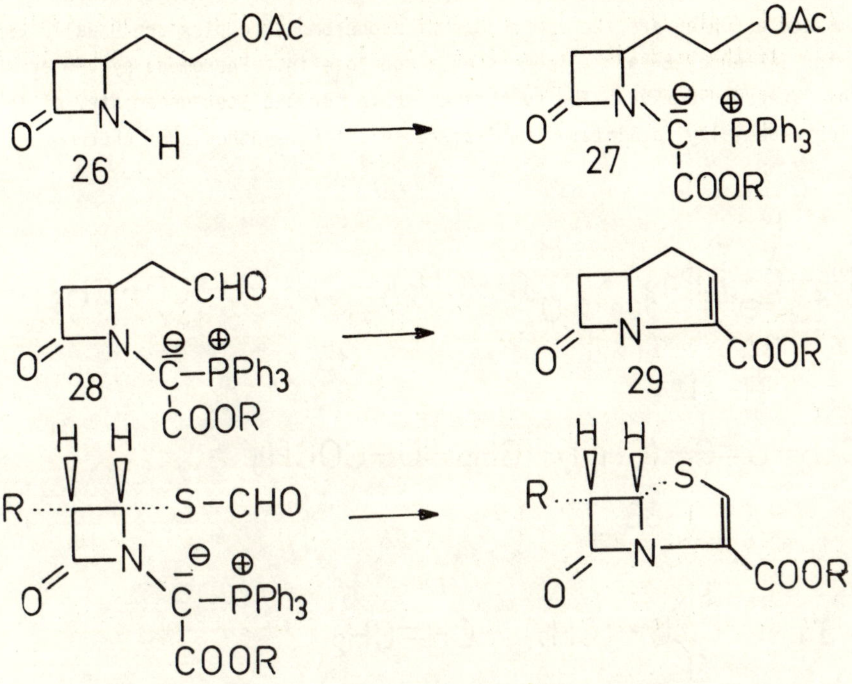

[Structures 26 → 27, 28 → 29, 30 → 31 shown]

The Wittig-reaction has become especially important in the synthesis of butterfly-pheromones.
[18)] The pheromones have the general structure 32 or 33, i.e. they are mono-olefins or bis-olefins with varying distances between the double bonds.

$$CH_3-(CH_2)_m-CH=CH-(CH_2)_n-A$$
32

$$CH_3-(CH_2)_m-CH=CH-(CH_2)_n-A$$
33

A = CH_2OH
$CH_2O-CO-CH_3$
CHO

Chemical transmission of information proceeds via multicomponent systems. The concentration ratios of the individual components influence the "message".[19] The structural differences between two molecules which are the components of a pheromone complex are usually very small. They can be, for instance, E- and Z-isomers of a mono-olefin. Pheromone syntheses must therefore be highly stereoselective. The rules given above for the stereochemistry of the Wittig-reaction especially facilitate the Z-stereoselective synthesis of pheromones.

$$Alk-CH_2-\underset{\ominus}{\overset{H}{C}}-\overset{\oplus}{P}Ph_3 \quad + \quad \underset{O}{\overset{H}{C}}-(CH_2)_7-CH_2-O-CO-CH_3 \longrightarrow$$
$$\text{34} \hspace{5cm} \text{35}$$

$$Alk-CH_2-\overset{H}{C}=\overset{H}{C}-(CH_2)_7-CH_2-O-COCH_3$$
$$\text{36}$$

$$34 + \quad \underset{O}{\overset{H}{C}}-(CH_2)_8-CH=CH_2$$
$$\text{37}$$

$$Alk-CH_2-\overset{H}{C}=\overset{H}{C}-(CH_2)_8-CH_2-CH_2OH \quad \xrightarrow[2)\ H_2O_2/OH^\ominus]{1)\ 9\text{-}BBN}$$
$$\text{38}$$

$$Alk-CH_2-\overset{H}{C}=\overset{H}{C}-(CH_2)_8-CH_2-CH_2OH$$
$$\text{39}$$

Ylids of type 34 (Alk = alkyl) undergo stereoselective Wittig-reactions with acetoxy-nonanal, 35, to give the pheromones, 36, with a Z-9-double bond (with 3-5% E-isomer).[20] The Z-11-dienes, 38, are prepared from 34 and undecenal, 37. Hydroboration with 9-bora-bicyclo [3,3,1] nonane (9-BBN) followed by H_2O_2 oxidation yields the alcohols, 39, with a Z-11-double bond. These alcohols can be converted to acetates and aldehydes, which are common in pheromone complexes, by the usual methods.[21]

These syntheses allow systematic structure changes to be made in known pheromones. Comparative electrophysiological testing of these modified pheromones has led to the discovery of the first rules concerning the relations between structure and activity of pheromone molecules,[22][23] and to the proposal of a dynamic model for the interaction between pheromones and the receptor regions of the cell membranes of dendrites.[23]

The E-stereoselective Wittig-reaction can be used in the following ways to synthesise E-olefinic pheromones.[24]

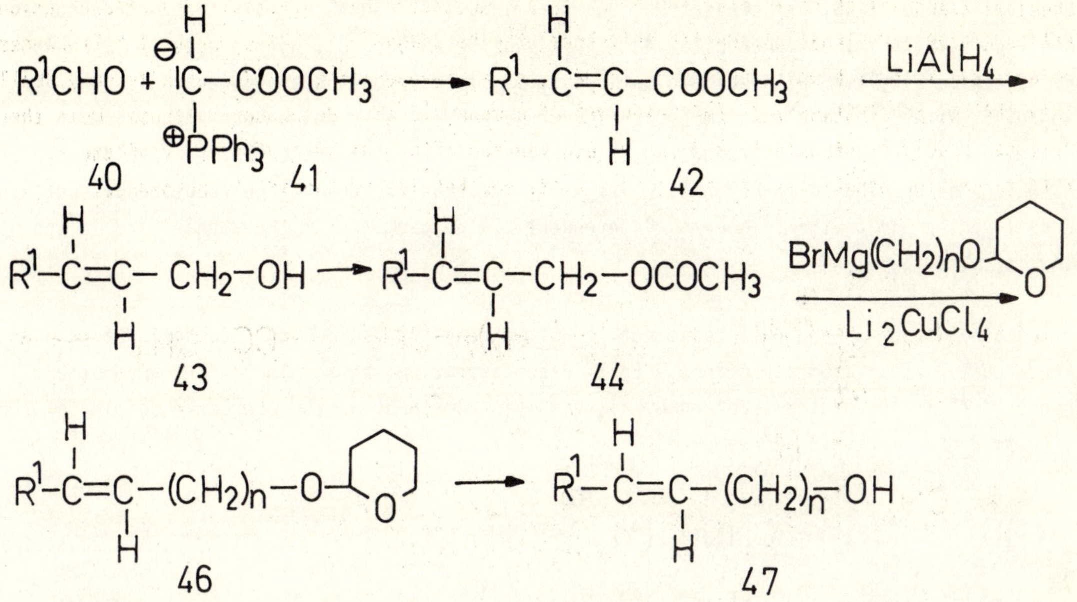

R^1CHO + $\overset{H}{\underset{\overset{\oplus}{P}Ph_3}{\overset{\ominus}{C}}}$–COOCH$_3$ ⟶ R^1–$\overset{H}{\underset{H}{C}}$=$\overset{}{\underset{}{C}}$–COOCH$_3$ $\xrightarrow{\text{LiAlH}_4}$

40   41   42

R^1–$\overset{H}{\underset{H}{C}}$=$\overset{}{\underset{}{C}}$–CH$_2$–OH ⟶ R^1–$\overset{H}{\underset{H}{C}}$=$\overset{}{\underset{}{C}}$–CH$_2$–OCOCH$_3$ $\xrightarrow[\text{Li}_2\text{CuCl}_4]{\text{BrMg(CH}_2)_n\text{O–THP}}$

     43   44

R^1–$\overset{H}{\underset{H}{C}}$=$\overset{}{\underset{}{C}}$–(CH$_2$)$_n$–O–THP ⟶ R^1–$\overset{H}{\underset{H}{C}}$=$\overset{}{\underset{}{C}}$–(CH$_2$)$_{\overline{n}}$–OH

      46   47

Aldehydes, 40, react with carbomethoxymethylenetriphenylphosphorane, 41, to give E-α,β-unsaturated esters, 42, as discussed above. The esters, 42, can be reduced to allyl alcohols, 43, with lithium aluminium hydride. After acetylation of 43 a Grignard coupling with 45 in the presence of Li$_2$CuCl$_4$ gives 46, which can be cleaved with acid to give the E-olefinc alcohol 47. The alcohol group of 47 can easily be further modified.

The following method for the preparation of E-olefins has proven very effective in our hands.[26]

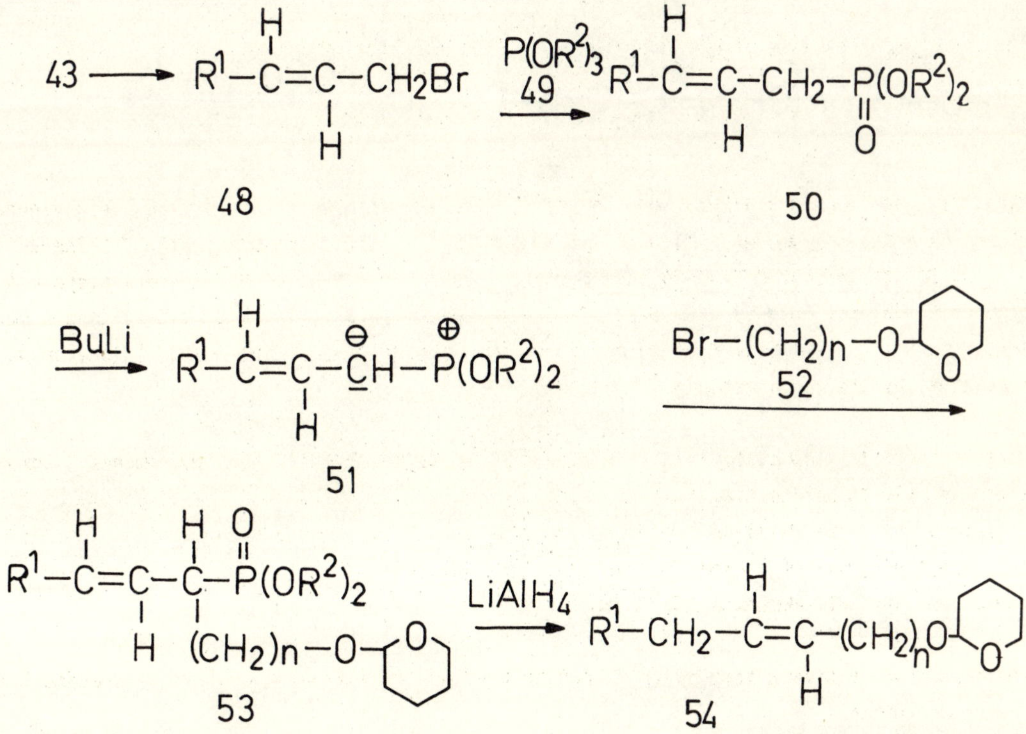

43 ⟶ R^1–$\overset{H}{\underset{H}{C}}$=$\overset{}{\underset{}{C}}$–CH$_2$Br $\xrightarrow[49]{\text{P(OR}^2)_3}$ R^1–$\overset{H}{\underset{H}{C}}$=$\overset{}{\underset{}{C}}$–CH$_2$–$\overset{O}{\underset{}{\overset{\|}{P}}}(OR^2)_2$

         48   50

$\xrightarrow{\text{BuLi}}$ R^1–$\overset{H}{\underset{H}{C}}$=$\overset{}{\underset{}{C}}$–$\overset{\ominus}{\underset{}{C}}H$–$\overset{\oplus}{P}(OR^2)_2$ $\xrightarrow{\text{Br–(CH}_2)_n\text{–O–THP} \atop 52}$

                51

R^1–$\overset{H}{\underset{H}{C}}$=$\overset{}{\underset{}{C}}$–$\overset{H}{\underset{(CH_2)_n-O-THP}{C}}$–$\overset{O}{\underset{}{\overset{\|}{P}}}(OR^2)_2$ $\xrightarrow{\text{LiAlH}_4}$ R^1–CH$_2$–$\overset{H}{\underset{H}{C}}$=$\overset{}{\underset{}{C}}$–(CH$_2$)$_n$–O–THP

        53   54

The allyl alcohols, 43, are converted to the bromides, 48, and then converted to the phosphonate, 50, using a trialkylphosphite, 49. 50 gives the anions, 51, with butyl lithium. The anions can be alkylated with halogen compounds, 52, to give the phosphonates, 53. The reduction of 53 with lithium aluminium hydride occurs via hydride attack on the γ-position to the P-atom, a shift of the double bond and elimination of the phosphonaterest.[27] E-Olefins, 54, which can be cleaved to alcohols of type 47, are thus produced. The reaction product contains less than 1% of the Z-isomers, independent of the geometry of the double bond in the starting material.

The combination of Z- and E-stereoselective Wittig-reactions allows the synthesis of conjugated doubly unsaturated pheromones with E, Z-configuration, as are found for many butterfly species.[24][28] The synthesis of bombykol, the main component of the pheromone complex of the silkworm Bombyx mori, is an example.

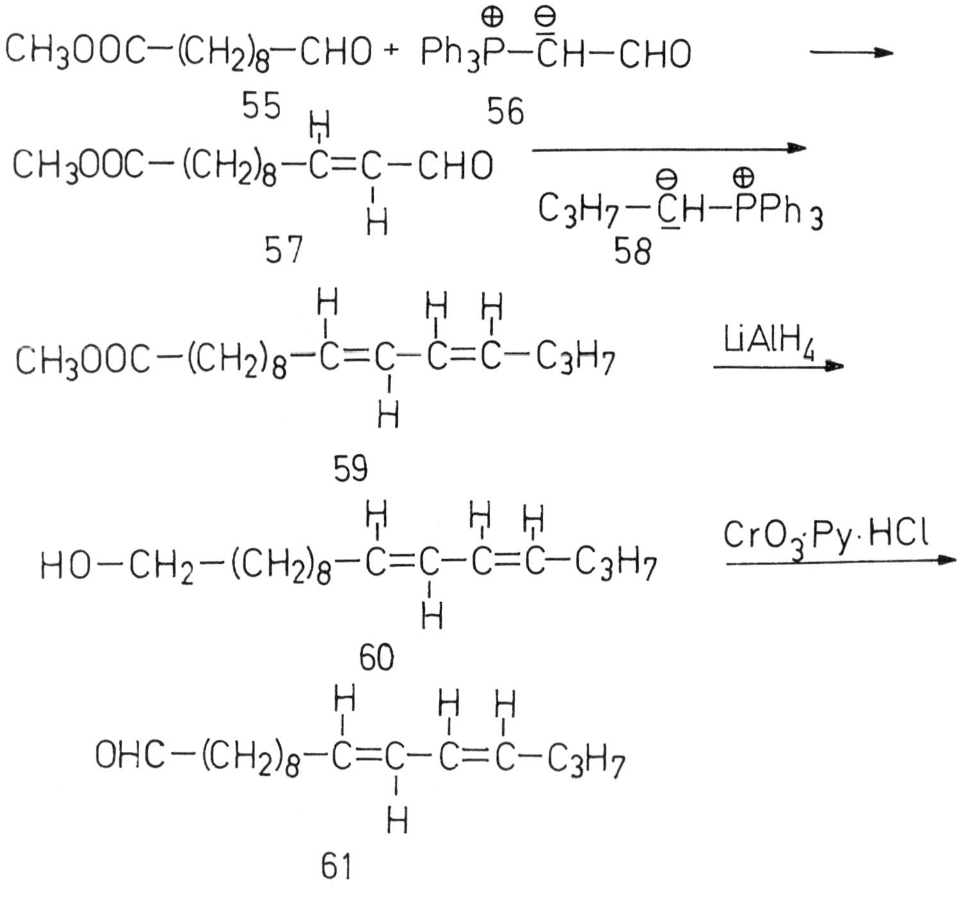

$$CH_3OOC-(CH_2)_8-CHO + Ph_3\overset{\oplus}{P}-\overset{\ominus}{C}H-CHO \longrightarrow$$

55   56

$$CH_3OOC-(CH_2)_8-\underset{H}{\overset{H}{C}}=\underset{H}{C}-CHO \longrightarrow$$

$$C_3H_7-\overset{\ominus}{C}H-\overset{\oplus}{P}Ph_3$$

57   58

$$CH_3OOC-(CH_2)_8-\underset{H}{\overset{H}{C}}=C-\overset{H}{C}=\overset{H}{C}-C_3H_7 \xrightarrow{LiAlH_4}$$

59

$$HO-CH_2-(CH_2)_8-\underset{H}{\overset{H}{C}}=C-\overset{H}{C}=\overset{H}{C}-C_3H_7 \xrightarrow{CrO_3 \cdot Py \cdot HCl}$$

60

$$OHC-(CH_2)_8-\underset{H}{\overset{H}{C}}=C-\overset{H}{C}=\overset{H}{C}-C_3H_7$$

61

The reaction of aldehyde, 55, with formylmethylenetriphenylphosphorane, 56, proceeds E-stereoselectively to the α,β-unsaturated aldehyde, 57, which in turn reacts Z-stereoselectively with the ylid, 58. The product is the E,Z-diene, 59, which can be reduced to bombykol, 60. Oxidation of 60 yields bombykal, 61, which we were able to identify as the second pheromone component of Bombyx mori.[29]

We were able to synthesise Z-9,E-12-tetradecadienylacetate, 66, which occurs as a pheromone in several Lepidoptera species, as follows.

Cyclopropylmethylcarbinol, 62, is converted with HBr to the E-olefinic bromide 63.[31] It is then reacted with triphenylphosphine, 10, to give the phosphonium salt, 64. The corresponding ylid, 65, reacts Z-stereospecifically with acetoxynonanal, 35, to give the pheromone, 66. We have employed a new method, which we recently discovered, to interrupt conjugated double bonds with a methylene group for the synthesis of the E,E-isomer.[32]

Sorbylbromide, 67, is converted to the phosphonate, 68, with triethylphosphite. The anion of 68 is then coupled with the bromide, 69. On reduction with lithium aluminium hydride the conjugated system is interrupted and 71 is formed. This can then easily be converted to 72.

After the discovery of the Wittig-reaction it was shown that phosphorus ylids can be compared with Grignard compounds in the multitude of their possible nucleophilic reactions with electrophilic substrates.[33] A series of natural products could be synthesised with these new reactions. The following synthesis was developed for the queen substance, which occurs in honey bee secretions.[34]

$$HOOC-(CH_2)_5-COOC_2H_5 \longrightarrow H_5C_2SCO-(CH_2)_5-COOC_2H_5$$
$$73 \qquad\qquad 74$$

$$\overset{\ominus}{C}H_2-\overset{\oplus}{P}Ph_3$$
$$\underset{75}{} \longrightarrow Ph_3\overset{\oplus}{P}-\overset{\ominus}{C}H-CO-(CH_2)_5-COOC_2H_5 \xrightarrow{\overset{\ominus}{O}H}$$
$$76$$

$$CH_3-CO-(CH_2)_5-COOC_2H_5 \xrightarrow[\substack{2)SOCl_2 \\ 3)NaSC_2H_5}]{1)\overset{\ominus}{O}H} CH_3-CO-(CH_2)_5-COSC_2H_5$$
$$77 \qquad\qquad\qquad\qquad\qquad\qquad 78$$

$$\text{Ra·Ni} \quad \overset{\ominus}{\underset{\oplus PPh_3}{\overset{|}{C}H}}-COOC_2H_5$$
$$\xrightarrow{79} CH_3CO-(CH_2)_5-\overset{H}{\underset{}{C}}=\overset{}{\underset{H}{C}}-COOC_2H_5 \longrightarrow$$
$$80$$

$$CH_3CO-(CH_2)_5-\overset{H}{\underset{}{C}}=\overset{}{\underset{H}{C}}-COOH$$
$$81$$

The pimelic acid mono ester, 73, is converted to the thioester 74. Acylating reaction with methylenetriphenylphosphorane,[35] 75, yields the acyl ylid, 76, which, upon saponification, gives the ketoester, 77. This is next converted to the thioester, 78, using a method which we developed for the preparation of bis-homologous carboxylic acids.[36] 78 is then treated with Raney nickel in the presence of methoxycarbonylmethylenetriphenylphosphorane, 79. The aldehyde which is first formed reacts in situ with 79 to give the α, β-unsaturated ester, 80, which is saponified to give the queen substance 9-oxo-(Z)-2-decenoic acid 81.

When ωω'-dihalogen compounds, 82, are reacted with 2 moles of triphenylphosphinemethylene, 75, the phosphonium salt, 83, is first formed via nucleophilic substitution. 83 reacts with a second molecule of 75 in a trans-ylidation equilibrium with the corresponding salt, 84, to give the ylid, 85. 85 cyclises via an intramolecular nucleophilic substitution to give the cyclic phosphonium salt, 86, which precipitates from the solution and is thus removed from the equilibrium.[37] One can obtain the cyclic ylid 87 from 86 and employ this in the well-known ylid reactions. We have used this generally applicable cyclisation method, in which a carbon atom unit is added to the dihalide starting material on cyclisation, for the synthesis of optically pure chinic- and shikimic acids starting from arabinose.[38]

D-1,5-ditosyl-2,3,4-O-benzylarabite, 88, is reacted with 3 moles of methylenetriphenylphosphorane, 75. As described above a cyclic phosphonium tosylate is first formed from 2 moles of 75 and 88. This then reacts with a further molecule of 75 in a transylidation to give ylid 89, which can be reacted with formaldehyde. Subsequent debenzylation gives exomethylene compound, 90. This is then acetylated and oxidised with OsO_4-$NaJO_4$. The optically pure Grewe ketone, 91, is obtained[39]. With HCN this gives the nitrile, 92, from which optically pure chinic acid can be obtained by treatment of the tetraacylated amide, 93, with $N_{2,3}$ and subsequent saponification.[40] The nitrile, 95, whose saponification yields shikimic acid is formed from 92 and $POCl_3$.

Many of the syntheses of biologically relevant molecules described above were only possible because of enthusiastic committment of my co-workers, who are named in the references. I should like here to express my sincerest thanks to them.

References:

1) H. J. Bestmann, Actes 1er Congrès International Comp. Phosphores, Rabat 1977, 519.
2) H. J. Bestmann, Pure and Applied Chemistry 51, 515 (1979).
3) H. J, Bestmann, Pure and Applied Chemistry, in press.
4) H. J. Bestmann, K. Roth, E. Wilhelm, R. Böhme and H. Burzlaff, Angew. Chem., 91, 945 (1979).
5) H. J. Bestmann, Angew. Chem., 77, 609 (1965)! Angew. Chem. Int. Ed., 4, 583 (1965).
6) H. J. Bestmann, O. Vostrowsky and W. Stransky, Chem. Ber., 109, 1964 (1976).
7) H. J. Bestmann, W. Downey, K. Geibel, I. Ugi, D. Marquarding and R. Kamel, unpublished.
8) H. Mayr, W. Bollag, R. Hänni and R. Rüegg, Experientia 34, 115 (1978).
9) H. Pommer, Angew. Chem., 89, 437 (1977); Angew. Chem. Int. Ed. Engl., 16, 413 (1977).
10) H. J. Bestmann and O. Kratzer, Chem. Ber., 96, 1899 (1963).
11) H. J. Bestmann, R. Armsen and H. Wagner, Chem. Ber., 102, 2259 (1969).
12) H. J. Bestmann, L. Kisielowski and W. Distler, Angew. Chem., 88, 297 (1976); Angew. Chem., Int. Ed. Engl., 15, 298 (1976).
13) H. Pommer, Angew. Chem. 72, 811 (1960).
14) H. J. Bestmann and O. Kratzer, Angew. Chem., 73, 757 (1961); H. J. Bestmann, O. Kratzer, R. Armsen and E. Maekawa, Liebigs Ann. Chem., 1973, 760.
15) A. Nürrenbach, I. Paust, H. Pommer, I. Schneider and B. Schulz, Liebigs Ann. Chem. 1977, 1046.
16) L. D. Cama and B. G. Christensen, J. Amer. Chem. Soc., 100, 8006 (1978).
17) I. Ernest, J. Gosteli, G. W. Greengrass, W. Holick, D. E. Jackman, H. R. Pfaendler and R. B. Woodward, J. Amer. Chem. Soc., 100, 8214 (1978).
18) H. J. Bestmann and O. Vostrowsky, Chem. Phys. Lipids, 24, 335 (1979).
19) H. J. Bestmann, Mitt. dtsch. Ges. angew. allgem. Entomol., 1, 147 (1978); H. J. Bestmann and O. Vostrowsky, Seifen-Öle-Fette-Wachse, 105, 443 (1979).
20) H. J. Bestmann, O. Vostrowsky, W. Stransky and P. Range, Chem. Ber., 108, 3582 (1975).
21) H. J. Bestmann, I- Kantardjew, P. Rösel, W. Stransky and O. Vostrowsky, Chem. Ber., 111, 248 (1978):
22) O. Vostrowsky, H. J. Bestmann and E. Priesner, Nachr. Chem. Techn., 23, 501 (1975); E. Priesner, M. Jacobson and H. J. Bestmann, Z. Naturforsch., 30c, 283 (1975); E. Priesner, H. J. Bestmann, O. Vostrowsky and P. Rösel, Z. Naturforsch., 32c, 979 (1977).
23) H. J. Bestmann, P. Rösel and O. Vostrowsky, Liebigs Ann. Chem., 1979, 1189.
24) H. J. Bestmann, J. Süß and O. Vostrowsky, Tetrahedr. Lett., 1979, 2407 and unpublished.
25) G. Fouquet and M. Schlosser, Angew. Chem., 86, 50 (1974); Angew. Chem. Int. Ed. Engl., 13, 82 (1974).
26) C. Canevet, Th. Röder, O. Vostrowsky and H. J. Bestmann, Chem. Ber., in press.
27) K. Kondo, A. Negeshi and D. Tunemoto, Angew. Chem., 86, 415 (1974); Angew. Chem. Int. Ed. Engl., 13, 407 (1974).
28) H. J. Bestmann, O. Vostrowsky, H. Paulus, W. Billmann and W. Stransky, Tetraher. Lett., 1977, 121.
29) G. Kasang, K. E. Kaißling, O. Vostrowsky and H. J. Bestmann, Angew. Chem., 90, 74 (1978); Angew. Chem. Int. Ed. Engl., 17, 60 (1978).
30) H. J. Bestmann, O. Vostrowsky and A. Plenchette, Tetrahedr. Lett., 1974, 779.
31) M. Julia, S. Julia, and Son-Yu Tchen, Bull. Soc. Chim. France, 1961, 1849.
32) H. J. Bestmann, J. Süß and O. Vostrowsky, Tetrahedr. Lett., 1979, 245

33) H. J. Bestmann, Angew. Chem., 77, 609, 651 (1965); Angew. Chem. Int. Ed., 4, 583, 645, 830 (1965); H. J. Bestmann and R. Zimmermann, Fortschr. chem. Forsch., (Topics in Curr. Chem.) 20, 1 (1971); H. J. Bestmann, Bull. Soc, Chim. France, 5, 1619 (1971); H. J. Bestmann and R. Zimmermann, Chemikerzeitung, 96, 649 (1972); H. J. Bestmann and R. Zimmermann in Kosolapoff-Maier, Phosphorus Organic Compounds Vol 3, 1, John Wiley and Sons, New York 1972; H. J. Bestmann, Angew. Chem., 89, 361 (1977); Angew. Chem. Int. Ed. Engl., 16, 349 (1977); H. J. Bestmann and R. Zimmermann, in Augustine, Carbon - Carbon-Bond Formation Vol 1, 353, Marcel Dekker, New York 1979.

34) H. J. Bestmann, R, Kunstmann and H. Schulz, Liebigs Ann. Chem., 699, 33 (1966).

35) H. J. Bestmann and B. Arnason, Chem. Ber., 95, 1513 (1962).

36) H. J. Bestmann, H. Schulz, R. Kunstmann and K. Rostock, Chem. Ber., 99, 1906 (1966).

37) H. J. Bestmann and E. Kranz, Chem. Ber., 102, 1802 (1969).

38) H. J. Bestmann and H. A. Heid, Angew. Chem., 83, 329 (1971); Angew. Chem. Int. Ed. Engl., 10, 336 (1971).

39) R. Grewe and E. Vangermain, Chem. Ber., 98, 104 (1965).

40) C. f. R. M. Baldwin, C. D. Snyder and H. Rapoport, J. Amer. Chem. Soc., 95, 276 (1973).

THE PHOSPHATE TRANSFER PROCESS

A. J. Kirby

University Chemical Laboratory, Cambridge CB2 1EW, U.K.

Abstract - Enzyme catalysed phosphate transfer is considered in the light of the detailed mechanisms revealed by studies on simple systems. An attempt is made to define the limits of the $S_N1(P)$ and $S_N2(P)$ mechanisms for phosphate transfer from monoester dianions, $ROPO_3^{2-}$, in terms of the basicity of the leaving group.

INTRODUCTION

We have been looking at the mechanisms of phosphate transfer reactions because of our interest in the enzyme catalysed process. In this paper I consider some of the factors involved in the simplest such reaction, the transfer of the unsubstituted phosphate group from a monosubstituted phosphate (**1**) to a nucleophile Nu.

$$RO-PO_3^{2-} + Nu \longrightarrow {}^{2-}O_3P-Nu + ROH$$
$$\mathbf{1}$$

This is one of the commonest biological reactions, and is catalysed by a wide range of enzymes (kinases, phosphatases, etc.). It is rarely profitable to attempt very detailed mechanistic studies on enzyme reactions, because of the technical difficulties involved, and the conventional approach is to study the reaction concerned in systems which are simple enough to understand in detail. Given a sufficiently detailed understanding of the basic reaction one can make informed predictions about the mechanism of the enzyme catalysed reaction, and use these as a basis for direct studies.

This is an exciting time for anyone interested in phosphate transfer, because of the current rush of information about the enzyme catalysed reactions, especially from structural and stereochemical studies. What I will do is look in some detail at what we know about the basic reaction from the work on simple systems, to define some of the problems a phosphate transfer enzyme has to solve: and then look briefly at some recent results with enzymes, which should show whether we are thinking along the right lines.

MECHANISMS IN SIMPLE SYSTEMS

The first generalisation I will rely on is that phosphate monoesters are the most reactive, and diesters the least, for any leaving group RO⁻ likely to be of biological interest (1). So mechanisms involving simple displacement (**2**) on monoester monoanions can probably be ruled out; because these species are known to react like diester monoanions (2).

$$RO-P\underset{\underset{\mathbf{2}}{O}}{\overset{OH(R')}{\overset{|}{-}}}O:Nu \longrightarrow ROH + \underset{O^-}{\overset{HO}{\overset{|}{-}}}O-P-Nu$$

The high reactivity of monoesters relative to more highly esterified phosphates depends, of course, on the mobile proton. When the leaving group (RO⁻) is very good (weakly basic) the dianion is the most reactive form, because the $S_N1(P)$ mechanism (**3**) becomes important.

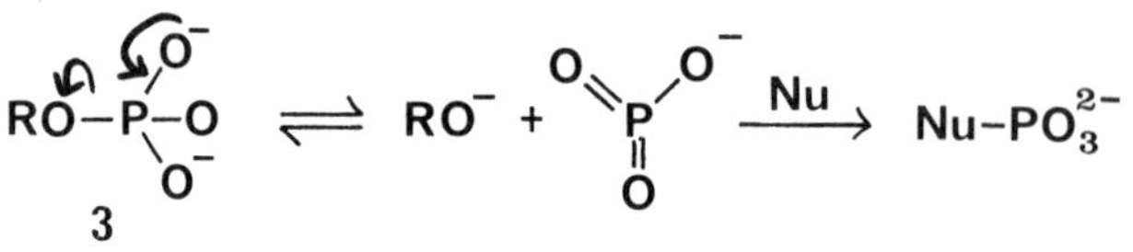

$$\text{RO-P(O)(O}^-\text{)-O}^- \rightleftharpoons \text{RO}^- + \text{PO}_3^- \xrightarrow{\text{Nu}} \text{Nu-PO}_3^{2-}$$
                              3

This reaction is exceedingly sensitive to the basicity of the leaving group, so that for most esters the monoanion is the reactive form: but the <u>mechanism</u> is still basically the same.

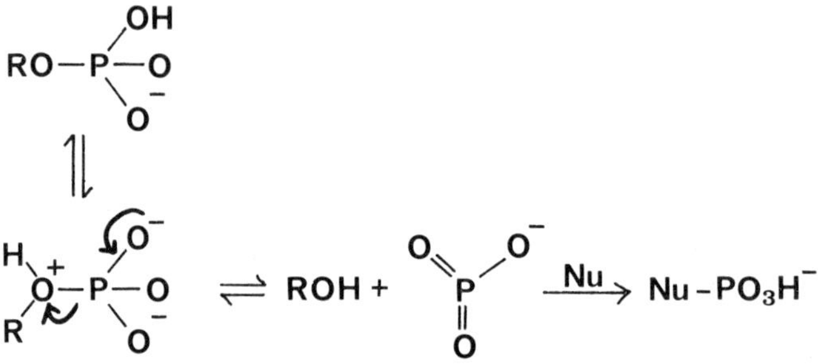

$$\text{RO-P(OH)(O)-O}^-$$
$$\updownarrow$$
$$\text{R-O(H)}^+\text{-P(O}^-\text{)-O}^- \rightleftharpoons \text{ROH} + \text{PO}_3^- \xrightarrow{\text{Nu}} \text{Nu-PO}_3\text{H}^-$$

We will be hearing more about metaphosphate from Professor Westheimer, but part of the evidence is that, at least under some conditions, phosphate transfer is non-selective (3). (This means that a species is formed which reacts indiscriminately with the first nucleophile it meets, and is therefore highly electrophilic.) Such behaviour is a characteristic of diffusion-controlled reactions, and I would like to look more closely at the mechanisms involved in the light of recent ideas on catalysis (4).

First the monoanion reaction. Under conditions where phosphate transfer is indiscriminate the reaction of metaphosphate with R^1OH presumably involves the rapid* addition of the nucleophile to the intermediate.

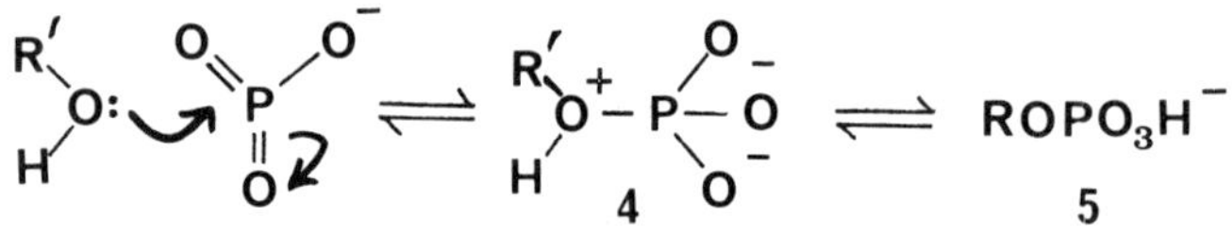

$$R'\text{O(H)}:\text{PO}_3^- \rightleftharpoons R'\text{-O(H)}^+\text{-P(O}^-\text{)-O}^- \rightleftharpoons \text{ROPO}_3\text{H}^-$$
$$\hspace{5em} \textbf{4} \hspace{6em} \textbf{5}$$

This generates the zwitterionic intermediate (**4**), which undergoes a final rapid prototropic shift to give the product in its thermodynamically stable form (**5**). If this picture is correct, and these two steps are indeed fast, we have defined the mechanism of the <u>forward</u> reaction, as the specific acid catalysed hydrolysis of the dianion (Scheme 1).

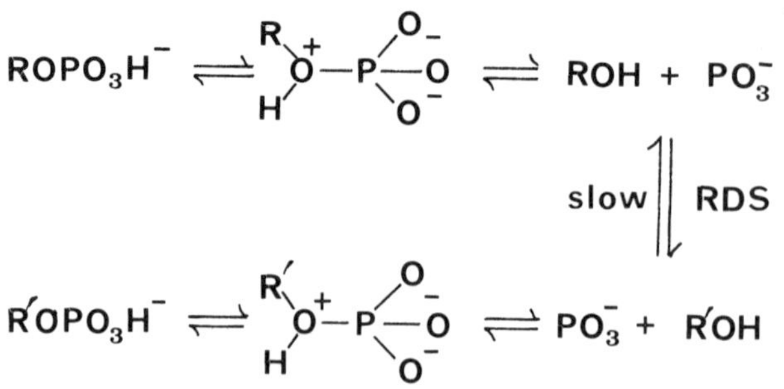

$$\text{ROPO}_3\text{H}^- \rightleftharpoons R\text{-O(H)}^+\text{-P(O}^-\text{)-O}^- \rightleftharpoons \text{ROH} + \text{PO}_3^-$$
$$\text{slow} \updownarrow \text{RDS}$$
$$R'\text{OPO}_3\text{H}^- \rightleftharpoons R'\text{-O(H)}^+\text{-P(O}^-\text{)-O}^- \rightleftharpoons \text{PO}_3^- + R'\text{OH}$$

Scheme 1

---

*There is good evidence (3) that the rate determining step is unimolecular, so this reaction must occur after the rate determining step, which must therefore be a diffusion or solvent-reorganisation process.

If even the hydrolysis of the monoanion is really a reaction of the dianion, it is to the properties of the dianion we should look if we want to understand phosphate transfer. We will start with reactions very close to that described in Scheme 1, where metaphosphate is generated in the presence of a weakly basic leaving group. Under these circumstances we expect metaphosphate to have a significant lifetime; and the situation is similar when it is generated from a dianion with a very good leaving group, like phosphate or dinitrophenoxide.

$$XOP(O^-)_2O \rightleftharpoons XO^- + PO_3^-$$

The reactions of amine nucleophiles with 2,4-dinitrophenyl phosphate in water are also indiscriminate (2), but now the reaction is clearly bimolecular. The rate determining step is therefore most likely the diffusion controlled formation of the encounter complex of nucleophile and reactive intermediate.

$$XOPO_3^{2-} \rightleftharpoons XO^- + PO_3^-$$
$$Nu \updownarrow \text{diffusion steps}$$
$$NuPO_3^{2-} \rightleftharpoons Nu + PO_3^-$$

This explains why the rate of the reaction is independent of the reactivity of the nucleophile, even though it is involved in the transition state.

Now consider what happens when the leaving group XO^- is less good. The lifetime, and hence the concentration of the intermediate are reduced, both because it is formed less easily and because the more basic XO^- reacts back more rapidly with metaphosphate. There must eventually come a point where this reverse reaction becomes faster than the rate of diffusion of the nucleophile through the medium. In that case phosphate transfer via metaphosphate can only occur if the nucleophile is already present in the encounter complex when it is formed (the so-called preassociation mechanism (4)). For still poorer leaving groups metaphosphate no longer has a significant lifetime: phosphate transfer can take place only if the nucleophile actually assists in the displacement of the leaving group, and only the concerted mechanism is possible.

We can construct a diagram (Scheme 2), of the sort used by Jencks (5) for carbonyl reactions, to show how the expected mechanism of phosphate transfer from $XOPO_3^{2-}$ depends on the basicity of the leaving group. For very good leaving groups (ROH, ArOH, where the pKa of the conjugate acid (ROH_2^+, etc.) is < -4) metaphosphate is formed rapidly, and the reaction is clearly stepwise. For very poor leaving groups like alkoxides (pKa of ROH > 10) only the concerted displacement is possible (though it is not actually observed in practice because reaction has become too slow). In between there must be a borderline region, and since reactivity in the concerted reaction depends on the nucleophile concerned, so too will the position of this borderline. For example, it might be expected to lie at lower pKa for solvolysis reactions, where only weak nucleophiles are involved, than for reactions with nucleophiles which are more reactive towards phosphorus.

We do not have enough information to define the different regions with any precision, except in the case of phosphate transfer to nitrogen. The reactions of amines with p-nitrophenyl phosphate dianion (pKa of ArOH = 7.1) depend weakly but distinctly on the basicity of the amine (6). This dependence increases for poorer leaving groups, and decreases when the leaving group is less basic; so that the rates of reactions with the 2,4-dinitrophenyl phosphate dianion are entirely independent of the basicity of the nucleophile (2). We can therefore place the borderline between concerted and stepwise mechanisms rather precisely for phosphate transfer to nitrogen, around pKa = 4 (Scheme 2).

Phosphate Transfer to Neutral Amine Nucleophiles

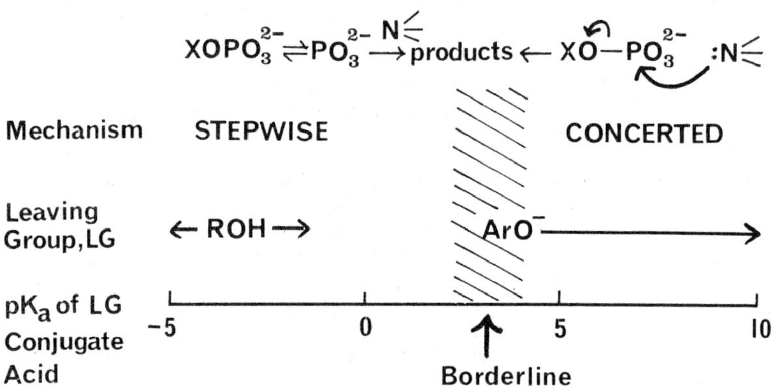

$$XOPO_3^{2-} \rightleftharpoons PO_3^{2-} \xrightarrow{N\leqslant} \text{products} \leftarrow X\overset{\curvearrowright}{O}-PO_3^{2-} :N\leqslant$$

Mechanism	STEPWISE	CONCERTED
Leaving Group, LG	← ROH →	ArO⁻ →
pKa of LG Conjugate Acid	−5 ..... 0	↑ 5 10
		Borderline

Scheme 2

ENZYME REACTIONS

Now let us consider an enzyme reaction involving phosphate transfer to nitrogen. A relevant example is creatine kinase, which catalyses the transfer of the terminal phosphate group of ATP to the guanidine nitrogen of creatine.

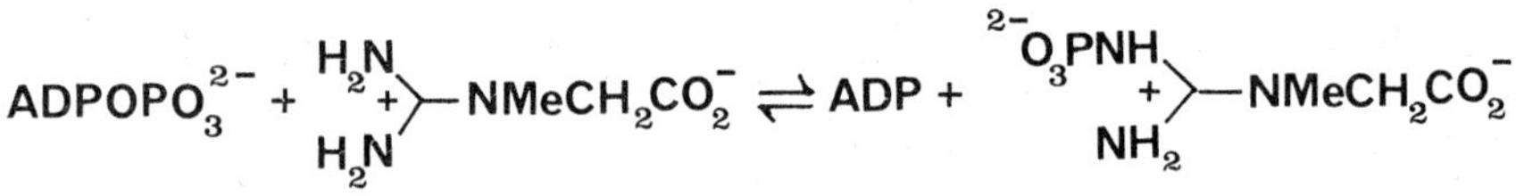

$$ADPOPO_3^{2-} + \overset{H_2N}{\underset{H_2N}{\gtrless}}-NMeCH_2CO_2^- \rightleftharpoons ADP + \overset{^{2-}O_3PNH}{\underset{NH_2}{\gtrless}}-NMeCH_2CO_2^-$$

The leaving group here is ADP, so the pKa of the conjugate acid of the leaving group (the second dissociation of the terminal phosphate group of ADP) is about 6; though this will no doubt be reduced by electrophilic catalysis by the metal cation known to be present in the active site. Inspection of Scheme 2 shows that the mechanism of a bimolecular reaction of this sort is expected to be concerted. And if we look at the beautiful picture (**6**) of the transition state for the enzyme reaction built by McLaughlin et al (7) from NMR experiments we find good evidence that the enzyme reaction is indeed concerted.

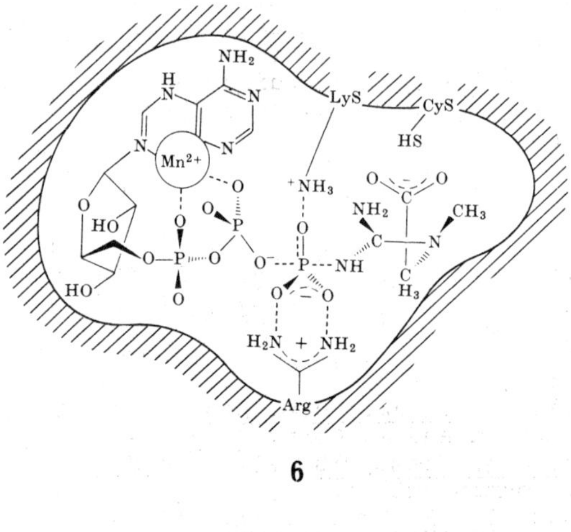

6

In fact Scheme 2 should be modified before it can be applied to enzyme catalysed reactions. In particular, reaction with a nucleophile in an enzyme active site is effectively an intramolecular reaction. As a result the reactivity of the nucleophile is likely to be substantially enhanced - in simple intramolecular reactions factors of $10^6 - 10^8$ are common (8), - so that the borderline between stepwise and concerted mechanisms (Scheme 2) will be shifted to much lower pKa for enzyme-catalysed phosphate transfer. All reactions of nitrogen nucleophiles with ATP, for example, should be well within the region of concerted mechanism.

It is difficult to extend this analysis with confidence to phosphate transfer to oxygen, but the solvolysis of p-nitrophenyl and 2,4-dinitrophenyl phosphate dianions in alcohol-water mixtures shows some selectivity at 39°, though none at 100° (3). This suggests that the borderline region for phosphate transfer to oxygen may not differ too much from the picture deduced for nitrogen (Scheme 2). In which case concerted mechanisms are to be expected for these reactions also. Here there is a good deal of recent evidence from stereochemical studies of reactions of substrates with chiral terminal phosphate groups, and the general conclusion at this stage, based on results with 9 different enzymes (9) is that all enzyme-catalysed phosphate transfer steps go with inversion at phosphorus. This is at least consistent with all-concerted mechanisms, though more searching probes of mechanism could modify this view (see, for example, 10).

We have recently been working with some acetals, which behave very much like phosphate monoester dianions. The reactivity towards neutral nucleophiles of methoxymethoxy-2,4-dinitrobenzene (7), for example, is closely similar - both qualitatively and quantitatively - to that of the 2,4-dinitrophenyl phosphate dianion (11). In both cases donor oxygen(s) provide a powerful driving force for S_N1- like mechanisms, but the instability of the intermediate (8 in the case of the acetal) limits the scope of the stepwise process, so that borderline concerted reactions are observed.

The important point is that these systems remain highly reactive because the 'push' from donor oxygen is still the most important factor in the displacement of the leaving group, even in the concerted reactions. This is because the transition state remains very close in structure to the intermediate (8 or metaphosphate). One consequence is that it may be very difficult in practice to distinguish this sort of concerted displacement from the stepwise mechanism. But the clear conclusion from our analysis is that enzyme catalysed phosphate transfers will generally, and perhaps exclusively, be concerted processes.

REFERENCES

1. S.A. Khan and A.J. Kirby, J. Chem. Soc. B, 1970, 1172.
2. A.G. Varvoglis and A.J. Kirby, J. Chem. Soc. B, 1968, 135.
3. A.G. Varvoglis and A.J. Kirby, J. Am. Chem. Soc., 89, 415 (1967).
4. W.P. Jencks, Acc. Chem. Res., 9, 425 (1976).
5. W.P. Jencks, 'Catalysis in Chemistry and Enzymology', McGraw-Hill, New York, 1969, p. 530.
6. A.J. Kirby and W.P. Jencks, J. Am. Chem. Soc., 87, 3209 (1965).
7. A.C. McLaughlin, J.S. Leigh and M. Cohn, J. Biol. Chem., 251, 2777, (1976).
8. A.J. Kirby, Adv. Phys. Org. Chem., 18, (in press).
9. J.R. Knowles, private communication.
10. G. Lowe and B.S. Sproat, J.C.S. Chem. Comm., 1978, 783.
11. G.-A. Craze and A.J. Kirby, J.C.S. Perkin 2, 1978, 357.

PHOSPHATE SYNTHESIS, EXCHANGE AND INTERACTION IN NUCLEOTIDES AND RELATED COMPOUNDS

N. J. Leonard

Roger Adams Laboratory, School of Chemical Sciences, University of Illinois, Urbana, Illinois 61801, U.S.A.

<u>Abstract</u> - Aspects of phosphorus chemistry directed towards biology that we have discussed include the following: modified nucleoside 3',5'-bisphosphates, substrates for T4 RNA ligase; direct synthesis of modified nucleoside 5'-monophosphates; enzymatic synthesis of nucleoside 5'-triphosphate analogues; interaction of phosphate(s) in adenine nucleotides and <u>linear</u>-benzoadenine nucleotides with the heteroaromatic moiety; oligophosphate inhibitors of adenylate kinase; oxidation of nucleoside and nucleotide analogues catalyzed by xanthine oxidase; and fluorescent analogues of guanine nucleotides.

INTRODUCTION

In the general area of synthesis of components of nucleic acids, with special reference to the phosphate moieties, we have set forth a number of questions of more or less fundamental interest to which we have sought to provide answers. It has been especially helpful to synthesize analogues of nucleotides and coenzymes to obtain comparative data for direct application to the behavior of the natural compounds.

RESULTS AND DISCUSSION

Modified nucleoside 3',5'-bisphosphates

For the examination of the substrate specificity of T4 RNA ligase with the purpose of establishing a convenient method of synthesis of oligoribonucleotides of defined sequence (1,2), we have synthesized various nucleoside and modified nucleoside 3',5'-bisphosphates as donor molecules in the ligase reaction (3). The bisphosphorylation of unprotected ribonucleosides with pyrophosphoryl chloride at low temperature, first introduced in 1963 (4), produces almost exclusively 3'(2'),5'-bisphosphates (3). These mixtures of pure bisphosphates can be used directly with the T4-induced RNA ligase since the 3',5' component of the mixture is a substrate and the 2',5' component is neither a substrate nor an inhibitor (2).

Generally, a 65 ± 5% to 35 ± 5% ratio of pN(3')p to pN(2')p is observed after rapid hydrolysis of the pyrophosphoryl chloride reaction mixture with ice and triethylammonium bicarbonate. The composition of the mixture is dictated either by stereoselective nucleophilic attack by water on a cyclic 2',3'-chlorophosphate 5'-dichlorophosphate intermediate (5,6) or by a regioselective phosphorylation step (7,8).

When complete relaxation of the NMR phosphorus signal is allowed, the ^{31}P relative chemical shifts are useful in structure assignment and in the quantitative estimation of proportions in mixtures such as those encountered in the bisphorylation reaction. Examination of the proton undecoupled ^{31}P spectra of a variety of nucleoside 5'-monophosphates and nucleoside 3'(2'),5'-bisphosphates reveals that the 5'-P signals experience very little change from an average value of δ 3.88 downfield from 85% H_3PO_4 among the compounds that we have examined (3). With the signal due to the 5'-P nearly constant, the 3'-P and 2'-P can be readily differentiated, especially when the ribonucleoside 3',5'- and 2',5'-bisphosphates are present in different proportions. The 2'-P signals are consistently upfield from the 5'-P resonance. The 3'-P signals are shifted downfield by ca. 0.20 ppm from the 5'-P average position, a result of the 3'-P environment and the equilibrium among the rotamers available to the 3'-phosphate group. From these results it is clear that one can distinguish readily and in general between 5'-, 3'-, and 2'-O-phosphates on ribonucleosides.

Modified nucleoside 5'-monophosphates
A variation of the pyrophosphoryl chloride reaction, namely, the use of this reagent with m-cresol at 0-5 °C, produces 5'-monophosphates from unprotected nucleosides (9). It is now clear that this method can be extended to the synthesis of 5'-phosphates in high yield from various analogues of the natural ribonucleosides. For example, we have introduced the concept of testing the dimensional restrictions of enzyme-active sites by synthesizing cofactor analogues (e.g., 1-4) which are related to the natural cofactors by defined dimensional changes in the molecules (10). The linear-benzopurine ribotides (1-4), defined by the formal insertion of a benzene

ring (actually four carbons) into the center of the purine ring system, are of special interest because the enzyme-binding characteristics of the terminal rings are preserved but are further separated by 2.4 Å while the potential for π interaction is increased.

The basic ring systems and the related ribonucleosides are entirely synthetic; accordingly, it is highly desirable that the first phosphorylation be efficient and specific and, if possible, that it proceed without the necessity of protection-deprotection steps. The pyrophosphoryl chloride/m-cresol method (9) is indeed efficient for the conversion of the unprotected lin-benzopurine ribosides to their corresponding 5'-monophosphates 1a-4a (11-13). In the case of lin-benzoadenosine, the product (1a) is characterized by high performance liquid chromatography (HPLC), electrophoresis, ^{31}P NMR spectroscopy, and microanalysis. The integrity of the 5'-phosphorylation is established by ^{31}P NMR and by the complete reversion of the ribonucleotide to lin-benzoadenosine on incubation with 5'-nucleotidase, which is a highly specific enzyme.

The same compound (1a) is obtained by incubation of the synthetic lin-benzoadenosine 3',5'-monophosphate with beef heart nucleotide 3',5'-phosphodiesterase. This enzyme is specific for 3',5'-monophosphates and plays an important role in regulating intracellular cAMP. The lin-benzo-cAMP is active not only with this enzyme but also maximally activates brain protein kinase and protein kinase from skeletal muscle, thus interacting with protein kinase in a manner similar to cAMP (14).

Nucleoside 5'-triphosphate analogues
The 5'-diphosphates, lin-benzo-ADP (1b) (11,12), lin-benzo-GDP (2b), lin-benzo-IDP (3b), and lin-benzo-XDP (4b) can be prepared by the general method of Moffatt and Khorana (15) from the 5'-phosphoromorpholidates. The most convenient route from the modified nucleoside 5'-diphosphates to the 5'-triphosphates (1c-4c) is an enzymatic one utilizing pyruvate kinase and phosphoenolpyruvate (PEP). Here we have taken advantage of the somewhat broad specificity of nucleotide substrate (16). The lin-benzopurine nucleotides 1b-4b as substrates are not phosphorylated as efficiently as the purine nucleotides ADP, GDP, and IDP, but they appear to be converted at least as well as pyrimidine nucleotides. The fact that compounds 1b-4b containing an extended tricyclic nucleus can be readily phosphorylated adds

further weight to the conclusion that the active site does not have very stringent steric requirements (16).

We found earlier that $1,N^6$-ethenoadenosine 5'-diphosphate (εADP), containing another fluorescent tricyclic nucleus, substitutes very well for ADP in the pyruvate kinase-Mg^{2+} or Mn^{2+}-phosphoenolpyruvate system (17). Moreover, by fluorescence depolarization measurements in the case of εADP, it is shown that εADP binds to the protein in such a way that the fluorophor portion (basic nucleus) is not strongly associated through multiple points of attachment.

The γ (terminal) phosphate radiolabeling of lin-benzo-ATP (1c) is accomplished by the ability of lin-benzo-ATP to phosphorylate 3-phosphoglyceric acid catalyzed by yeast 3-phosphoglycerate kinase, which permits equilibration of the γ phosphate with $^{32}PO_4^{3-}$ (12), comparable to the preparation of [γ-^{32}P]ATP, [γ-^{32}P]εATP, and [γ-^{32}P]εCTP. A hinge-bending action of the enzyme, proposed for bringing the ATP site and the phosphoglycerate site together (18), will apparently accommodate the wider nuclei represented by lin-benzo-ATP (1c) and εATP.

Interaction of phosphate(s) in adenine nucleotides and lin-benzoadenine nucleotides with the heteroaromatic moiety.
X-ray analysis of crystalline disodium ATP (19) displays two distinguishable molecules of ATP in the asymmetric unit. In the A form the side chain is folded so that the γ phosphate is proximate to the furanose oxygen and the 8-H of the adenine ring, while folding in the B form places the γ phosphate in the vicinity of the 3'-H. Solutions of ATP, examined by nuclear magnetic resonance with paramagnetic metal ions as probes of phosphate geometry, appear to adopt side-chain folding corresponding to a conformational average resembling the A form of the crystal (20-22). In an examination of the conformational possibilities for metal-nucleotide interaction, Sundaralingam (23) discounted the 6-NH_2 as a complexing site (24) for ATP combinations with transition metal ions: e.g., Mn^{2+}, Co^{2+}, Ni^{2+}. Instead, it was concluded that the triphosphate group assumes an anti-gauche conformation, allowing metal ions to bind simultaneously to 7-N and two or three phosphates (ATP) or to 7-N and two phosphates (ADP), with ribose in the anti conformation. There is general agreement on the preferred anti conformation for the ribotyl substituent on the 9-N at neutral or basic pH (25,26) and the line broadening of the 2-H in the NMR spectrum of AMP-Mn^{2+} at pH 8.5 indicates a similar conformation at the monophosphate level (27,28). For ADP and ATP, titrimetric data obtained in aqueous solution have been used as evidence both for and against interaction between the adenine ring and the phosphate side chain (24,29).

Among the analogous lin-benzoadenine nucleotides, the spectroscopically determined pK_a values of the series 1a, b, c in aqueous solution give a unique response to the presence and conformation of the phosphate side chain (11,30). When no intramolecular interaction can occur between phosphate and base, as in lin-benzoadenosine and its 3',5'-cyclic monophosphate, the N^+-H pK_a values are identical: 5.6. By contrast, the pK_a values of 1a, b, c, which are 7.6, 7.3, 7.1, respectively, indicate that the phosphates in these molecules are involved in base protonation-deprotonation. The decrease in the pK_a values of lin-benzo-ADP (1b) and lin-benzo-ATP to 6.9 and 6.6, respectively, in the presence of 5 mM Mg^{2+} is parallel to, but greater than, that observed for ADP and ATP (31) and indicates the formation of Mg^{2+} complexes with the pyrophosphate unit (32) of 1b and 1c. The pK_a values for 1b and 1c are also lowered in the presence of Mn^{2+} and Co^{2+}. Association of lin-benzo-ATP with quaternary ammonium micelles reduces the N^+-H pK_a value to that of lin-benzoadenosine by breaking the interaction between the phosphate and the base. The rise in fluorescence polarization of lin-benzo-ATP in the presence of $CetNMe_3Cl$ micelles confirms the electrostatic binding.

It is the fluorescence of the lin-benzoadenine nucleotides that enables us to determine their association constants with Mg^{2+}, Mn^{2+}, and Co^{2+}. The fluorescence quenching effect of a paramagnetic metal ion is shown in Fig. 1 in the fluorometric titration of lin-benzo-ATP (1c) with $CoCl_2$, followed by restoration of fluorescence with $MgSO_4$ in much higher concentration. In addition to the measurement of fluorescence quantum yields, the measurement of fluorescence lifetimes of lin-benzoadenine nucleotides by phase and modulation as a function of Co^{2+} concentration permits determination of the

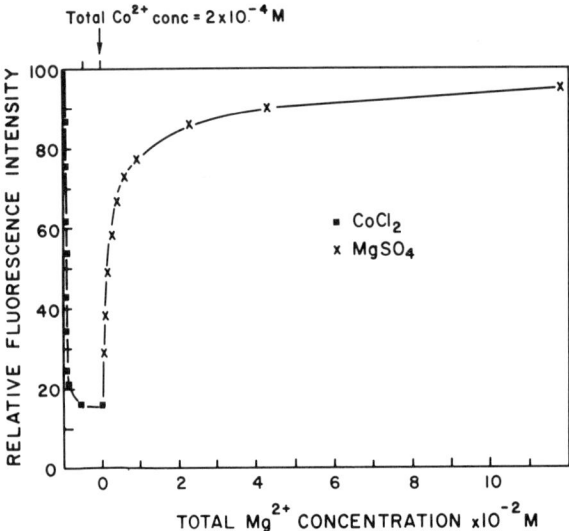

Fig. 1. Fluorometric titration of < 10 μM lin-benzo-ATP (accurately determined) with $CoCl_2$, followed by $MgSO_4$, in aqueous solution at pH 8.5 and 23°. Total Co^{2+} = 0.2 mM. ■, $CoCl_2$; x, $MgSO_4$ (28).

static component of the quenching due to intramolecular complex formation. Although there is some collisional quenching of the heteroaromatic ring by the bound Co^{2+}, the major quenching effect is due to the formation of a static complex. The association constants of paramagnetic metal ions, such as Co^{2+} and Mn^{2+}, to the lin-benzoadenine nucleotides are determined by the

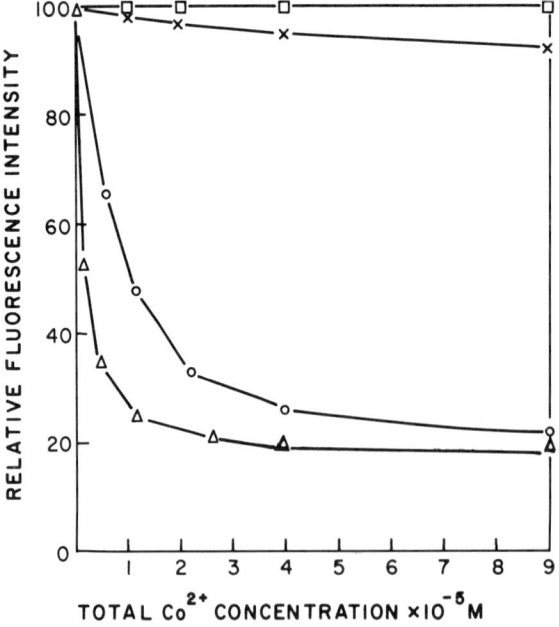

Fig. 2. Change in fluorescence intensity with increasing Co^{2+} concentration in aqueous solution at pH 8.5 and 23°. Derivatives present at < 10 μM. □, lin-Benzoadenosine; x, lin-benzo-AMP; O, lin-benzo-ADP; Δ, lin-benzo-ATP (28).

decrease in fluorescence intensity of the lin-benzoadenine moiety with increasing concentration of the metal ion (Fig. 2). The association constants of diamagnetic metal ions such as Mg^{2+}, which is not a fluorescence quencher, are obtained by competition studies. Similar values are obtained for Co^{2+} and Mn^{2+}, greater than those for Mg^{2+} for the series and

greater than the association constants observed for the corresponding
adenine nucleotides. The association constants are in the order: lin-benzo-ATP > lin-benzo-ADP > lin-benzo-AMP.

The paramagnetic quenching effect of Mn^{2+} on the fluorescent ATP analogue,
$1,N^6$-ethenoadenosine triphosphate or εATP, has been used to determine the
proximity of the nucleotide binding site to the high affinity divalent
cation binding site on actin, the fibrous protein constituent of muscle
(33). The degree of fluorescence quenching is identical for the
εATP-Mn^{2+} complex and the εATP-Mn^{2+}-actin complex, implying that the
nucleotide and divalent cation bind to actin as a metal-nucleotide complex.

Additional evidence for the interaction of Co^{2+} and the heteroaromatic
ring of lin-benzo-ATP in aqueous solution at pH 8.5 is obtained by means
of Fourier transform ^{1}H NMR. When lin-benzo-ATP (1c) is titrated with
Co^{2+}, a progressive broadening of the aromatic proton signals is observed
with the 2-H (5) (corresponding to the 8-H of ATP) showing the greatest
broadening. lin-Benzo-cAMP, serving as a control, shows no significant
broadening of the ^{1}H signals even at higher Co^{2+}/nucleotide concentration
ratios than for the triphosphate 1c. It can therefore be concluded

additionally that the phosphate-bound Co^{2+} or Mn^{2+} interacts with 1-N of
the ring system in the anti conformation of 1c and 1b. The positioning
of the ring and the conformation of the phosphate chain of lin-benzo-ATP
chelated with paramagnetic ions involve 1-N just as ATP can involve the
equivalent 7-N of the natural ring system.

In the adenine nucleotides, e.g., 8-bromo-5'-AMP, the NMR signal of the
2-H does not detect the presence of the syn conformation (25). In the
fluorescent lin-benzoadenine nucleotides (1a-c), we have the advantage of
a 4-H on the central ring that can detect the syn conformation. It is
necessary to assign unequivocally the proton magnetic resonances of the
heteroaromatic nucleus in order to make safe conclusions about intra- and
intermolecular interactions or about the site of protonation of the various
lin-benzoadenine nucleotides (1).

The 2-H in the series can be exchanged with deuterium (and 2-D with 2-H) by
heating the nucleotides in D_2O (H_2O) for several hours (34), in accord
with the behavior of most benzimidazoles. Identification of the chemical
shift corresponding to the 2-H was important in coming to the conformational
conclusions described earlier. The 4-H and 9-H can be exchanged
unequivocally and to varying extents in the common synthetic precursors of
lin-benzoadenosine, the 5'-phosphate series 1a-c, lin-benzo-cAMP, and a
dimeric monitor of intramolecular complexation, lin-benzo-A(5')p_2(5')lin-benzo-A (34). The controlled incorporation of deuterium to set
different extents in the three positions 2, 4, and 9, along with complete
retention of hydrogen at position 6, permits convenient assignment of the
^{1}H NMR signals throughout since their integration corresponds to the
percent H (versus D) at each position. This technique greatly reduces
the number of experiments and variables involved and prevents misassignments
of the signals upon changing solution concentration and the multiplicity
of phosphate groups.

Since the analogues possess a proximal aromatic proton on each side of the
ribotyl-N (2-H and 4-H in 5), the assigned NMR chemical shifts are sensitive
to a phosphate environment and are indicative of N-C_1, anti or syn
conformations under different conditions. The 6-H chemical shift is a
monitor of charge; protonation occurs on the pyrimidine ring. The 6-H has

similar chemical shifts for all the lin-benzoadenine nucleotides at pD 8.5 and infinite dilution, and similar downfield shifts are observed upon protonation of the nucleus (Table 1). The absence of a downfield shift

TABLE 1. Differential chemical shifts ($\delta_{pD\ 8.5} - \delta_{pD\ 4.0}$) for the lin-benzoadenine nucleotides at infinite dilution.[a]

Compound	2-H	4-H	6-H	9-H	1'-H
lin-Benzo-cAMP	0.00	+0.05	-0.25	-0.05	-0.15
lin-Benzo-AMP (1a)	+0.20	-0.60	-0.15	+0.10	0.00
lin-Benzo-ADP (1b)	+0.25	-0.60	-0.15	+0.15	0.00
lin-Benzo-ATP (1c)	+0.30	-0.45	-0.15	+0.20	0.00

[a]Values are within a precision of ±0.03 ppm (34).

for 2-H upon protonation provides additional evidence that at pD 4.0 the base is protonated on the pyrimidine ring. Acid does not change the chemical shifts for 4-H in lin-benzo-cAMP appreciably, whereas it produces a substantial shielding of 4-H in 1a-c. In the titration of lin-benzo-AMP (pK_a = 7.6), the 4-H first experiences a dramatic downfield shift, $\sim$ 0.85 ppm at 3 mM, upon changing the pD from 8.5 to the range 6.5-6.0. The two negative charges on the α phosphate produce a deshielding effect on 4-H that is not observed in 1b and 1c at similar pD. When the pD is lowered to 4.0, protonation of the secondary phosphate occurs and an upfield shift (0.25 ppm) is observed. The net downfield shift for the 4-H of lin-benzo-AMP in going from pD 8.5 to 4.0 is 0.60 ppm at 3 mM (Table 1, cf. 1b, c). These NMR results indicate that in the N-protonated form the syn conformation is preferred in comparison with the anti conformation in the neutral form. In the N-protonated syn conformation, a better stabilization of the positive charge on the base results from intramolecular coulombic interaction between first ionized phosphate and protonated base at low pH.

Oligophosphate inhibitors of adenylate kinase

Adenylate kinase, the widely occurring enzyme that catalyzes the reaction: ATP-Mg^{2+} + AMP $\rightleftarrows$ ADP-Mg^{2+} + ADP, is important in maintaining equilibrium among the several species in the adenine nucleotide pool (35). In the design of inhibitors of adenylate kinase, the thousand-fold greater inhibition of A(5')p_5(5')A compared with A(5')p_4(5')A (36-40) suggests that the additional phosphate of the former, which lengthens the phosphate chain by about 2.7 Å (based on sodium triphosphate, $Na_5P_3O_{10}$, as a model (41,42)) is crucial for strong inhibition.

We have asked the question whether there will be a similar inhibition-enhancing effect of stretching one of the terminal adenines by 2.4 Å (43,44). lin-Benzo-A(5')p_4(5')A (6, n = 3) is an excellent candidate for examining inhibitor interaction with adenylate kinase since association constants can be obtained by both kinetics and fluorescence experiments.

6

lin-Benzo-A(5')p$_5$(5')A (6, n = 4) and lin-benzo-A(5')p$_4$(5')A (6, n = 3) are synthesized by reaction of lin-benzoadenosine 5'-monophosphoromorpholidate with adenosine 5'-tetraphosphate (tri-n-octylammonium salt) (40). Degradation of adenosine tetraphosphate during the course of the reaction results in the formation of both desired compounds and other combinations. Under identical conditions, the substitution of ATP for adenosine tetraphosphate results in the formation of lin-benzo-A(5')p$_4$(5')A in higher yield. Before fractionation of the reaction mixture on a DEAE cellulose column, all terminal phosphates are hydrolyzed by alkaline phosphatase. The elimination of these terminal phosphates simplifies the column chromatography. lin-Benzo-A(5')p$_5$(5')A and lin-benzo-A(5')p$_4$(5')A are identified on the basis of the following evidence. Both compounds contain identical amounts of adenine and lin-benzoadenine since the absorbance ratio of 259 to 331 nm observed on hydrolysis is that of an equimolar mixture of AMP and lin-benzo-AMP. Total phosphate digestion of lin-benzo-A(5')p$_5$(5')A and lin-benzo-A(5')p$_4$(5')A by snake venom phosphodiesterase and alkaline phosphatase yields ratios of phosphate to lin-benzoadenosine residue of 4.8 and 3.9, respectively. No terminal phosphates are present as judged by both ^{31}P NMR spectra and by treatment of 6 (n = 3, 4) with alkaline phosphatase and determination of inorganic phosphate. The ^{31}P chemical shifts and coupling constants of the lin-benzo-A(5')p$_4$(5')A and lin-benzo-A(5')p$_5$(5')A are also consistent with these identifications, employing A(5')p$_4$(5')A and A(5')p$_5$(5')A as model compounds.

Compounds A(5')p$_4$(5')A and A(5')p$_5$(5')A are inhibitors of adenylate kinase (Table 2) and are competitive with respect to ATP and AMP. The association

TABLE 2. Association constants of dinucleoside oligophosphates with adenylate kinase (43,44).

Compound	$K(M^{-1})^a$
A(5')p$_4$(5')A^b	8.0 X 10^4
A(5')p$_5$(5')A^b	8.0 X 10^7
lin-Benzo-A(5')p$_4$(5')A^b	2.2 X 10^6
lin-Benzo-A(5')p$_4$(5')A^c	2.1 X 10^6
lin-Benzo-A(5')p$_4$(5')A^d	1.0 X 10^6
lin-Benzo-A(5')p$_5$(5')A^b	2.1 X 10^5
lin-Benzo-A(5')p$_5$(5')A^c	2.0 X 10^5
lin-Benzo-A(5')p$_5$(5')A^d	1.7 X 10^5

^{a}Reciprocals of these association constants are the kinetic inhibition constants. ^{b}Determined by kinetic inhibition with varying ATP concentration. ^{c}Determined by kinetic inhibition with varying AMP concentration. ^{d}Determined using fluorescence to monitor equilibrium binding to adenylate kinase.

constants are within the range of the values previously reported for these compounds as competitive inhibitors of adenylate kinase with respect to both ATP and AMP (37-39). Adenylate kinase has a random bi bi mechanism (i.e., there is no order in which substrates must add or products must leave the enzyme (39)), and the inhibition constant of any inhibitor competitive with either of the two substrates will be identical with a dissociation constant determined by means other than kinetics. That binding of these inhibitors occurs at the active site and not at a non-catalytic substrate site is consistent with what is known about adenylate kinase, namely that a non-catalytic site for substrate or substrate analogues has never been indicated (35).

Although A(5')p$_5$(5')A is a better inhibitor (ca. 1000 X) than A(5')p$_4$(5')A, the reverse order holds for lin-benzo-A(5')p$_4$(5')A and lin-benzo-A(5')p$_5$(5')A, the former being a better inhibitor (ca. 10 X) than the latter. The finding that the lin-benzo-A(5')p$_4$(5')A inhibition constant is intermediate between the values for A(5')p$_4$(5')A and A(5')p$_5$(5')A suggests that the changes in inhibition constants, as a partial function of the distance between the two pyrimidine rings, are qualitatively similar whether the phosphate chain is lengthened (ca. 2.7 Å from p$_4$ to p$_5$) or one terminal base is widened (2.4 Å from A to lin-benzo-A) (6, n = 3). Furthermore, the pyrimidine-to-pyrimidine distance in lin-benzo-A(5')p$_4$(5')A is longer than that of A(5')p$_4$(5')A only if the lin-benzoadenosine moiety is in the anti conformation as it is for the lin-benzoadenine nucleotides. A tentative conclusion is thereby reached that this is the preferred conformation of the lin-benzoadenosine portion of lin-benzo-A(5')p$_4$(5')A bound to adenylate kinase. It has been demonstrated previously that AMP is bound to adenylate kinase in the anti conformation (45). Since lin-benzo-ATP but not lin-benzo-AMP is a substrate (12), it is reasonable to conclude that the ATP site of adenylate kinase accepts its substrates in the anti conformation.

lin-Benzo-A(5')p$_4$(5')A and lin-benzo-A(5')p$_5$(5')A (6, n = 3, 4) have substantially lower fluorescence quantum yields and shorter lifetimes than lin-benzo-AMP, a result of intramolecular stacking interaction between the lin-benzoadenine and adenine portions of these molecules in dilute solution (43,44). The increases in the fluorescence quantum yields of lin-benzo-A(5')p$_4$(5')A and lin-benzo-A(5')p$_5$(5')A when bound to adenylate kinase result from the breaking of this intramolecular stacking and implicate inhibitor binding to the enzyme in an "open" or "extended" form of the oligophosphate chain.

Oxidation of nucleoside and nucleotide analogues catalyzed by xanthine oxidase
Milk xanthine oxidase catalyzes the oxidation of a very wide variety of heterocyclic bases (46,47) and xanthosine at one-twentieth the rate of xanthine. Other ribonucleosides are not substrates, nor have any ribonucleotides been described as substrates.

In this laboratory we find that lin-benzohypoxanthine (3, R = H) undergoes xanthine-oxidase catalyzed oxidation to lin-benzoxanthine (4, R = H) and thence to lin-benzouric acid at an apparently faster rate than hypoxanthine (48). The reaction is not completely unexpected due to the broad substrate specificity attributed to this enzyme, but the relative rate and degree of oxidation of the tricyclic substrate in comparison with hypoxanthine does provide additional definition of the geometrical adaptivity of the active site(s). What is unusual is the enzymatic oxidation of the modified nucleoside lin-benzoinosine (3, R = ribosyl) at a rate equivalent to that of hypoxanthine, whereas inosine is not oxidized appreciably. Evidently, the ribosyl appendage on the tricyclic nucleus at the 3 position does not interfere with enzyme binding and/or oxidative attack, whereas the ribosyl at the 9 position of hypoxanthine, closer to the pyrimidine end of the molecule, provides steric interference.

A more dramatic test of the tolerance of bulk at the 3 position of the tricyclic nucleus lies in a procedure that has been developed for the synthesis of lin-benzo-XMP (4a). The xanthine-catalyzed oxidation of lin-benzo-IMP (3a) gives lin-benzo-XMP (4a) in 98% yield (13,49). The relative rate of oxidation has not been determined. Since the major goal was a satisfactory conversion, the reaction was not hastened but was allowed to run its course. This is the first reported enzymatic oxidation of a ribonucleotide to the corresponding dioxo product. It is, in fact, of interest that the oxidation does not proceed further. In this regard, it bears similarity to the mono-oxidation (in the pyrimidine ring) of distal-benzohypoxanthine (48) with xanthine oxidase. The observation in the case of 3a suggests that the distance of the ribose 5'-phosphate moiety from the pyrimidine ring permits oxidation to occur at the 6 carbon (see 4 for numbering) while maintaining steric protection of the imidazole 2 carbon.

Fluorescent analogues of guanine nucleotides
Although fluorescent derivatives of guanine have been reported, none to date possesses unaltered terminal rings and most are modifications of guanine nuclei preexistent in the system under investigation. We have been able to synthesize lin-benzoguanosine 5'-mono-, di-, and triphosphates

(2a-c), laterally extended versions of the natural nucleotides in which the substituted pyrimidine ring is not further modified (13,49). The keys to the synthesis of lin-benzoguanosine are in the introduction of a carbon and three nitrogens in a 1,2,4,5 pattern on a central benzene nucleus, with little or no regioisomer formation; introduction of at least one solubilizing group; and unequivocal positioning of an entering ribofuranosyl moiety at the desired position in a common intermediate (49,50). Then, by a divergent synthetic pathway we have synthesized lin-benzoguanosine (2, R = ribosyl), lin-benzoinosine (3, R = ribosyl), and lin-benzoxanthosine (4, R = ribosyl).

The monophosphates (2-4a) are obtained by 5'-monophosphorylation using pyrophosphoryl chloride and m-cresol, converted to the 5'-diphosphates (2-4b) chemically, and these are converted to the 5'-triphosphates (2-4c) enzymatically with pyruvate kinase and phosphoenolpyruvate, all as mentioned earlier.

These nucleotides of lin-benzoguanine, lin-benzoxanthosine, and lin-benzoinosine have fluorescence quantum yields in aqueous solution of 0.39, 0.55, and 0.04 and fluorescence lifetimes of 6, 9, and ≈ 1.5 nsec, respectively (13). Application of these as dimensional probes, and particularly of the fluorescent analogue of GTP (2c), in biological systems are anticipated that will utilize both the defined dimensional change from the natural GTP and the diagnostic value of the fluorescence.

Acknowledgment - This work was supported by Research Grant GM-05829 from the National Institutes of Health, U.S. Public Health Service.

REFERENCES

1. T. E. England, R. I. Gumport, and O. C. Uhlenbeck, Proc. Natl. Acad. Sci. U.S.A. 74, 4839 (1977).
2. T. E. England and O. C. Uhlenbeck, Biochemistry 17, 2069 (1978).
3. J. R. Barrio, M. del C. G. Barrio, N. J. Leonard, T. E. England, and O. C. Uhlenbeck, Biochemistry 17, 2077 (1978).
4. M. Honjo, K. Imai, Y. Furukawa, H. Morijama, K. Yasumatsu, and A. Imada, Takeda Kenyuskko Nempo 22, 47 (1963).
5. F. H. Westheimer, Acc. Chem. Res. 1, 70 (1968).
6. J. Tomasz and A. Simoncsits, J. Carbohydr. Nucleosides Nucleotides 2, 315 (1975).
7. F. Lynen, Proc. 4th. Int. Congr. Biochem. 11, 230 (1958).
8. W. Gruber and F. Lynen, Justus Liebigs Ann. Chem. 659, 139 (1962).
9. K. Imai, S. Fujii, K. Takanohashi, Y. Furukawa, T. Masuda, and M. Honjo, J. Org. Chem. 34, 1547 (1969).
10. N. J. Leonard, Heterocycles 12, 129 (1979).
11. D. I. C. Scopes, J. R. Barrio, and N. J. Leonard, Science 195, 296 (1977).
12. N. J. Leonard, D. I. C. Scopes, P. VanDerLijn, and J. R. Barrio, Biochemistry 17, 3677 (1978).
13. N. J. Leonard and G. E. Keyser, Proc. Natl. Acad. Sci. U.S.A., in press.
14. M. J. Schmidt, L. L. Truex, N. J. Leonard, D. I. Scopes, and J. R. Barrio, J. Cyclic Nucleotide Res. 4, 201 (1978).
15. J. G. Moffatt and H. G. Khorana, J. Am. Chem. Soc. 83, 649 (1961).
16. F. J. Kayne, The Enzymes 8, 353 (1973).
17. J. R. Barrio, J. A. Secrist III, Y.-H. Chien, P. J. Taylor, J. L. Robinson, and N. J. Leonard, FEBS Lett. 29, 215 (1973).
18. R. D. Banks, C. C. F. Blake, P. R. Evans, R. Haser, D. W. Rice, G. W. Hardy, M. Merrett, and A. W. Phillips, Nature 279, 773 (1979).
19. O. Kennard, N. W. Isaacs, J. C. Coppola, A. J. Kirby, S. Warren, W. D. S. Motherwell, D. G. Watson, D. L. Wampler, D. H. Chenery, A. C. Larson, K. Ann Kerr, and L. R. Di Sanseverino, Nature 225, 333 (1970).
20. T. A. Glassman, C. Cooper, L. W. Harrison, and T. J. Swift, Biochemistry 10, 843 (1971).
21. P. Tanswell, J. M. Thornton, A. V. Korda, and R. J. P. Williams, Eur. J. Biochem. 57, 135 (1975).
22. P. A. Hart, J. Am. Chem. Soc. 98, 3735 (1976).
23. M. Sundaralingam, Biopolymers 7, 821 (1969).

24. R. J. Labotka, T. Glonek, and T. C. Myers, J. Am. Chem. Soc. 98, 3699 (1976).
25. F. E. Evans and R. H. Sarma, FEBS Lett. 41, 253 (1974).
26. M. P. Schweizer, A. D. Broom, P. O. P. Ts'O, and D. P. Hollis, J. Am. Chem. Soc. 90, 1042 (1968).
27. S. I. Chan and J. H. Nelson, J. Am. Chem. Soc. 91, 168 (1969).
28. R. A. Dwek, Nuclear Magnetic Resonance in Biochemistry, Clarendon Press, Oxford, pp. 275-278 (1973).
29. R. Phillips, P. Eisenberg, P. George, and R. J. Rutman, J. Biol. Chem. 240, 4393 (1965).
30. P. VanDerLijn, J. R. Barrio, and N. J. Leonard, Proc. Natl. Acad. Sci. U.S.A. 75, 4204 (1978).
31. R. M. Izatt, J. J. Christensen, and J. H. Rytting, Chem. Rev. 71, 439 (1971).
32. M. Cohn and T. R. Hughes, J. Biol. Chem. 237, 176 (1962).
33. J. Loscalzo and G. H. Reed, Biochemistry 15, 5407 (1976).
34. J. R. Barrio, F.-T. Liu, G. E. Keyser, P. VanDerLijn, and N. J. Leonard, J. Am. Chem. Soc. 101, 1564 (1979).
35. L. Noda, The Enzymes 8, 279 (1973).
36. C. M. Anderson, F. H. Zucker, and T. A. Steitz, Science 204, 375 (1979).
37. P. Feldhaus, T. Fröhlich, R. S. Goody, M. Isakov, and R. H. Schirmer, Europ. J. Biochem. 57, 197 (1975).
38. G. E. Lienhard and I. I. Secemski, J. Biol. Chem. 248, 1121 (1973).
39. D. L. Purich and H. J. Fromm, Biochim. Biophys. Acta 276, 563 (1972).
40. J. R. Reiss and J. G. Moffatt, J. Org. Chem. 30, 3381 (1965).
41. D. E. C. Corbridge, Acta Crystallogr. 13, 263 (1960).
42. D. R. Davies and D. E. C. Corbridge, Acta Crystallogr. 11, 315 (1958).
43. P. J. VanDerLijn, Ph.D. Thesis, University of Illinois, 1979.
44. P. VanDerLijn, J. R. Barrio, and N. J. Leonard, Biochemistry, in press.
45. A. Hampton, P. J. Harper, and T. Sasaki, Biochemistry 11, 4965 (1972).
46. T. A. Krenitsky, S. M. Niel, G. B. Elion, and G. H. Hitchings, Arch. Biochem. Biophys. 150, 585 (1972).
47. R. C. Bray, The Enzymes 12, 209 (1975), and references therein.
48. N. J. Leonard, M. A. Sprecker, and A. G. Morrice, J. Am. Chem. Soc. 98, 3987 (1976).
49. G. E. Keyser, Ph.D. Thesis, University of Illinois, 1979.
50. G. E. Keyser and N. J. Leonard, J. Org. Chem., in press.

CHIRAL ANTICANCER OXAZAPHOSPHORINANES - STEREOSPECIFIC SYNTHESIS, CONFIGURATIONAL ASSIGNMENTS AND BIOLOGICAL EVALUATIONS

W. J. Stec

Polish Academy of Sciences, Centre of Molecular and Macromolecular Studies, 90-362 Łódź, Boczna 5, Poland

<u>Abstract</u> - *The chirality of Cyclophosphamide (<u>1</u>) and its analogs <u>2</u>, <u>3</u> and <u>4</u> necessitates consideration of stereochemical factors within the mechanism of action of these drugs, and moreover, the chemical significance thereof. The syntheses of enantiomeric forms of Cyclophosphamide (<u>1</u>), Isophosphamide (<u>2</u>), Trophosphamide (<u>3</u>) and Sulphosphamide (<u>4</u>) are described. Optically active 1-methylbenzylamine has been used as a substrate for the synthesis of intermediary diastereomeric N-substituted 1,3,2-oxazaphosphorinanes. Their separation and hydrogenolytic removal of 1-methylbenzyl group led to desired products. The synthesis of some major chiral metabolites of <u>1</u> and <u>2</u> is described as well. The results of preliminary biological studies on stereodifferentiated metabolism and mode of action of chiral anticancer oxazaphosphorinanes are presented.*

The importance of stereochemistry as a determinant of pharmacological activity is widely recognized. Often, when a drug molecule possesses one or more chiral centres, its biological activity resides largely or even exclusively in one enantiomer or diastereomer. Such differences in activity could result from the differences in the rate or profile of metabolism, absorption or tissue distribution, or a combination thereof.
Cyclophosphamide{2-|bis(2-chloroethyl)amino|tetrahydro-2H-1,3,2-oxazaphosphorine 2-oxide <u>1</u>} has been one of the most widely used antitumour agents since its discovery in 1958 (Ref.1). Despite the wealth of chemical and experimental knowledge generated during that time one feature of this compound has not been explored which relates to the structure of the molecule itself. Cyclophosphamide is a chiral molecule by virtue of its asymmetric phosphorus atom, and optical isomers are therefore possible. Hitherto the drug has been administered clinically as the racemate.
It was of great interest to undertake efforts to synthesize the enantiomeric forms of <u>1</u>, and if successful, to perform the comparative studies on their metabolism and therapeutic efficacies thereof.
There are precedents for supposing that asymmetry at phosphorus can elicit therapeutic differences. As an example the greater enzyme inhibition by the S-(-) isomer of some anticholinesterases, the alkylmethylphosphonothioates compared with their R-(+) antipodes may be cited (Ref.2). The enantiomers had also different toxicities *in vivo* (rat), although the (S)-form was not always the more toxic one. Comparative metabolic studies on the enantiomers and racemate of O-p-cyanophenyl O-ethyl phenylphosphonothioate have shown that metabolism can also be stereoselective (Ref.3). Besides the practical aspects of metabolic studies on the enantiomers of <u>1</u>, such as the more effective use of existing drugs, there is   another aspect of these studies namely the use of stereochemistry as a investigative tool for the mode of action and metabolism profile of a family of known anticancer drugs based on the 1,3,2-oxazaphorine skeleton (<u>1-4</u>). These are chiral by virtue of an asymmetric phosphorus atom:

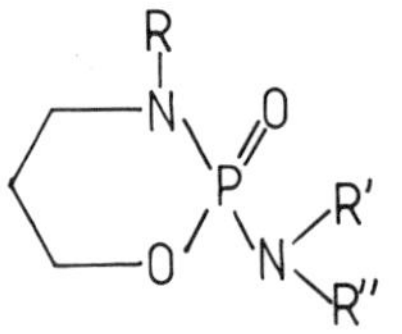

1. Cyclophosphamide R=H, R'=R''=-CH$_2$CH$_2$Cl
2. Isophosphamide R=R'=-CH$_2$CH$_2$Cl, R''=H
3. Trophosphamide R=R'=R''=-CH$_2$CH$_2$Cl
4. Sulphosphamide R=-CH$_2$CH$_2$Cl, R'=-CH$_2$CH$_2$OSO$_2$CH$_3$
   R''=H

Our first attempt to synthesize the enantiomers of 1 failed. 2-Methoxy-2-seleno-1,3,2-oxazaphosphorinane did not react with (+) strychnine. Therefore the potential chiral precursor 5, was not obtained. This precursor after Se-methylation could have been converted to 1 (Ref.4).

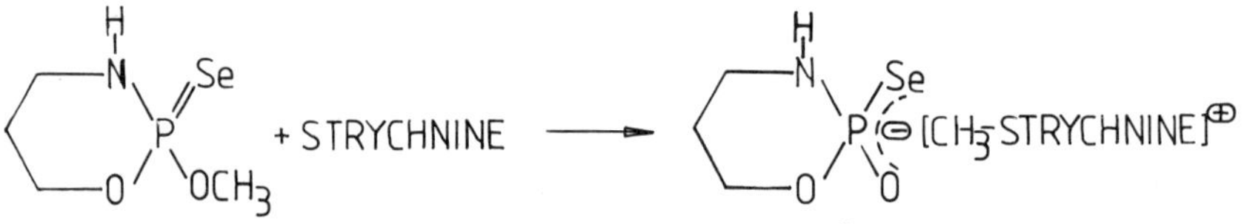

Thus, we took a different route to synthesize the enantiomers of 1, using an approach so far not explored in the synthesis of optically active phosphoramidates. The strategy is shown in Scheme 1. Condensation of chiral racemic phosphochloridate with the optically active 1-methyl benzylamine

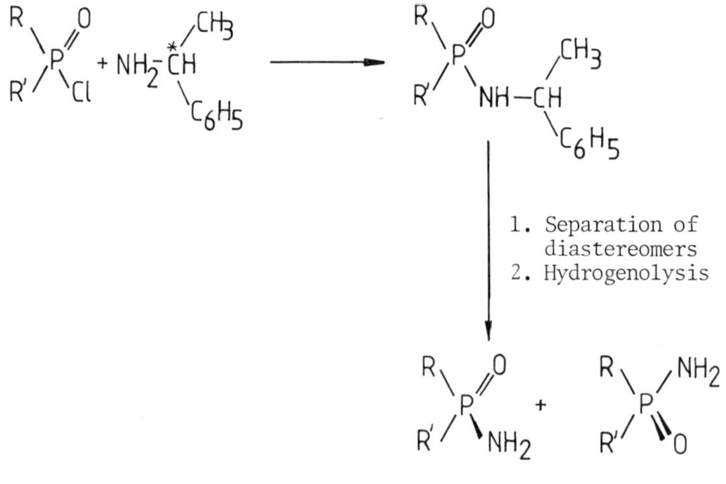

SCHEME 1.

should lead to two separable covalent diastereoisomeric phospho-1-methyl-benzylamidates. Removal of the chiral 1-methylbenzyl moiety by means of hydrogenolytic cleavage of the N-C bond should give the desired optically active phosphoamidate. The effectiveness of this synthetic plan was proved by the synthesis of both enantiomers of 1 (Ref.5). Condensation of (-)-S-1-methyl-benzylamine (6) with 3-chloropropan-1-ol gave (-)-S-N-(3-hydroxypropyl)-1-methylbenzylamine (7). Its condensation with N,N-bis(2-chloroethyl)phosphoramidic dichloride (8) led to diastereomeric N-(1'-methylbenzyl)cyclophosphamides 9, that were separated by means of column chromatography. The 1-methylbenzyl group was removed by catalytic hydrogenolysis. The overal procedure is depicted in Scheme 2.

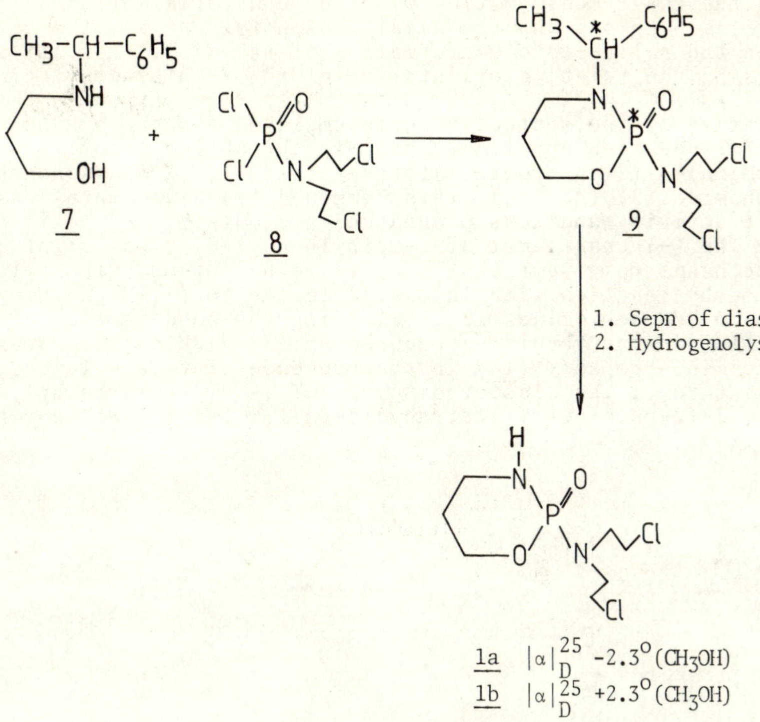

SCHEME 2.

A similar synthesis of 9 was independently published by Zon (Ref.6), but his synthesis of enantiomeric 1 (Ref.7) appeared two years after the publication of our work (Ref.5).
In our synthetic procedure optically pure 6 was employed and the effective separation of the diastereomers of 9 proved by ^{31}P NMR. Assuming that catalytic debenzylation did not cause racemization of 9 nor 1 we considered that the optical rotations values $|\alpha|_D^{25}$ -2.3° for 1a and ∓2.3° (1b) corresponded to optically pure compounds. This has been confirmed by Zon (Ref.7) and independently by Verkade who prepared both 1a and 1b in the different way (Scheme 3), using α-naphtyl-methyl-phenyl-chlorosilane for N-silylation of the racemic anhydrous 1 and subsequent resolution of N-silylated diastereomers of 11 (Ref.8).

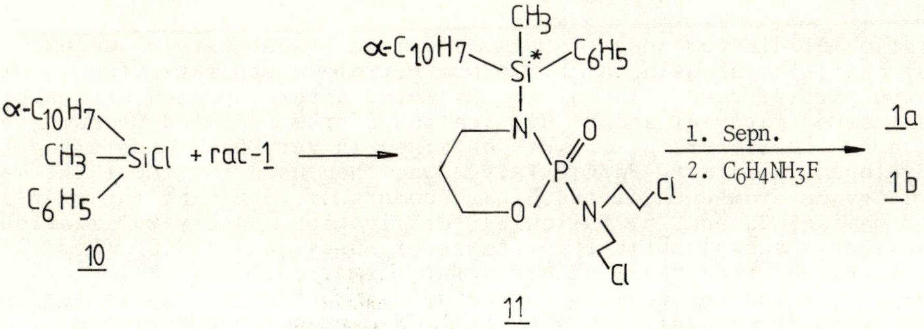

SCHEME 3.

The absolute configuration of 1a was established by means of X-ray analysis as (-)-S (Ref.9). Independently, American colleagues have assessed for 1b the absolute configuration (+)-R (Ref.10). Each of the enantiomers of 1 exists in solid state in the chair conformation with the equatorial orientation of bis-(N-chloroethyl)amino group and the axial phosphoryl oxygen. The only difference between the solid-state conformation of racemic 1 (Ref.11) and its enantiomers exists in the relative orientation of the 2-chloroethyl groups of bis(2-chloroethylamino) moiety.

The key feature of the synthetic route employed for the synthesis of the enantiomers of 1 has been applied to prepare the optical isomers of 2-(2-chloroethylamino)-3-(2-chloroethyl)tetrahydro-2H-1,3,2-oxazaphosphorine 2-oxide (Isophosphamide, 2) (Ref.12). This compound (as a racemate) has shown better cumulative activity and less pronounced accumulation of toxic activity than 1 (Ref. 13). The (-) enantiomer of 1-methylbenzylamine was first condensed with 2-chloroethanol to give (-)-N-(2-hydroxyethyl)-1-methylbenzylamine which was converted by reaction with thionyl chloride to (-)-N-(2-chloroethyl)--1-methylbenzylamine hydrochloride (12). Following the condensation of 12 with phosphoryl chloride the resulting phosphoramidic dichloride 13 was treated with ethyleneimino-propanol (14) in the presence of triethylamine, yielding directly a mixture of the diastereoisomers of 2-N-(2-chloroethyl)-N--(1'-methylbenzyl)-3-(2-chloroethyl)tetrahydro-2H-1,3,2-oxazaphosphorine 2-oxide (15).

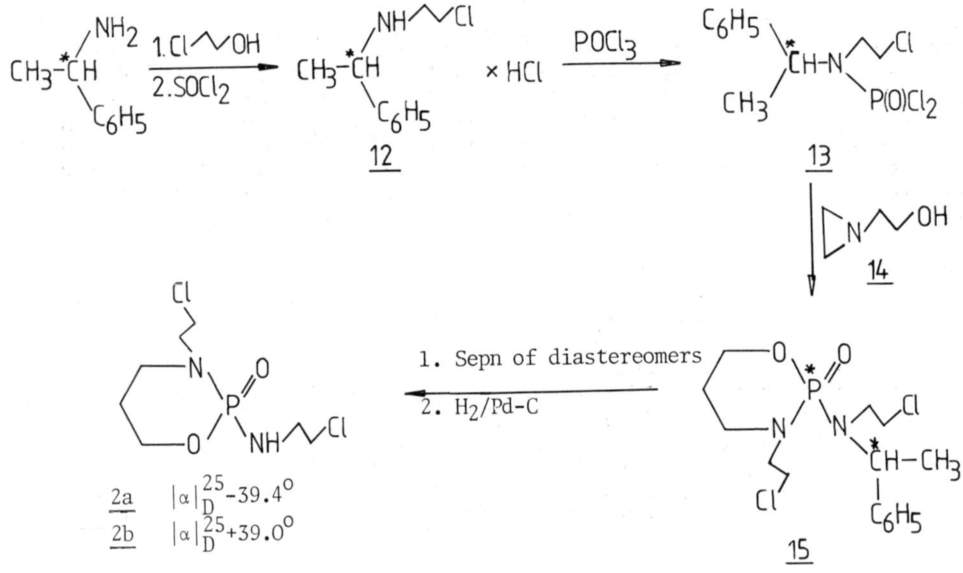

SCHEME 4.

The separation of diastereomers 15 proved difficult but careful column chromatography on silicagel using a chloroform-petroleum ether-t-butanol solvent mixture gave partial resolution of the isomers. After repeated chromatography of the unresolved material and repurification of the isolated products by chromatography the isomers of 15 were obtained in very high diastereomeric purity. Hydrogenolysis with Pd/C catalyst was then used to remove the N-(1-methylbenzyl) group from each of the diastereomers of 15 giving the enantiomers of Isophosphamide, 2a and 2b. Following purification and crystallization these enantiomers showed optical rotations of -39.4° and +39.0° (Ref. 12), considerably higher than these observed for isomeric 1 (Ref. 5).

During the further development of our synthetic procedures leading to the enantiomeric forms of 1 it has been noticed that condensation of (+)-R-7 with phosphoryl chloride gives the mixture of the diastereomeric phosphorochloridates 16a,b in the ratio 80:20. Such an observation indicated that synthesis directed towards the predominate formation of one enantiomer of 1 could be possible. However, the direct condensation of 16 with bis(2-chloroethyl)amine, which should lead to 9, has failed independently of the reaction conditions.

Fortunately the mixture of 16a,b reacted with aziridine and the mixture of the aziridides 17 was obtained in the ratio 20:80. The predominant 17a was

separated via crystallization and converted to 18a under treatment with dry HCl.
Independently, the mixture 17a,b after treatment with HCl gave 18a,b which was separated into the pure diastereomers 18a and 18b:

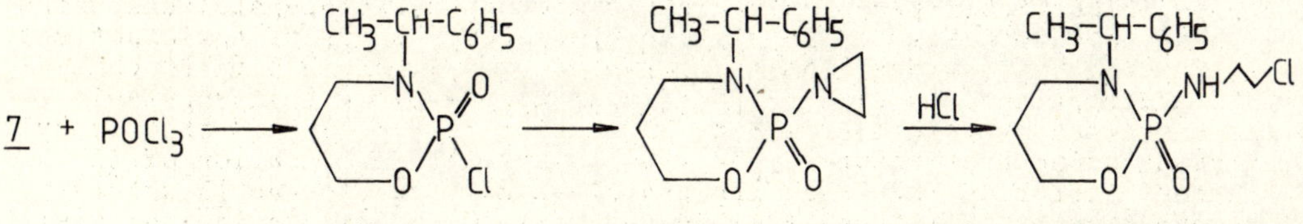

SCHEME 5.

It should be pointed out that the use of (-)-S-7 led to 16a,b with the inverted proportions 16a/16b, which in the light of potentially better therapeutic efficacy of one enantiomer of 1 than that of the racemate, created a very attractive aspect of this synthetic approach to the "desired" enantiomer of 1.
Any attempts of direct 2-chloroethylation of the exocyclic nitrogen atom of 18, were unsuccessful. Treatment of the diastereomers 18 with sodium hydride led to 17. Neither the use of 1-chloro-2-iodo-ethane nor of 2-chloroethyl tosylate gave the desired 9. However, treatment of 18 with chloroacetyl chloride gave 19:

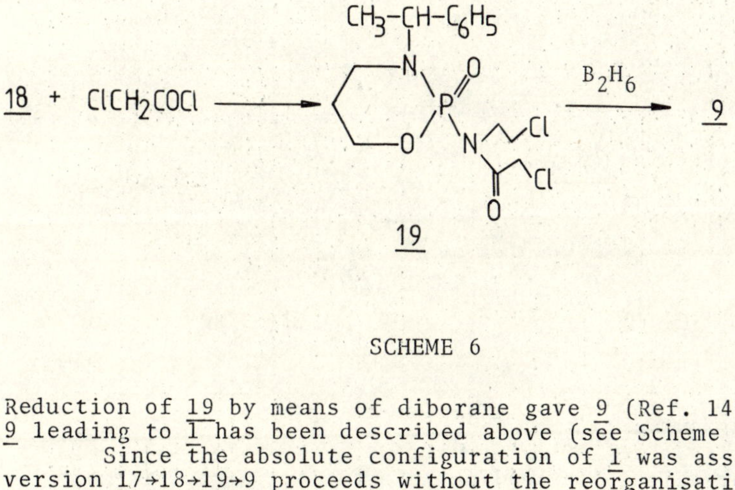

SCHEME 6

Reduction of 19 by means of diborane gave 9 (Ref. 14). The hydrogenolysis of 9 leading to 1 has been described above (see Scheme 2).
Since the absolute configuration of 1 was assigned earlier and the conversion 17→18→19→9 proceeds without the reorganisation of bonds attached to phosphorus, we were able to assign the absolute configuration of 17, 18, 19 and 9. This data is given in the Table below:

No. of cpd	Absol.conf.	$\|\alpha\|_D^{25}$		δ_{31_P} (ppm)*	
9	(R,R_p)	+65.0°	(benzene)	10.2	(benzene)
19	(R,S_p)	+64.0°	(MeOH)	0.1	(MeOH)
18	(R,R_p)	+72.5°	(MeOH)	9.9	(benzene)
17	(R,R_p)	+28.7°	(MeOH)	16.8	(benzene)

* Positive values are assigned for compounds absorbing at lower field than H_3PO_4.

In light of the information that compounds related to 1, such as 2, 3 and 4, constitute valuable drugs against human neoplasms and autoimmune deseases, we undertook additional efforts for their preparation based on the reactions and by products shown in Schemes 5 and 6.

Trophosphamide |3, 2-bis(2-chloroethyl)amino-3-(2-chloroethyl)tetrahydro-2H-1,3,2-oxazaphosphorine 2-oxide| which is more effective than 1 against Ehrlich carcinoma and sarcoma 37 (both resistant to 1, Ref. 15) was obtained in enantiomeric forms from 1 by its chloroacetylation and reduction of 3-N-chloroacetyl-cyclophosphamide (20) with diborane (Ref.14):

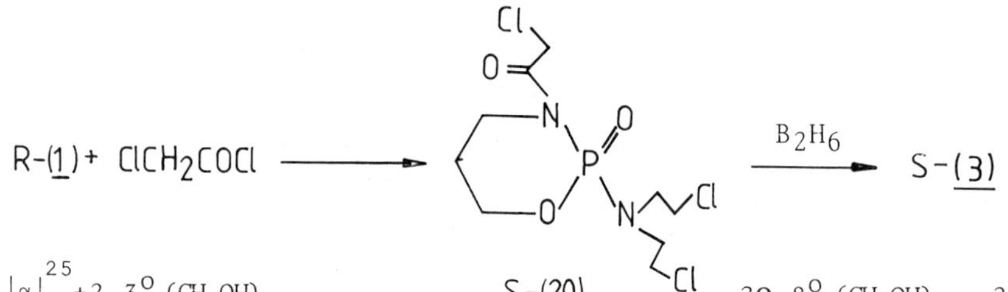

R-(1) + ClCH$_2$COCl $\longrightarrow$ S-(20) $\xrightarrow{B_2H_6}$ S-(3)

$|\alpha|_D^{25}$ +2.3° (CH$_3$OH)   −20.8° (CH$_3$OH)   −28.6° (CH$_3$OH)
(Ref.62)

SCHEME 7.

Diastereomers of 18 were exploited for the preparation of enantiomers of Isophosphamide 2. Catalytic cleavage of the C-N bond led to enantiomeric 2-(2-chloroethylamino)tetrahydro-2H-1,3,2-oxazaphosphorine 2-oxide (21)(Ref.16). Chloroacetylation of 21, which is fully regioselective, gave compound 22. Its treatment with diborane gave the desired 2 (Ref. 14).

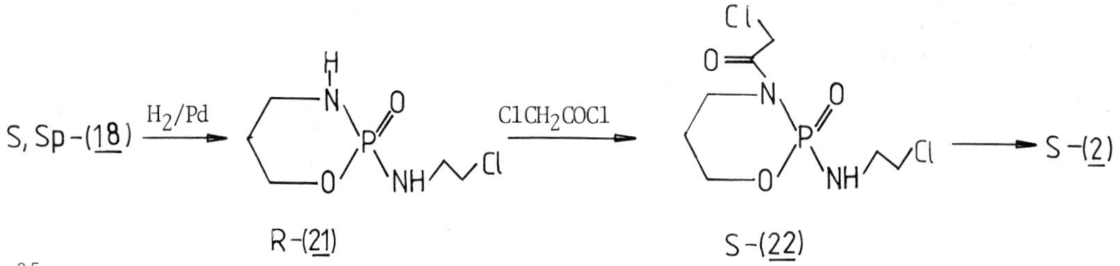

S,Sp-(18) $\xrightarrow{H_2/Pd}$ R-(21) $\xrightarrow{ClCH_2COCl}$ S-(22) $\longrightarrow$ S-(2)

$|\alpha|_D^{25}$ −72.6°(CH$_3$OH)   −15.1°(CH$_3$OH)   −45.4°(CH$_3$OH)   −39.0° (CH$_3$OH)

SCHEME 8.

It must be emphasized, that this route leading to enantiomers of 2 is much more efficient than that presented in Scheme 4.

Access to the enantiomers of 2 opened the way to the enantiomers of 2-(2-mesyloxyethylamino)-3-(2-chloroethyl)tetrahydro-2H-1,3,2-oxazaphosphorine 2-oxide (Sulphosphamide, 4), which , as a racemate has shown promising effects in the induction of both humoral and cellular immunotolerance (Refs. 17 and 18).

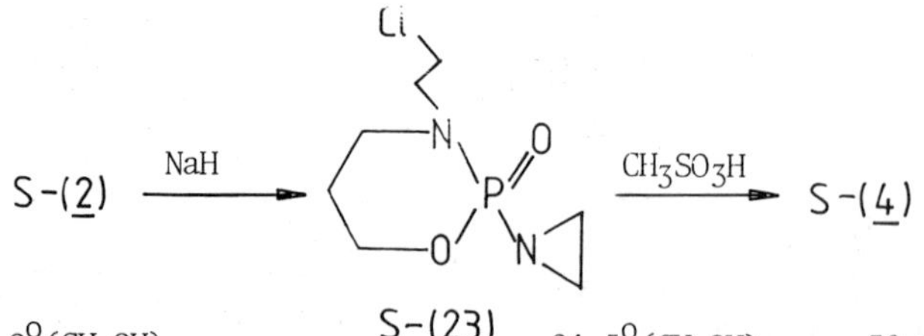

S-(2) $\xrightarrow{NaH}$ S-(23) $\xrightarrow{CH_3SO_3H}$ S-(4)

$|\alpha|_D^{25}$ −39.0°(CH$_3$OH)   −24.5°(CH$_3$OH)   −38.0°(CH$_3$OH)

SCHEME 9.

Treatment of 2 with NaH leads exclusively to 23. There was no indication (^{31}P NMR assay) that compound 24 is formed in this process:

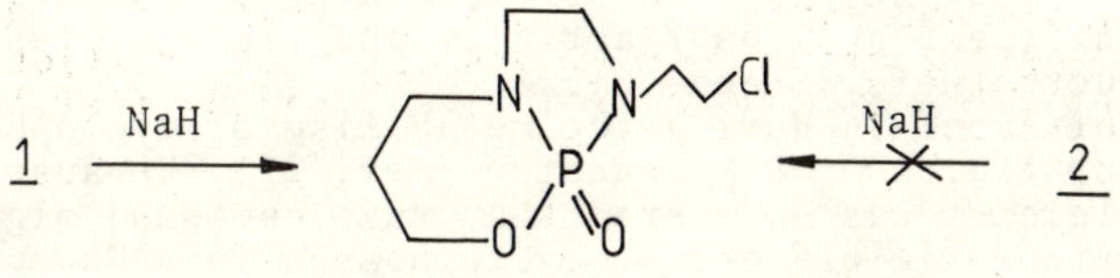

$$1 \xrightarrow{NaH} \underset{24}{\text{[structure]}} \xcancel{\xleftarrow{NaH}} 2$$

The reaction of compound S-(23) with methanesulphonic acid gave S-(4) in high yield (Ref.14).

Before presenting the preliminary biological results it seems wise to review the metabolism profile of cyclophosphamide. Cyclophosphamide proved to be inactive *in vitro* against normal cells and tumor cells, even though it was both cytotoxic and therapeutically effective in the whole animal (Refs. 1 and 19). Foley and co-workers (Ref. 20) demonstrated in 1960 that the liver is the principal site of cyclophosphamide activation, and subsequent studies by Brock and Hohorst (Ref. 21) indicated that the activating enzymes are located in the microsomal fraction of the liver and that the activation process is oxidation requiring the presence of NADPH and oxygen. Further studies by Cohen and Jao (Ref. 22) confirmed this observation and provided evidence that hepatic mixed function oxidases are responsible for the activation.

Although the basic requirements for metabolic activation and the chemical details of the reaction were understood by 1963, it is only in the last few years that the intermediates in the metabolism of cyclophosphamide have actually been characterized. From the evidence accumulated in recent years it seems likely that the metabolism of cyclophosphamide proceeds as shown in Scheme 10. The initial oxidative activation of the parent compound occurs at C-4, the atom adjacent to the ring nitrogen, to produce the cyclic hemiaminal 4-hydroxycyclophosphamide (25). This compound either undergoes further oxidation (detoxification) to 4-ketocyclophosphamide (26) or tautomerizes to the open-chain aldehyde aldophosphamide (27). Aldophosphamide, in turn, may be oxidized to carboxyphosphamide (28), a detoxification reaction, or may undergo spontaneous β-elimination of acrolein to form phosphoramide mustard (29), a toxification reaction. This compound is believed to be the biologically active alkylating agent derived from cyclophosphamide.

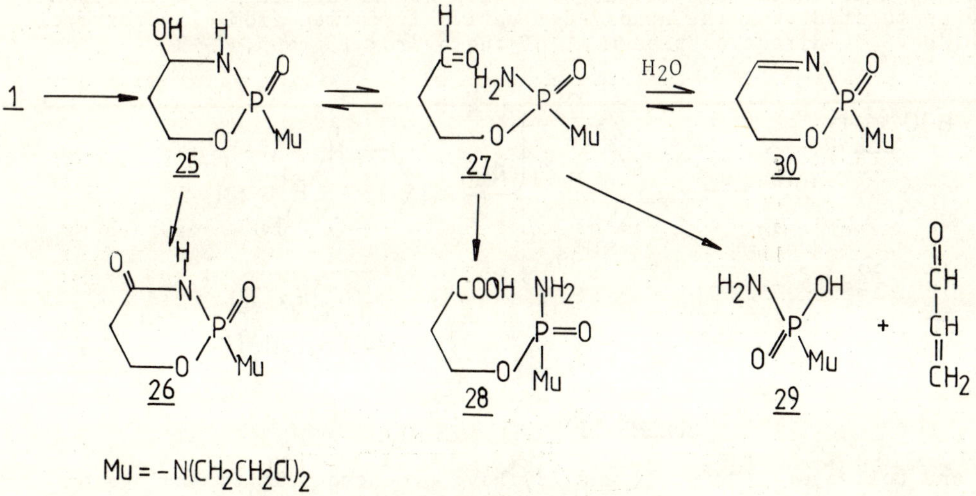

Mu = -N(CH$_2$CH$_2$Cl)$_2$

SCHEME 10.

The major urinary metabolites carboxyphosphamide (28) and 4-ketocyclophosphamide (26) were discovered independently in 1971 by Struck and associates in dog and man (Ref. 23) and by Bakke and co-workers (Ref. 24) in sheep. The urinary excretion of carboxyphosphamide 28 accounts for about 50% of the administrated dose of cyclophosphamide, whereas 4-ketocyclophosphamide accounts for about 15% of the dose (Refs. 23 and 25). However, these compounds do not exhibit significant antitumor activity and probably represent inactivated excretion products (Refs. 23 and 26).

Production of acrolein during the metabolism of cyclophosphamide has been demonstrated by Alarcon and Meienhofer (Ref. 27). Because of the known toxicities of acrolein it was suggested that this compound might play a role in the toxic or therapeutic effects of cyclophosphamide. However, the metabolites of cyclophosphamide have been shown to be cytotoxic at levels considerably below the level of acrolein required for cytotoxicity (Ref. 28). That intracellular acrolein release is unlikely to contribute significantly to the action of cyclophosphamide is also inferred from the fact that compound 31 readily produces acrolein on oxidation (Ref. 29) but is relatively noncytotoxic and has little antitumor effect in the whole animal (Ref. 30). Very recently it has been shown that acrolein is responsible for urotoxic side effects (Ref. 31).

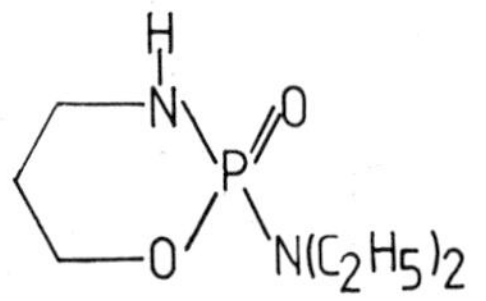

31

Two key cyclophosphamide metabolites are 4-hydroxycyclophosphamide (25) and aldophosphamide (27). These two compounds were initially postulated (Ref. 23) as the putative precursors for the oxidation products 4-ketocyclophosphamide (26) and carboxyphosphamide (28), respectively, but have proved difficult to synthesize or isolate from biologic sources. Takamizawa and co-workers (Ref. 32) originally synthesized 4-hydroxycyclophosphamide by reduction of 4-hydroperoxycyclophosphamide (32) with triphenylphosphine. 4-Hydroxycyclophosphamide is also generated spontaneously from 4-hydroperoxycyclophosphamide in aqueous solution (Ref. 33); although it has been found to be cytotoxic *in vitro* at considerably lower concentrations than those of phosphoramide mustard, it lacks alkylating activity (Refs. 34 and 35).

Connors and co-workers (Ref. 36) have identified 4-ethoxycyclophosphamide (33) in microsomal incubations of cyclophosphamide treated with ethanol and have postulated that the ethoxy-derivative is formed from 4-hydroxycyclophosphamide 25 by direct displacement of the hydroxyl group.

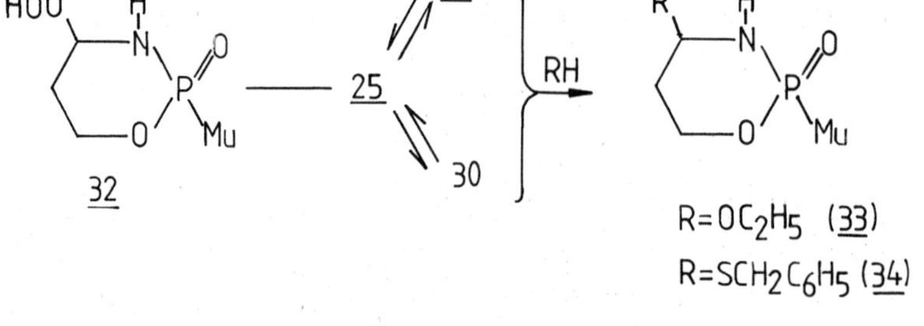

$R = OC_2H_5$ (33)
$R = SCH_2C_6H_5$ (34)

SCHEME 11

Hohorst and colleagues (Refs. 37 and 38) have produced sulphur analogs of 4-hydroxycyclophosphamide by treating either 4-hydroxycyclophosphamide or metabolites of cyclophosphamide in human body fluids with thiols. The authors have suggested that these compounds (e.g., 34) are formed *via* a hemithioacetal derivative of aldophosphamide.

Another potential explanation for the formation of derivatives such as 33 and 34 is reaction of the appropriate nucleophile with the iminophosphamide 30 (Ref. 39). By analogy with nicotine metabolism, one might expect such an intermediate (Ref. 40). Evidence for its existence has recently been obtained by treatment of synthetic 4-hydroxycyclophosphamide with sodium borodeuteride to produce cyclophosphamide which is monodeuterated at C-4 (Ref. 39). The nitrogen-carbon double bond should be very polar, and addition of either an alcohol or a thiol across the bond would be expected to occur readily. Studies on the metabolism of Isophosphamide 2 have shown that its metabolism profile is the same as that of 1 (Refs. 41 and 42). However, as recently shown by several Laboratories, initial hydroxylation does not occur exclusively at the endocyclic carbon atom α to the ring nitrogen and two other metabolites are formed in the moderate yields (Refs. 43 and 44) resulting from hydroxylation at the α-carbon of both 2-chloroethyl groups.

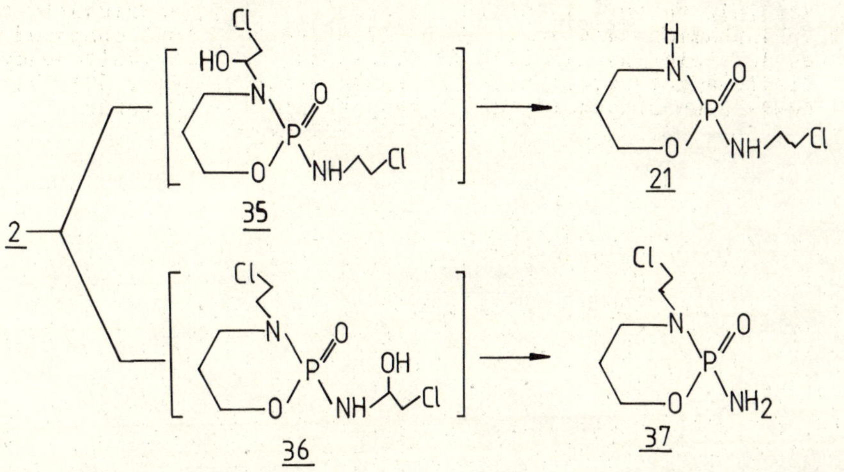

SCHEME 12

The description presented above of the metabolism profiles of 1 and 2 clearly shows that, besides phosphoramide mustards, all other phosphorus-containing metabolites are chiral by virtue of an asymmetric phosphorus atom. Microsomal activation of 1 creates a new chiral centre in the molecule and a pair of diastereomeric 4-hydroxycyclophosphamides 25 may be expected. The stereoselectivity question of the initial hydroxylation step is still open and our studies on the preparation of diastereomers of 25 are in progress. However, for further evaluation of stereodifferentiation in the metabolism of enantiomers of 1 optically active 4-ketocyclophosphamide 26 was recently obtained by oxidation of enantiomeric 1 with Fenton's Reagent (Ref. 45):

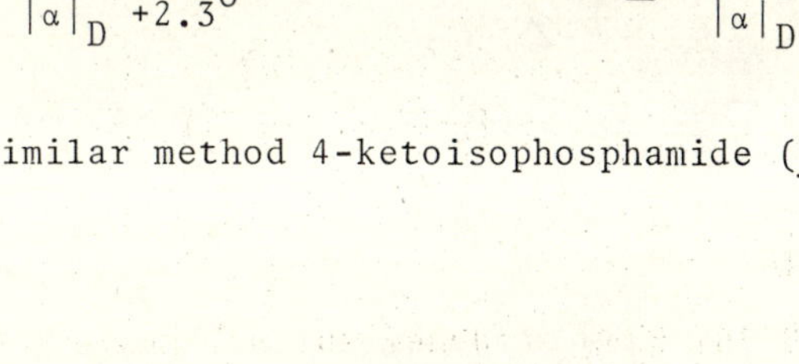

(+)R-(1) ⟶ (-)R (26)

$|\alpha|_D^{25}$ +2.3° $|\alpha|_D^{25}$ -30.0° (c 1.2, CH$_3$OH)
                                (100% o.p.)

By a similar method 4-ketoisophosphamide (38) was obtained (Ref. 46):

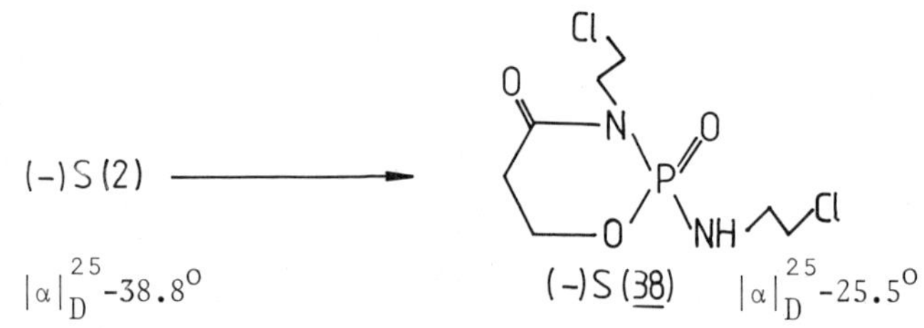

$(-)S(2) \longrightarrow$ (-)S (**38**) $|\alpha|_D^{25} -25.5°$

$|\alpha|_D^{25} -38.8°$

(contaminated with ca. 5% of unknown compound)

The metabolites of 2, compounds 21 and 37, were also obtained in optically pure form. The synthesis of 21 is presented above (Scheme 8). Synthesis of the enantiomers of 37 was performed in the following way:

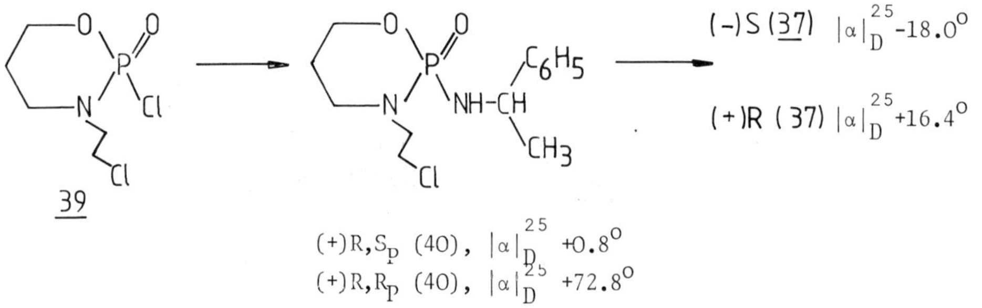

39 → → (-)S (**37**) $|\alpha|_D^{25} -18.0°$

(+)R (**37**) $|\alpha|_D^{25} +16.4°$

(+)R,S$_p$ (**40**), $|\alpha|_D^{25} +0.8°$
(+)R,R$_p$ (**40**), $|\alpha|_D^{25} +72.8°$

SCHEME 13.

Condensation of N-(2-chloroethyl)-3-hydroxypropylamine with phosphoryl chloride gave compound 39 which when treated with (+)R-1-methylbenzylamine have 2-|(R)-1'-methylbenzylamino|-3-(2-chloroethyl)tetrahydro-2H-1,3,2-oxazaphosphorine 2-oxides, 40 (R,Sp) and 40 (R,Rp).
Separation of the diastereomers and their catalytic hydrogenation led to 2-(R)-amino-3-(2-chloroethyl)tetrahydro-2H-1,3,2-oxazaphosphorine 2-oxides, (-)S-37 and its enantiomer (+)R-37, respectively (Ref. 45).
Because the metabolic activation of 1 occurs at the endocyclic carbon atom adjacent to the ring nitrogen, it was of interest to know in what way monomethyl-substitution at this atom influences the biological activity and metabolism profile of C-4 substituted cyclophosphamide derivatives (Refs. 47 and 48). Introduction of a substituent at C-4 prevents further oxidative metabolism of the 4-hydroxyderivative and additionally it creates a new centre of chirality in the molecule: therefore four optically active forms of 2-|bis-(2-chloroethyl)amino|-4-methyl-tetrahydro-2H-1,3,2-oxazaphosphorine 2-oxide (4-methylcyclophosphamide, 44) are possible.
The synthesis of the four optically active forms of 4-methylcyclophosphamide (44) was based on (+)- and (-)-3-aminobutan-1-ol (41) as depicted in Scheme 14.

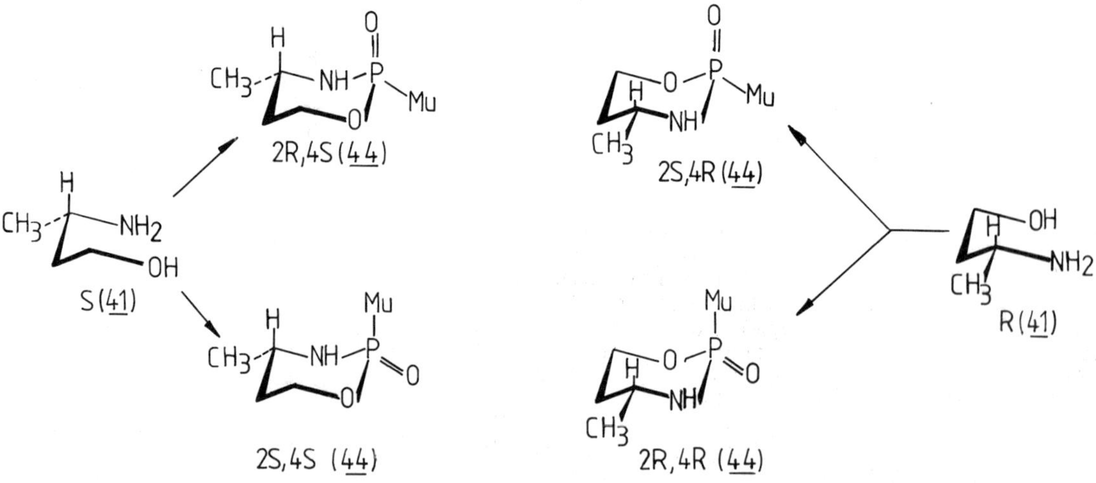

2R,4S (**44**)   2S,4R (**44**)

S(**41**)   R(**41**)

2S,4S (**44**)   2R,4R (**44**)

SCHEME 14.

Treatment of ethyl crotonate with (-)-1-(S)-phenylethylamine gave the separable diastereomers of ethyl 3-N-|1(S)-methylbenzyl|aminobutyrate (42) having readily determinable absolute configurations at the newly introduced asymmetric carbon atom (C-3) (Ref. 49.)
These were reduced in two stages to give the requisite enantiomers of 3-aminobutan-1-ol (41). Condensation of each of these enantiomers with N,N-bis(2-chloroethyl)phosphoramidic dichloride gave in each case a mixture of two separable isomers of 4-methylcyclophosphamide, designated "fast" and "slow" according to their relative mobilities on TLC. Since the absolute configuration of the starting materials 41 is known, the four enantiomeric 4-methylcyclophosphamides have predetermined configurations at C-4.

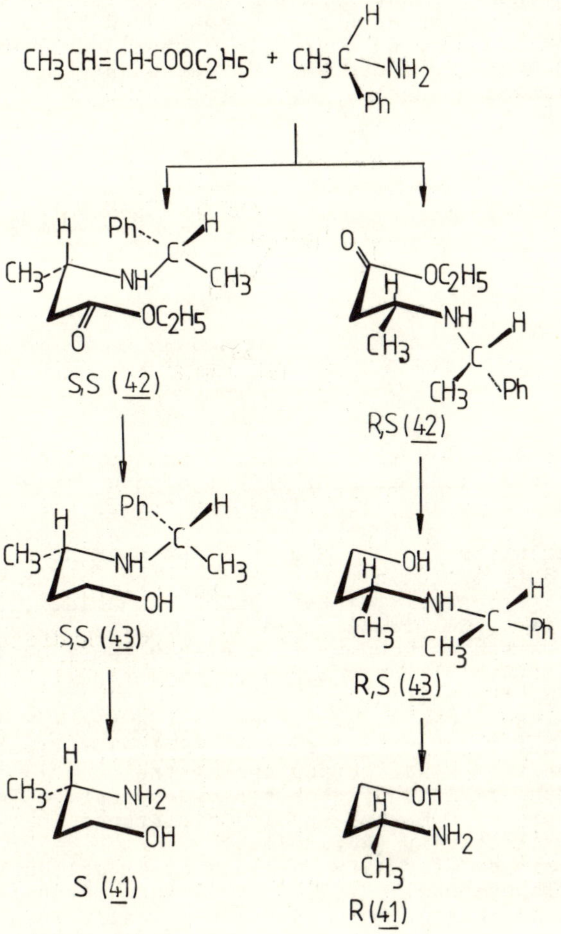

SCHEME 15.

Determination of the absolute configurations at phosphorus for the pair of diastereomers 44 for which the absolute configuration at C-4 is known, is equivalent to the assignment of *cis* and *trans* geometry to the "fast" and the "slow" isomers 44. Contrary to the assignments made by Struck et al. (Ref. 48) for the faster migrating racemic 44 (mp 72-74°C) and the slower migrating racemic 44 (mp 102°C) we assigned, taking the example of the fast and the slow products derived from 3-(R)-aminobutan-1-ol, the spatial arrangement 4-Me$_{eq}$-2NR$_{2eq}$ (*trans*) to the slower migrating 44 (and hence the configuration 2S, 4R) and 4-Me$_{eq}$-2NR$_{2ax}$ ⇌ 4-Me$_{ax}$-2NR$_{2eq}$ (*cis*) to the faster migrating 2R,4R-(44) on the basis of the following arguments: (1) 2S,4R-(44) absorbs at lower field in its ^{31}P NMR spectrum (δ_{31P} 13.5 ppm, external H$_3$PO$_4$) than 2R,4R-(44) (δ_{31P} 11.0 ppm) (Ref. 50); (2) 2S,4R-(44) exhibits a lower ν_{PO} value (1218 cm^{-1} in CCl$_4$) than 2R,4R-(44) (1231 cm^{-1})(Refs. 51 and 52); (3) The value (2.9 Hz) of $^4J_{PNCCH_3}$ (^{1}H NMR) for 2S,4R-(44) is higher than that (1.9 Hz) (Ref. 53) for 2R,4R-(44).

The conformational stability of 2S,4R-(44) was established on the basis of the following data: $^3J_{POCH_{eq}}$=22.75 (^{1}H NMR), $^3J_{PNCC_5}$=3.2 and $^3J_{PNCCH_3}$=12.1 Hz (^{13}C NMR). Although analysis of the ^{1}H NMR spectrum of 2R,4R-(44) was not possible because of its complexity, the ^{13}C NMR spectrum gave the data $^3J_{PNCC_5}=7.0$ and $^3J_{PNCCH_3}=7.6$ Hz, which suggested rapid equilibrium of two or more conformations of 2R,4R-(44). The assignment 4-Me$_{eq}$-2NR$_{2eq}$ to 2S,4R-(44) and its conformational stability is consistent with the known equatorial preference of the 4-methyl and 2-dialkylamino groups in dioxa- and oxazaphosphorine ring systems.

Arguments analogous to these applied to the products from 3-(R)-aminobutan-1-ol enabled the assignment of absolute configuration to the products from the 3-(S) isomer, namely, 2S,4S-(44) and 2R,4S-(44) (Ref. 49). Which kind of biological information was obtained from the compounds so far described in this lecture?

1. The differential metabolism by three species (mouse, rat and rabbit) of (R)- and (S)-cyclophosphamide has been monitored by mass spectrometry by following the metabolism of pseudoracemates comprized of either the unlabeled (S)- and the tetradeuterated ($^2H_4$-labeled) (R)-enantiomer or of the $^2H_4$-labeled (S)- (Ref. 54) and unlabeled (R)- form.

Metabolism by microsomes from rabbit livers, but not those from rat and mouse, was markedly stereoselective, the percentage of drug metabolized during incubation for 15 min at 37° being 2.5-4 fold greater for the (S)-isomer than for the (R)-antipode.

Following administration of the cyclophosphamide pseudoracemates, enantiomer ratios were determined in rat and rabbit urine for the unchanged drug and for two metabolites, namely, 4-ketocyclophosphamide (26) (rat and rabbit) and carboxyphosphamide (28) (rabbit) and the results compared with those for mouse. There was a marked enrichment in one enantiomer of 4-ketocyclophosphamide (26) from mouse (R):(S) ratio 2.3-3.0:1 and rabbit (S):(R) ratio 3.8-11.1 but not from the rat.
Carboxyphosphamide (28) excreted by the mouse and the rabbit was enriched in the (S)-enantiomer but to a smaller extent |(S):(R) ratio 2:1|. The results of this study demonstrate that metabolism of the enantiomers of cyclophosphamide and its intermediary metabolites is stereoselective in mouse, rat and rabbit, and that there is a marked species difference in the extent and direction of this stereoselectivity (Ref. 55).

2. The *in vitro* cytotoxicity in bioassay or cell culture assays of (-)-*cis*-4-methylcyclophosphamide |2R,4R-(44)| when metabolized by microsomes from the livers of male Wistar rats was significantly less than that of metabolized (+)-*cis*, (+)-*trans*-, and (-)-*trans*-4-methylcyclophosphamide. However, metabolism of the individual stereoisomers by liver microsomes from female BALB/c mice yielded metabolite mixtures of approximately equal cytotoxicity.

The extent of metabolism by rat liver microsomes of the four stereoisomers was similar when assessed by substrate disappearance, but the yield of alkylating metabolites (expressed as phosphoramide mustard equivalents) was comparable in the case of three of the isomers but substantially less for the 2R,4R-(44) isomer.

A new metabolite was isolated following the incubation of (-)-*cis*-4-methylcyclophosphamide |2R,4R-(44)| with rat liver microsomes. It was identified by its breakdown products and its electron impact and chemical ionization mass spectra, especially by the presence in the latter of a pseudomolecular ion. Its stucture was 2-|(2-chloroethyl)(2-chloro-1-hydroxyethyl)amino|-4-methyl-tetrahydro-2H-1,3,2-oxazaphosphorine 2-oxide (45), an intermediate in dechloroethylation.

Species variation (mouse, rat and rabbit) in the formation of this metabolite of low cytotoxicity (1% of that of 4-hydroxy-4-methylcyclophosphamide) was noted and was in good accord with the observed species differences in cytotoxicity assays (Ref.56).

3. The enantiomers of cyclophosphamide were tested (two-fold dose spacing) against the ADJ/PC6 plasma cell tumour in mice (Ref. 57).

PC6 Tumour Test Results

	LD_{50} (mg/Kg)	ID_{90} (mg/Kg)	Therapeutic index (LD_{50}/ID_{90})
(+)-cyclophosphamide	365	5.3	68.9
(-)-cyclophosphamide	365	2.85	128.1
Racemic cyclophosphamide	335	3.6	93.0

The (-)-isomer is considerably more effective (lower ID_{90}) in killing the tumour cells in this test system than is the (+)-isomer. Preliminary experiments using a bioassay technique showed that the (-)-isomer had greater toxicity towards tumour cells in tests involving *in vitro* incubation of TLX 5 cells with the drug in the presence of a microsomal activating system and subsequent injection into mice. Measuring the inhibition of growth of Walker ascites cells in the culture after microsomal activation of the drugs showed that when metabolism of the racemate and the (+)- and (-)- forms was 98% complete, the metabolic products were of equal toxicity.

The metabolism of racemic, (+)- and (-)-, cyclophosphamide was monitored by quantitating the unchanged drug by stable isotope dilution-mass spectrometry using racemic cyclophosphamide-4,5,6-d_6 as an internal standard. The initial rates for the three forms were virtually identical, as was the extent of metabolism (45%) after 15 minutes.

4. The comparative metabolism of the enantiomers of cyclophosphamide and of the racemate has been studied in man (Ref. 58). Four patients (squamous cell carcinoma of the lung) were each given sequentially the racemate, the (+)-enantiomer and its (-)-antipode. The plasma levels of the parent drug and the urinary output (24 hr) of unchanged drug and of two enzymatically produced metabolites, 4-ketocyclophosphamide (26) and carboxyphosphamide (28) were determined using mass spectrometry-stable isotope dilution. There was no significant difference between the three forms of cyclophosphamide in respect to the plasma half-life (β-phase) or in the urinary outputs of the drug or of carboxyphosphamide. The output of 4-ketocyclophosphamide after administration of (+)-cyclophosphamide was significantly greater than that produced from the racemate. Cyclophosphamide recovered from the urine of patients given the racemate was either racemic, or only slightly enriched in the (-)-enantiomer. The two enantiomers were almost equally bound to plasma protein.

Based on these metabolic studies alone, there is little reason to predict that the enantiomers will differ from each other or from the racemate in their therapeutic effects in man, but there are other factors, e.g. stereoselective uptake of the intermediary 4-hydroxylated metabolites by neoplastic cells, which could elicit such differences.

5. Independently from our efforts biological studies on stereodifferentiated metabolism of enantiomers of 1 were performed by Zon et al. (Ref. 59). Animal screening data (test/control percentages) for (+)-, (-)- and (±)-1 activity against mouse L-1210 lymphoid leukemia showed no outstanding differences in therapeutic value. Collectively, these various experimental probes suggest that there is an unusually low degree of biological stereoselectivity associated with the metabolism of enantiomers of 1.

6. Preliminary studies on the metabolism of Isophosphamide 2 have shown that enantiomers of Isophosphamide give, in ADJ/PC-6 plasma cell tumour in mice tests similar results to that described above for the enantiomer of 1. (-)-S-(2) possesses two-fold higher activity than its enantiomer. In the urine of patients treated with racemic 2 no trace of 37 has been found (Ref. 60).

7. The influence of cyclophosphamide chirality on the sensitivity of lymphoid cells was studied (Ref. 61). It has been shown that (+)-1 inhibits the formation of PFC in BALB/c mice more effectively than (-)-1 and (±)-1. The cytolytic activity of (+)-1 (spleen indices, mice) is lower than that of racemic and (-)-1. However, skin transplant survival time in mice was the same in groups of animals treated separately with (+)-1, (-)-1 and (±)-1. Elimination of T suppressor cells was most effective after treatment of guinea pigs immunized by DNCB with (+)-1. The eventual consequences of stereodifferentiated immunosuppressive activity of the enantiomeric forms of cyclophosphamide in the therapy of "autoimmune" and malignant diseases should be consi-

dered.
   Further studies on stereodifferentiated cytolytic, immunosuppressive, and teratogenic effects of enantiomers of 1, 2, 3 and 4 are in progress.

Acknowledgment - The work that I have described here has been supported by grants PR6/1703 to the Polish Academy of Sciences from the National Cancer Program, Poland, and G 973/788 to the Institute of Cancer Research; Royal Cancer Hospital, Chester Beatty Research Institute (London) from the Medical Research Council, England. I thank Dr. K.Pankiewicz, Dr. R.Kinas, Mr K.Misiura and Mrs K.Chorzewska for their contributions to the development of our projects. I owe a special debt of gratitude to my British colleagues, Professor A.B.Foster, Dr.P.J.Cox, Dr. P.B.Farmer and Dr. M.Jarman of the Chester Beatty Research Institute for their joint development of problems described in this paper.

NOTES and REFERENCES

1. H.Arnold, F.Bourseau and N.Brock, Naturwiss., 45, 64 (1958); For the recent revue on cyclophosphamide see paper by O.M.Friedman, A.Myles and M.Colvin in "Advances in Cancer Chemotherapy", Ed. by A. Rosowsky, 1979, Marcel Dekker, Inc., p. 143
2. R.C.Hall, T.D.Inch, R.H.Inns, A.N.Muir, D.J.Sellers and A.P.Smith, J.Pharm.Pharmacol., 29, 574 (1977)
3. H.Ohkawa, N.Mikami and J.Miyamoto, Agr.Biol.Chem., 41, 369 (1977)
4. K.Pankiewicz - unpublished results
5. R.Kinas, K.Pankiewicz and W.J.Stec, Bull.Acad.Polon.Sci., Ser.sci.chim., 23, 981 (1975)
6. G.Zon, Tetrahedron Letters, 3139 (1975)
7. G.Zon, J.A.Brandt and W.Egan, J.Natn.Cancer Inst.,58, 1117 (1977)
8. T.Kawashima, R.D.Kroshefsky, R.A.Kok and J.Verkade, J.Org.Chem., 43, 1111 (1978)
9. D.A.Adamiak, W.Saenger, R.Kinas and W.J.Stec, Angew.Chem., 89, 337 (1977); D.A.Adamiak, W.Saenger, R.Kinas and W.J.Stec, Z.Naturforsch., 32C, 672 (1977)
10. I.L.Karle, J.M.Karle, W.Egan, G.Zon and J.A.Brandt, J.Am.Chem.Soc., 99, 4803 (1977)
11. S.Garcia-Blanco and A.Perales, Acta Cryst., B28, 2647 (1972)
12. R.W.Kinas, K.Pankiewicz, W.J.Stec, P.B.Farmer, A.B.Foster and M.Jarman, Bull.Acad.Polon.Sci., Ser.sci.chim., 24, 39 (1978)
13. P.J.Creaven, L.H.Allen, M.H.Cohen and R.L.Nelson, Cancer Treat.Rep., 60, 445 (1976); G.Falkson and H.C.Falkson, ibid., 60, 955 (1976); H.Burket, H.C.Voigt, Proceedings International Holoxan Symposium, Düsseldorf, March 21-23, 1977; J.Schnitker, N.Brock, H.Burket and E.Fichtner, Arzneim.-Forsch., 26, 1783 (1976)
14. K.Pankiewicz, R.Kinas, W.J.Stec, A.B.Foster, M.Jarman and Jan M.S. van Maanen, J.Am.Chem.Soc., - accepted for publication
15. N.Brock, Med.Monatssch., 27, 290, 294 (1973)
16. Compound 21 of unspecified enantiomers composition was isolated from urine of patients treated with racemic 2 - unpublished results
17. N.Brock and I.Kuhlman, Arzneim.-Forsch., 24, 1139 (1974)
18. N.Brock and J.Potel, Arzneim.-Forsch., 24, 1149 (1974); V.Botzenhard and E.M.Lemeel, ibid., 24, 1167 (1974)
19. N.Brock, Arzneim.-Forsch., 8, 1 (1958)
20. G.E.Foley, O.M.Friedman and B.R.Drolet, Proc.Amer.Assoc.Cancer Res., 3, 111 (1960)
21. N.Brock and H.J.Hohorst, Arzneim.-Forsch., 13, 1021 (1963)
22. J.L.Cohen and J.Y.Jao, J.Pharmacol.Exp.Ther., 174, 206 (1970)
23. F.R.Struck, M.C.Kirk, L.B.Mellet, S.El.Dareer and D.L.Hill, Mol.Pharmacol., 7, 519 (1971)
24. J.E.Bakke, V.J.Feil and R.G.Zaylskie, J.Agr.Food Chem., 19, 788 (1971)
25. J.E.Bakke, V.J.Feil, C.E.Fjelstul and E.J.Thacker, J.Agr.Food.Chem., 20, 384 (1972)
26. A.Takamizawa, Y.Tochino, Y.Hamashima and T.Iwata, Chem.Pharm.Bull., 20, 1612 (1972)
27. R.A.Alarcon and J.Meienhofer, Nature (New Biol.), 233, 250 (1971)
28. N.E.Sladek, Cancer Res., 33, 1150 (1973)
29. R.A.Alarcon, J.Meienhofer and E.Atherton, Cancer Res., 32, 2519 (1972)
30. H.Arnold, F.Borseaux and N.Brock, Arzneim.-Forsch., 11, 143 (1961)

31. N.Brock, J.Stekar, J.Pohl, U.Niemeyer and G.Scheffler, Arzneim.-Forsch., 29, 659 (1979)
32. A.Takamizawa, S.Matsumoto, T.Iwata, K.Katagiri, Y.Tochino and K.Yamaguchi, J.Am.Chem.Soc., 95, 985 (1973)
33. J. van der Steen, E.C.Timmer, J.G.Westra and C.Benckhuysen, J.Am.Chem. Soc., 95, 7535 (1973)
34. H.J.Hohorst, U.Draeger, G.Peter and C.Voelcker, Cancer Treat.Rep., 60, 309 (1976)
35. R.F.Struck, M.C.Kirk, M.H.Witt and R.W.Laster., Jr., Biomed.Mass Spectrom., 2, 46 (1975)
36. T.A.Connors, P.J.Cox, P.B.Farmer, A.B.Foster and M.Jarman, Biochem. Pharmacol., 23, 115 (1974)
37. G.Peter, T.Wagner and H.J.Hohorst, Cancer Treat.Rep., 60, 429 (1976)
38. T.Wagner, G.Peter, G.Voelcker and H.J.Hohorst, Cancer Res., 37, 2592 (1977)
39. M.Colvin, O.M.Friedman and C.Fenselau, Proc.Amer.Assoc.Cancer Res., 18, 27 (1977)
40. P.J.Murphy, J.Biol.Chem., 248, 2796 (1973)
41. K.Norpoth, Cancer Treat.Rep., 60, 437 (1976)
42. D.L.Hill, W.R.Laster, Jr., M.C.Kirk, S.El Dareer and R.F.Struck, Cancer Res., 33, 1016 (1973)
43. K.Norpoth, G.Müller and H.Raidt, Arzneim.-Forsch., 26, 1376 (1976)
44. A.Takamizawa, S.Matsumoto, T.Iwata and I.Makino, Chem.Pharm.Bull., 25, 1891 (1977)
45. K.Misiura, K.Pankiewicz, W.J.Stec and M.Jarman - submitted for publication
46. K.Pankiewicz, K.Misiura, W.J.Stec - unpublished results
47. P.J.Cox, P.B.Farmer and M.Jarman, Biochem.Pharmacol., 24, 599 (1975)
48. R.F.Struck, M.C.Thorpe, W.C.Coburn, Jr., and M.C.Kirk, Cancer Res., 35, 3160 (1975)
49. R.Kinas, K.Pankiewicz, W.J.Stec, P.B.Farmer, A.B.Foster and M.Jarman, J.Org.Chem., 42, 1650 (1977)
50. W.G.Bentrude, Han-Wan Tan and V.C.Yee, J.Am.Chem.Soc., 97, 573 (1973)
51. J.M.Harrison, T.D.Inch and G.L.Lewis, J.Chem.Soc., Perkin I, 1892 (1975)
52. J.P.Majoral and J.Navech, Spectrochim.Acta, Part A, 28, 2247 (1972)
53. L.O.Hall and R.B.Malcolm, Chem.Ind. (London), 92 (1968)
54. P.J.Cox, P.B.Farmer, A.B.Foster, L.J.Griggs, M.Jarman, R.Kinas, K.Pankiewicz and W.J.Stec, Biomed.Mass Spectrom., 4, 371 (1977)
55. P.J.Cox, P.B.Farmer, M.Jarman, R.W.Kinas and W.J.Stec, Drug Metabolism Disp., 6, 617 (1978)
56. G.Abel, P.J.Cox, P.B.Farmer, N.J.Haskins, M.Jarman, K.Merai and W.J.Stec, Cancer Res., 38, 2592 (1978)
57. P.J.Cox, P.B.Farmer, M.Jarman, M.Jones, W.J.Stec and R.Kinas, Biochem. Pharmacol., 25, 993 (1979)
58. M.Jarman, R.A.V.Milsted, J.F.Smyth, R.W.Kinas, K.Pankiewicz and W.J.Stec, Cancer Res., 39, 2762 (1979)
59. F.P.Tsui, J.A.Brandt and G.Zon, Biochem.Pharmacol., 28, 367 (1979)
60. K.Pankiewicz, R.Kinas, K.Misiura, W.J.Stec, A.B.Foster, M.Jarman and Jan M.S. van Maanen - manuscript in preparation
61. H.Tchorzewski, W.Soszynska, K.Zeman, K.Pankiewicz and W.J.Stec - manuscript in preparation.
62. Recently the elegant synthesis of enantiomers $\underline{3}$ {$|\alpha|_D^{25}$ +30.1° (MeOH) and $|\alpha|_D^{25}$ -29.7° (MeOH)} has been accomplished: A.E.Wroblewski and J.G.Verkade, J.Am.Chem.Soc. - in press.

v

SYNTHETIC POLYNUCLEOTIDES AS INTERFERON INDUCERS

D. W. Hutchinson

Department of Chemistry and Molecular Sciences, University of Warwick, Coventry CV4 7AL, U.K.

Abstract - The induction of the antiviral glycoprotein interferon has been studied using a very sensitive line of human cells and polynucleotides of high specific radioactivity. Contrary to earlier observations, polynucleotides can enter cells with ease at 4° C whether they are interferon inducers or not. Thus, the stringent structural requirements of interferon inducers must be required at a subsequent step in the induction process. Studies with cross-linked double-stranded nucleic acids and polynucleotides indicate that strand separation is not an important factor during interferon induction.

INTRODUCTION

Interferon, the antiviral glycoprotein, is showing considerable promise in chemical trials against both viruses and certain types of cancer. Consequently, considerable efforts are being made throughout the world to develop inexpensive, large-scale methods of interferon production. Susceptible cells can be made to synthesise interferon by exposing them to viruses or certain double-stranded synthetic polynucleotides[1], and at the University of Warwick we are investigating the induction of interferon by synthetic polynucleotides. This is to try to establish which structural features are necessary for a polynucleotide to be a good inducer, whether the polynucleotide has to enter the cell for the induction process to occur and whether the two strands of a double-stranded polynucleotide come apart during the induction process.

STRUCTURE-FUNCTION RELATIONSHIPS IN INTERFERON INDUCERS

Once it had been established[2] that synthetic polynucleotides could act as interferon inducers, much effort has been expended to discover what, if any, structure-function relationships exist and which of many synthetic polynucleotides was the best inducer. This work has been summarised[1,3] and the main structural features required are as follows.

  (a) Double-strandedness of the polynucleotide is important - although the two strands need not be added at the same time.

  (b) A highly ordered structure is required (usually an A nucleic acid structure[4]) with a melting temperature $\geq 60°$.

  (c) The presence of 2'-hydroxyl groups on the sugar residues in both polynucleotide chains was thought to be essential but recent work[5] has shown that this is not necessarily the case.

In addition, stability to nucleases and low toxicity are also important for *in vivo* work. It has been found that poly(rI).poly(rC) is probably the best synthetic inducer so far discovered on account of its ready availability, high activity, melting temperature of *ca.* 60° in 0.1 M salt solution, and comparatively good stability towards nucleases.

The distinction should be made, however, between polynucleotides which are interferon inducers and those which are antiviral agents without the mediation of interferon. Some years ago, we prepared poly(5-hydroxycytidylic acid) by the polymerisation of 5-hydroxycytidine diphosphate with polynucleotide phosphorylase[6]. The diphosphate (1) can be obtained in high yield by treating CDP with bromine in the presence of 2,4,6-collidine.

To our disappointment, poly(ho^{5}C) did not form a double-stranded hybrid with poly(I) or any other purine-containing polynucleotide. Presumably, ionic interactions between hydroxyl groups on the cytidine rings and the phosphate groups in the polynucleotide chain cause the polymer to adopt a conformation unsuitable for hybrid formation. However, in mice single-stranded poly(ho⁵C) is an active antiviral in the absence or presence of poly(I) although in the latter case enhanced interferon production is not noticed[7].

We are now extending this work by investigating poly(5-methoxycytidylic acid) which we have synthesised by the following route[8-12].

(i) HCOOEt/NaOEt
(ii) NH_2CONH_2
(iii) 1-Acetyl-2,3,5-tribenzoyl-β-D-ribofuranoside
(iv) P_2S_5
(v) NH_3/MeOH
(vi) $POCl_3$/(MeO)$_3$PO
(vii) NH_4OH
(viii) $H_2PO_4^-$
(ix) Polynucleotide phosphorylase

poly(5-methoxycytidylic) acid

We have not yet tested the antiviral properties of poly(meo⁵C) but expect it to form a hybrid with poly(I) and to behave like poly(br⁵C) or poly(cl⁵C) as it lacks the ionisable hydroxyl group at C(5).

MUST THE POLYNUCLEOTIDE ENTER THE CELL FOR INTERFERON INDUCTION TO OCCUR?

It is important commercially to know whether a double-stranded polynucleotide has to enter the cell for interferon induction to occur. For example, if the polynucleotide acts as an external trigger like a hormone, then insolubilised inducers could be used in the large-scale production of interferon. Alternatively, if the polynucleotide must enter the cell before induction can occur, then methods can be developed for the improvement of the uptake of polynucleotide. A number of observations indicate that the polynucleotide must enter the cell for induction to occur. Firstly, when immobilised polynucleotides are used as interferon inducers, there is always some loss of polynucleotide from the insoluble support before an antiviral effect is noticed[3]. Secondly, cells which are grown under conditions which render their plasma membranes permeable tend to be more sensitive to poly(I).poly(C) than native cells[13]. We have recently obtained further evidence to support the hypothesis that uptake of poly(rI).poly(rC) must occur before a cell starts to synthesise interferon. Early work[14] using insensitive cells and poly(rI).poly(rC) of low specific radioactivity distinguished between two steps in the induction process, the binding of the polynucleotide to the cell which occurred at 4° and the formation of interferon which required the cells to be brought to 37°. Most of the polynucleotide which bound at 4° could be removed by washing with salt[15]. The antiviral effect in the insensitive cells could be destroyed by treatment of the polynucleotide-cell complex with ribonucleases without warming to 37°C. If the cells were warmed to 37° C for any appreciable time, all the poly(rI).poly(rC) could not be removed by ribonuclease treatment and the antiviral activity was unaffected by such treatment. These observations were interpreted as the poly(rI).poly(rC) binding at 4° followed by internalisation and interferon induction at 37°. Distinction could not be made between external triggering followed by polynucleotide-independent interferon formation or uptake of the polynucleotide as an essential prerequisite for interferon formation. Using poly(rI).poly(rC) of high specific radioactivity and a very sensitive line of human cells we have shown[16] that small amounts of polynucleotide are taken up at 4° and this small amount cannot be removed by salt washing or destroyed by treatment with ribonuclease. Furthermore, these treatments do not now destroy the interferon synthesising capability of our sensitive cells. The uptake process does not appear to be the stage at which stringent structural requirements are necessary as both colloidal gold and a polynucleotide which is a non-inducer of interferon (poly(dI).poly(rC)) are taken up in a similar manner to poly(rI).poly(rC). There are precedents for viral nucleic acids entering cells at 4° and we believe that the summation of the above evidence suggests that the polynucleotide must enter a cell before interferon synthesis can start and that the structural requirements of an interferon inducer, outlined earlier, are important in subsequent step(s).

DO THE STRANDS COME APART DURING INTERFERON INDUCTION?

Since double-stranded nucleic acids are so much better inducers of interferon than single-stranded nucleic acids, it may be essential for the two strands to stay together throughout the induction process. This hypothesis can be tested by chemically cross-linking dsRNA and we have carried this out using both bifunctional alkylating agents and the furocoumarin 8-methoxy psoralen. Bifunctional alkylating agents such as mustard gas (2) or its nitrogen analogue methylbis(β-chloroethylamine) (3) are known to cross-link DNA and RNA through the N(7)-atoms of suitably placed guanine residues to give structures resembling (5)[17]. The alkylating agents react with inosine residues in a similar manner.

$ClCH_2CH_2SCH_2CH_2Cl$ $ClCH_2CH_2N(Me)CH_2CH_2Cl$ $Me_2NCH_2CH_2Cl$

(2)   (3)   (4)

(5)

Our initial experiments were carried out with a group of naturally occurring RNAs from the myophage *Penicillium stoloniferum*. These RNAs which were active interferon inducers had a G+C content of approximately 50% and hence should cross-link readily with our bifunctional agents. Cross-linking was detected by a change in the melting profiles of the nucleic acids. Natural RNA or those treated with monofunctional alkylating agents showed pronounced hysteresis of hyperchromicity when the melted RNAs were cooled. On the other hand, the strands of cross-linked RNA reformed the duplex rapidly on cooling and no hysteresis of hyperchromicity was observed when the melted RNAs were cooled.

We have found that the cross-linked RNAs were active inducers of interferon provided that the extent of alkylation was low ($\leq 10\%$). In some cases they were better inducers than the unalkylated material or the RNA alkylated with the monofunctional reagent (4). At high levels of alkylation either with (3) or (4) the secondary structure of the RNAs was destroyed and these highly alkylated nucleic acids were not interferon inducers. At low alkylation levels the melting characteristics of the nucleic acids were similar to the native material. As the alkylation levels rose above 10%, the melting profiles became broad and there was little hyperchromicity.

TABLE: Effect of alkylation on interferon induction by nucleic acids

Nucleic Acid	Alkylating Agent	% Alkylation	Minimal protective dose (μg/ml)[b]
RNA[a]	Nil	Nil	0.05
	(3)	5.4	0.01
	(3)	8.0	0.005
	(3)	12.7	0.075
	(4)	5.5	0.075
	(4)	8.25	0.075
poly(rG-C)	Nil	Nil	0.03
	(3)	5.2	0.03
	(4)	7.0	0.05

[a] RNA from *P. stoloniferum*
[b] Minimal protective dose (μg/ml) to cause complete protection of 10^6 human fibroblast cells against Sindbis virus.

As there were at least five species of RNA in the material we obtained from *P. stoloniferum*, we turned our attention to synthetic polynucleotides and cross-linked poly(rG-C) with alternating G and C residues in the two strands. Here again the cross-linked material was an active interferon inducer at low ($\leqslant$ 10%) levels of alkylation.

Recently, we have studied a second class of cross-linking agents, the furocoumarins, e.g. 8-methoxypsoralen (6) which are used to treat psoriasis and similar complaints.

(6)   (7)   (8)

The furocoumarins intercalate between the bases of DNA and on irradiation form cycloaddition products, e.g. (7)[18] with the heterocyclic bases, particularly the pyrimidines. We find that irradiation of a solution of 8-methoxypsoralen and poly(rI-C) leads to cross-linking of the polynucleotide but poly(rI).poly(rC) is not cross-linked under our conditions presumably owing to the low reactivity of the purine rings. The change in hysteresis on melting and cooling cannot be used for synthetic double stranded polynucleotides and we checked for cross-linking by chromatography on hydroxylapatite following denaturation with formaldehyde[19]. At levels of alkylation of 1 8 MOP per 50 base pairs we find our material has activity as an inducer of interferon. Our results can be compared with those of Hochkeppel and Gordon[20] who cross-linked both 16 s rRNA and poly(rI).poly(rC) using a different furocoumarin (8). While no chemical studies on the cross-linking reaction were carried out, they suggest that their furocoumarin may have peculiar advantages as the presence of the aminomethyl group may lead to better intercalation which might explain why poly(rI).poly(rC) is cross-linked. They state without details that the cross-linked poly(rI).poly(rC) is still a good interferon inducer.

Thus from the above evidence we suggest that the strands of the double-stranded RNA stay together for the crucial triggering of the synthesis of interferon.

I would like to thank my able collaborators who have made the work possible, particularly Mr. C. F. Hui and Drs. T. K. Bradshaw, I. L. Cartwright and D. Eaton.

REFERENCES

1. P. F. Torrence and E. De Clercq, *Pharmacology and Therapeutics, Part A* **2**, 1-88 (1977)
2. A. K. Field, A. A. Tytell, G. P. Lampson and M. R. Hilleman, *Proc. Nat. Acad. Sci. U.S.A.* **58**, 1004-1009 (1967)
3. P. M. Pitha and D. W. Hutchinson, *Interferon and their Actions* ed. W. E. Stewart II p.13, C.R.C. Press, Boca Raton, Florida, E.S.A. (1977)
4. S. Arnott, D. W. L. Hukins, S. D. Dover, W. Fuller and A. R. Hodgson, *J. Mol. Biol.* **81**, 107-122 (1973)

5. E. De Clercq, P. F. Torrence, B. D. Stollar, J. B. Hobbs, T. Fukui, N. Kakiuchi and M. Ikehara, *European J. Biochem.* 88, 341-349 (1978)
6. M. A. W. Eaton and D. W. Hutchinson, *Biochim. Biophys. Acta* 319, 281-287 (1973)
7. N. Stebbing, C. A. Grantham, I. J. D. Lindley, M. A. W. Eaton and N. H. Carey, *Ann. N. Y. Acad. Sci.* 284, 682-696 (1977)
8. J. H. Chesterfield, J. F. W. McOmie and M. S. Tute, *J. Chem. Soc.* 4590-4596 (1960)
9. H. Vorbrüggen and B. Bennua, *Tetrahedron Letters* 1339-1342 (1978)
10. P. E. Garret, *Synthetic Procedures in Nucleic Acid Chemistry* ed. W. W. Zorbach and R. S. Tipson, Vol. I, p.439-440 (1968)
11. J. J. Fox, D. van Praag, I. Wempen, I. L. Doerr, L. Cheong, J. E. Knoll, M. L. Eidinoff, A. Bendich and G. B. Brown, *J. Amer. Chem. Soc.* 81, 178-187 (1959)
12. J. Tomasz, A. Simoncsits, M. Kajtar, R. M. King and A. J. Shatkin, *Nucleic Acids Res.* 5, 2945-2957 (1978)
13. E. C. Borden and P. H. Leonhardt, *Antimicrobial Agents and Chemotherapy* 9, 551-553 (1976)
14. G. H. Bausek and T. C. Merigan, *Virology* 39, 491-498 (1969)
15. M. D. Johnston, K. T. Atherton, D. W. Hutchinson and D. C. Burke, *Biochim. Biophys. Acta* 435, 69-75 (1976)
16. T. K. Bradshaw, I. L. Cartwright, D. W. Hutchinson and D. C. Burke submitted to *Biochim. Biophys. Acta*
17. P. Brookes and P. D. Lawley, *J. Chem. Soc.* 3923-3928 (1961)
18. L. Musajo, F. Bordin and R. Bevilacqua, *Photochem. Photobiol.* 6, 927-931 (1967)
19. G. Bernardi, *Biochim. Biophys. Acta* 174, 449-457 (1969)
20. H. Hochkeppel and J. Gordon, *Biochemistry* 1979 (in press)

THE CASE FOR MONOMERIC METAPHOSPHATE

A. Satterthwait and F. H. Westheimer

Department of Chemistry, Harvard University, Cambridge, MA 02138, U.S.A.

Abstract - Monomeric methyl metaphosphate, CH_3OPO_2, can be generated by the fragmentation of a β-bromophosphonate. It reacts with methylaniline to yield the products of electrophilic substitution in the aromatic ring, as well as the expected phosphoramidate. In the presence of acetophenone and a strong non-nucleophilic base, the methyl ester of the enol phosphate of acetophenone is formed; in the presence of ethyl benzoate and aniline, the product is N-phenyl-O-ethylbenzimidate. These latter reactions, which mimic enzymic processes, presumably take place by attack of the monomeric metaphosphate on the carbonyl oxygen atom of acetophenone and ethyl benzoate respectively. They suggest the possibility that ATP has a kinetic as well as thermodynamic role in metabolism.

INTRODUCTION

Since many intermediary metabolites are monoesters of phosphoric acid, since genetic material is comprised of diesters of phosphoric acid, and since pyrophosphates and enol phosphates provide energy for all living systems, the importance of organic phosphates can scarcely be exaggerated. Two major pathways have been adduced for the hydrolyses of phosphate esters: In the first (Ref. 1,2), water or another nucleophile adds to the tetracovalent phosphorus atom of a phosphate ester to yield a pentacovalent intermediate that then decomposes to product. In the second, the phosphate ester loses one of its substituents, to yield a monomeric metaphosphate, a strongly electrophilic intermediate that can then react with unshared electron pairs on nearby molecules to yield product. Whereas nitrate ion is stable, and orthonitric acid (H_3NO_4) is unknown, orthophosphate is stable and monomeric metaphosphate (PO_3^-) has never been isolated.

Considerable evidence has nevertheless accumulated that metaphosphates function as unstable intermediates. PO_3^- is isoelectronic with SO_3, and would be expected to resemble SO_3 as an electrophile. If it is formed in the neighborhood of an alcohol or amine, it could then serve as a phosphorylating agent. Evidence from our laboratory for just such reactions is presented below; important contributions to metaphosphate chemistry have come from many other laboratories, including those of Niel Hamer, W. P. Jencks, Anthony Kirby, Julius Rebek, and others (Ref. 3). This paper is devoted to a discussion of the new evidence for, and the new chemistry of, monomeric metaphosphates; the information serves as background to discussions of the mechanisms of chemical and enzymic reactions at phosphorus.

The hydrolysis of monoesters of phosphoric acid probably takes place by way of the monomeric metaphosphate ion, PO_3^-. In the 1940's, Bailly and Desjobert (Ref. 4) showed that these hydrolyses proceed most rapidly at pH 4, where the monoester monoanion, $ROPO_3H^-$, is present in highest concentration. In 1955, C. A. Bunton and his coworkers at University College, London, and Walter Butcher and one of us, at the University of Chicago, suggested that the rate maximum could be explained by a monomeric metaphosphate mechanism (Ref. 5).

$$ROPO_3H^- \rightleftharpoons R\overset{+}{\underset{H}{O}}-PO_3^{=} \longrightarrow CH_3OH + [PO_3^-] \tag{1}$$

The postulated PO_3^- ion presumably reacts with solvent:

$$[PO_3^-] + H_2O \longrightarrow H_2PO_4^- \tag{2}$$

ELECTROPHILIC SUBSTITUTION IN AROMATIC RINGS

A few years ago, Charles Clapp and one of us (Ref. 6) devised a scheme to prepare monomeric methyl metaphosphate. Clapp pyrolysed methylbutenylphosphonate in the gas phase; the products of the reaction were butadiene and, apparently, the desired CH_3OPO_2. At any rate, when the gas stream from the pyrolysis was introduced into a stirred solution of N-methylaniline in butylbenzene as solvent, the corresponding phosphoramidate was formed; when the reactant was N,N-diethylaniline, the product was that of aromatic phosphorylation (see next paragraphs).

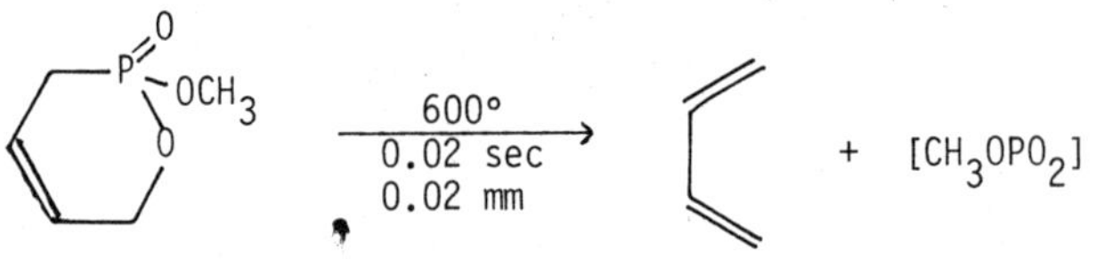

$$\text{(cyclic phosphonate)} \xrightarrow[\substack{0.02 \text{ sec} \\ 0.02 \text{ mm}}]{600°} \text{butadiene} + [CH_3OPO_2] \qquad (3)$$

$$[CH_3OPO_2] + C_6H_5NHCH_3 \longrightarrow C_6H_5 - \overset{+}{\underset{|}{N}}H\underset{|}{-}\overset{CH_3}{\underset{}{}} \overset{OCH_3}{\underset{|}{PO_2^-}} \qquad (4)$$

The fragmentation of β-halophosphonates was discovered by J. B. Conant and his collaborators in the 1920's; the reaction was rediscovered by J. Maynard and J. Swan in 1963 (Ref. 7). The fragmentation is stereospecific (Ref. 8) and <u>trans</u>; presumably it proceeds by way of monomeric metaphosphate as an intermediate. In continuing this work (Ref. 9), we have prepared what appears to be monomeric methyl metaphosphate by the fragmentation shown below.

$$\begin{array}{c} C_6H_5 - CBr - CHBr - CH_3 \\ | \\ CH_3O - PO_2^- \\ \mathbf{I} \end{array} \longrightarrow \begin{array}{c} C_6H_5 - CBr = CHCH_3 \\ + [CH_3OPO_2^-] + Br^- \end{array} \qquad (5)$$

The reaction has been carried out in various solvents at 70° in the presence of 2,2,6,6-tetramethylpiperidine, B, as a sterically hindered base. The base is needed since the anion, but not the free acid of the phosphonate undergoes cleavage; further, the phosphorylations generate acid, which must be neutralized. When the solvent for the reaction is N-methylaniline, the products are those shown below, where the combined yield of aromatic substitution products is 35%. When, however, the reaction mixture is diluted with other solvents, the yield of aromatic substitution falls, and that of phosphoramidate increases. Some pertinent data are shown in Table 1.

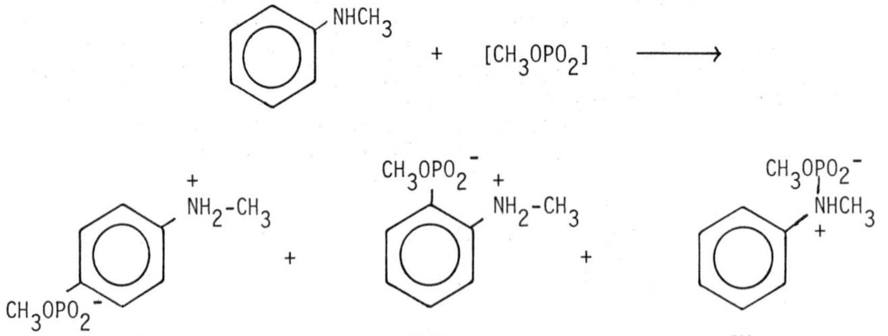

$$C_6H_5NHCH_3 + [CH_3OPO_2] \longrightarrow \qquad (6)$$

II: ortho-substituted aniline with $NH_2^+-CH_3$ and $CH_3OPO_2^-$
III: para-substituted aniline with $CH_3OPO_2^-$ and $NH_2^+-CH_3$
IV: N-substituted with $CH_3OPO_2^-$ and $\overset{+}{N}HCH_3$

TABLE 1. Aromatic substitution by "monomeric methyl metaphosphate"

Diluent 9/1,v/v	%Aromatic Substitution	%Nitrogen Addition
None	35	50
$CDCl_3$	29	67
CD_3CN	3	85
Dioxane	≤2	85

One can infer from these data that dioxane and acetonitrile strongly suppress the electrophilic character of the reactive intermediate, so that it is no longer sufficiently active to carry out aromatic substitution, whereas chloroform is much less destructive of the reagent. These findings are most easily interpreted by the assumption that monomeric methyl metaphosphate is indeed formed in the fragmentation, and can react with N-methylaniline either at the nitrogen atom or at an activated position of the ring. Since, however, the electrophile can react with any Lewis base, it will add readily to the unshared electron pairs of dioxane or acetonitrile. The reaction of monomeric methyl metaphosphate with dioxane presumably parallels the known reaction of SO_3 with that basic ether (Ref. 10).

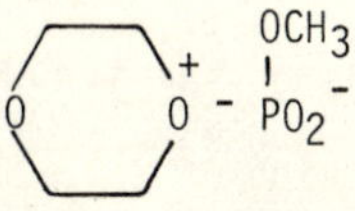

$$\underset{V}{\overset{}{\bigcirc}}\text{O}^+\text{—O}^-\text{—PO}_2^-\text{—OCH}_3 \qquad \underset{VI}{CH_3-C\equiv\overset{+}{N}-\overset{OCH_3}{\underset{}{PO_2^-}}} \qquad \underset{VII}{\overset{}{\bigcirc}}\text{O}^+\text{—O}^-\text{—SO}_3^-$$

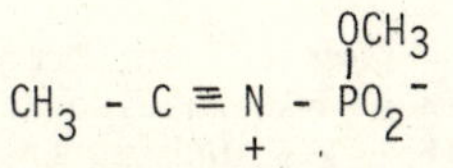

The probable reaction products are shown above. The zwitterionic phosphates are presumably sufficiently active as phosphorylating agents to react with a secondary amine, and yield a phosphoramide, but not sufficiently electrophilic to effect aromatic substitution, even with such an activated molecule as N-methylaniline.

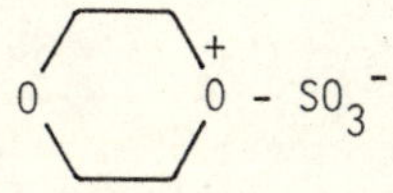

ATTACK ON CARBONYL GROUPS

The most interesting reactions that we have discovered for monomeric methyl metaphosphate are those with carbonyl compounds. In acetophenone as solvent, monomeric methyl metaphosphate, generated as shown in equation 5 above, gives an excellent yield of the enol phosphate, identical with that produced by an alternative pathway, and identified by analysis and NMR spectroscopy. Presumably the monomeric methyl metaphosphate directly attacks the carbonyl group of the ketone, according to equation 7.

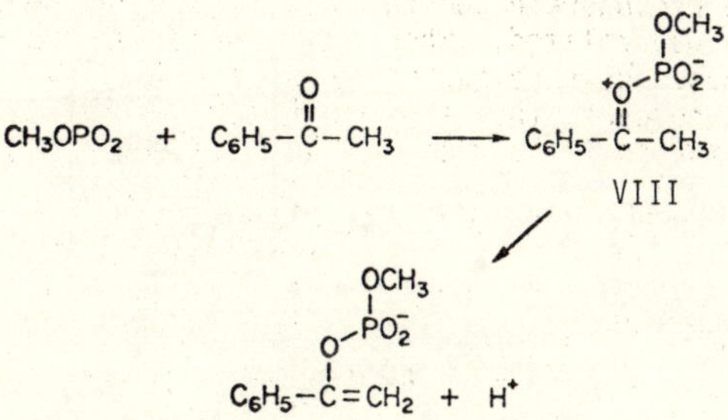

$$CH_3OPO_2 + C_6H_5-\overset{O}{\underset{}{C}}-CH_3 \longrightarrow \underset{VIII}{C_6H_5-\overset{\overset{+O-PO_2^-}{|}}{\underset{}{C}}-CH_3} \qquad (7)$$

$$\underset{IX}{C_6H_5-\overset{\overset{O-PO_2^-}{|}\overset{OCH_3}{}}{\underset{}{C}}=CH_2} + H^+$$

However, when aniline as well as acetophenone is present, the products are not the enol phosphate, but a mixture of the Schiff base of acetophenone and methyl dihydrogen phosphate. We postulate that the reaction takes place thru the same intermediate, VIII, formed in the preparation of the enol phosphate.

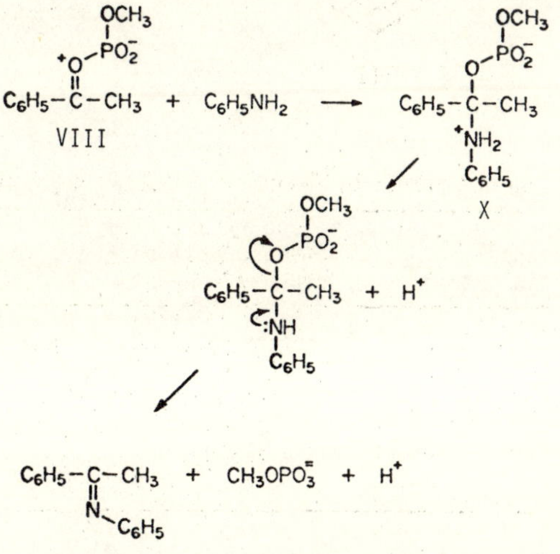

$$\underset{VIII}{C_6H_5-\overset{\overset{+O-PO_2^-}{|}\overset{OCH_3}{}}{\underset{}{C}}-CH_3} + C_6H_5NH_2 \longrightarrow \underset{X}{C_6H_5-\overset{\overset{O-PO_2^-}{|}\overset{OCH_3}{}}{\underset{\overset{+NH_2}{|}\underset{C_6H_5}{}}{C}}-CH_3}$$

$$\underset{}{C_6H_5-\overset{\overset{O-PO_2^-}{|}\overset{OCH_3}{}}{\underset{\overset{:NH}{|}\underset{C_6H_5}{}}{C}}-CH_3} + H^+ \qquad (8)$$

$$\underset{XI}{C_6H_5-\overset{}{\underset{\overset{||}{N_{C_6H_5}}}{C}}-CH_3} + CH_3OPO_3^= + H^+$$

If this pathway is correct, then the initial reaction here, as in the formation of the enol phosphate, is the attack of methyl metaphosphate on the carbonyl oxygen atom of acetophenone; the intermediate behaves like a protonated ketone, and is activated both for the abstraction of a proton from the methyl group to yield the enol phosphate and for attack by a nucleophile at the activated carbon atom of the carbonyl group. For this mechanism to be valid, the attack of aniline at the activated carbonyl group must be rapid relative to the attack of the strong but hindered base, 2,2,6,6-tetramethylpiperidine, at one of the hydrogen atoms of the methyl group.

Although we believe that this mechanism is correct, that conclusion is not at all obvious since the products can be accounted for in an alternative way. Aniline reacts with acetophenone in the presence of a mild acid catalyst, such as the hydrobromide of 2,2,6,6-tetramethylpiperidine, to yield the Schiff base and water; the methyl phosphate that was obtained could have arisen by the attack of monomeric methyl metaphosphate on the water. Further, when the fragmentation reaction is carried out in the presence of acetophenone and aniline, the yield of Schiff base is far in excess of the yield of methyl phosphate, or of the yield calculated on the basis of the phosphonate ester originally introduced into the reaction mixture. When, however, the reaction is repeated in the presence of o-trifluoromethylaniline instead of with aniline, the results proved unambiguous. Here the direct condensation of amine with acetophenone is far to slow to compete with the process promoted by monomeric methyl metaphosphate; only one mole of Schiff base is formed, and it is formed at the same rate as that for the formation of methyl dihydrogen phosphate.

In an even more significant process, ethyl benzoate reacts with aniline, when the mixture is treated with monomeric methyl metaphosphate, to yield N-phenyl-O-ethylbenzimidate. Since the direct action of aniline on ethyl benzoate leads to the formation of benzanilide, the actual process necessarily follows a special pathway; presumably the chemistry is that shown in equation 9. Here again, the initial attack by the metaphosphate is on the carbonyl oxygen atom, to yield a cationic intermediate that is activated at the carbonyl carbon atom.

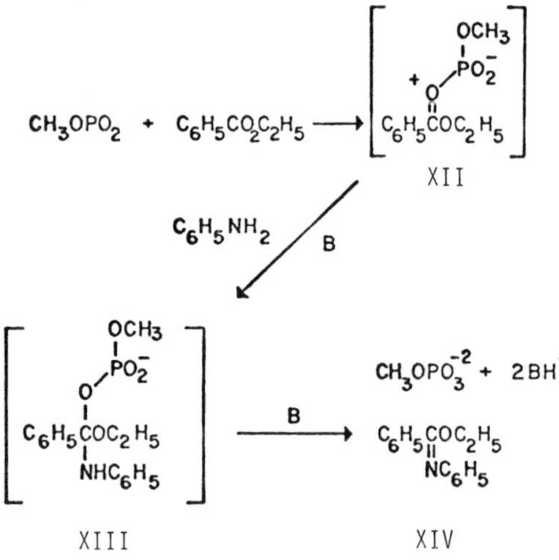

$$CH_3OPO_2 + C_6H_5COC_2H_5 \longrightarrow \begin{bmatrix} & & OCH_3 \\ & & | \\ & & PO_2^- \\ & +O & \nearrow \\ & \| & \\ & C_6H_5COC_2H_5 \end{bmatrix} \quad \text{XII}$$

$$\xrightarrow[B]{C_6H_5NH_2} \begin{bmatrix} & OCH_3 \\ & | \\ & PO_2^- \\ O & \nearrow \\ | & \\ C_6H_5COC_2H_5 \\ | \\ NHC_6H_5 \end{bmatrix} \xrightarrow{B} \begin{array}{c} CH_3OPO_3^{-2} + 2BH^+ \\ C_6H_5\underset{\|}{C}OC_2H_5 \\ NC_6H_5 \end{array} \quad (9)$$

XIII   XIV

FORMATION OF PHOSPHONATES

One of the minor products that can be isolated from the fragmentation of the methyl ester of 1-phenyl-1,2-dibromopropylphosphonate, I, in the presence of N-methylaniline is the methyl ester of 1-benzoylethylphosphonic acid. Two reasonable pathways for the process are shown below; one involves monomeric methyl metaphosphate, whereas the other requires the direct rearrangement of I to XV.

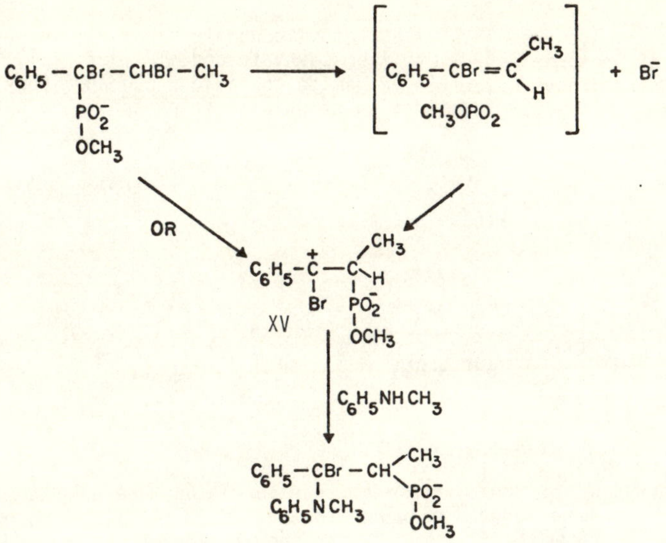

$$\text{(10)}$$

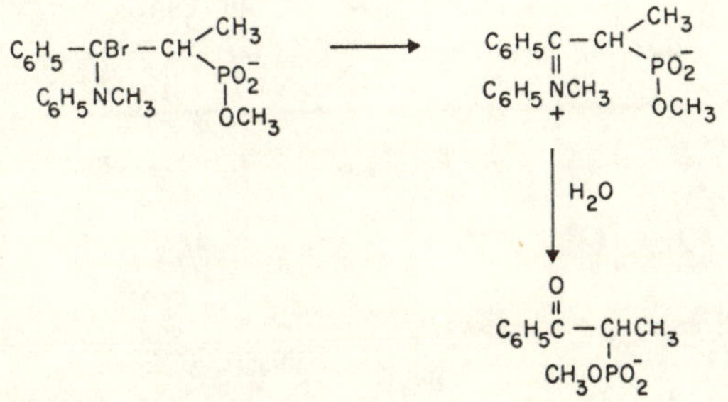

$$\text{(11)}$$

Although the formation of a phosphorus to carbon bond suggests the intervention of monomeric metaphosphate, no decision between the alternatives shown above can be made on the basis of information now available. In any event, the synthesis of a phosphonic acid parallels that in the enzymic formation of aminoethylphosphonate from phosphoenolpyruvate (Ref. 11); the pyruvylphosphonic acid in equation 12 below has been identified as an intermediate in the overall transformation.

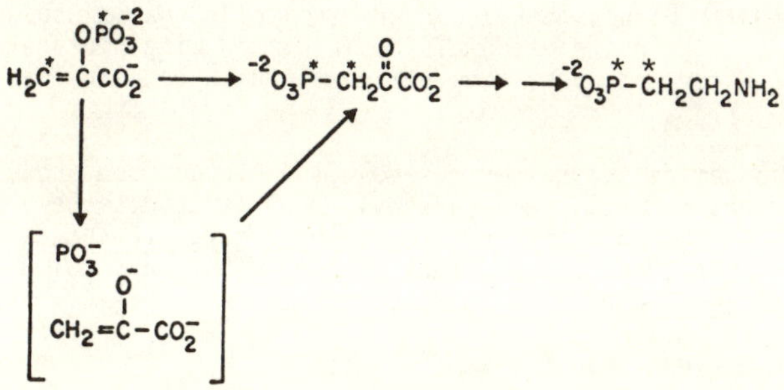

$$\text{(12)}$$

OTHER ANALOGIES TO METABOLIC PROCESSES

The processes that involve acetophenone and ethyl benzoate parallel, in a formal sense, the enzymic formation of phosphoenolpyruvate from pyruvate and ATP (Ref. 12), the enzymic amidation of 5-phosphoribosyl-N-formylglycinamide to yield 5-phosphoribosyl-N-formylglycinamidine (Ref. 13), and the enzymic amidation of uridine triphosphate to cytidine triphosphate (Ref. 14).

$$CH_3COCO_2^- + ATP \xrightleftharpoons[\text{kinase}]{\text{Pyruvate}} CH_2 = C(OPO_3^=)(CO_2^-) + ADP \qquad (13)$$

$$\text{(structure)} \xrightarrow[\text{GLUTAMINE}]{ATP/Mg^{++}} \text{(structure)} + ADP + P_i \qquad (14)$$

In detail, the mechanisms of the biochemical transformations are unknown. In particular, Rose and his coworkers have adduced evidence that the formation of enolpyruvate occurs by way of the enolization of pyruvate, followed by phosphorylation of the enol (Ref. 12). Further, the amidations could occur either by phosphorylation of the carbonyl groups, followed by attack of ammonia, or alternatively by attack of ammonia on the substrate to yield a carbinolamine, which is subsequently phosphorylated; the pathways for the synthesis of cytidine triphosphate are outlined in equation 15.

$$\text{(scheme)} \qquad (15)$$

Evidence for the transfer of a carbonyl oxygen atom to inorganic phosphate has been presented by Lieberman (Ref. 15), in the reaction of inosinic acid with glutamate to yield adenylosuccinate; oxygen transfer, however, does not distinguish between the two pathways of equation 15.

A number of other biochemical processes resemble those described above. Perhaps ATP functions in these processes, too, to activate a carbonyl group, and so serve a kinetic as well as a thermodynamic role. These reactions include, among others, that catalyzed by 5-formyltetrahydofolate cyclodehydrase (Ref. 16) and that catalyzed by 5-oxyprolinase (Ref. 17); the latter is a rare example of a hydrolytic reaction promoted by ATP. Further, a tetrahedral intermediate similar to that postulated in equation 15 has been observed in the reaction catalyzed by adenylosuccinate synthetase (Ref. 18).

Even if the attack on the carbonyl group provides the first step in the reactions such as that shown in equation 15, a question must arise as to whether ATP phosphorylates the carbonyl group by direct attack or with prior dissociation of monomeric metaphosphate. Two studies that bear upon these questions have been published by Lowe and Sproat (Ref. 19), and by Knowles and his collaborators (Ref. 20). The former showed that pyruvate kinase in the absence of pyruvate acts to scramble the oxygen atoms attached to the β-phosphorus atom of ATP, as shown in equation 16; this fact strongly suggests a monomeric metaphosphate mechanism.

$$Ad-O-\underset{\underset{O}{\|}}{\overset{\overset{O}{\|}}{P}}-\overset{*}{O}-\underset{\underset{*O}{\|}}{\overset{\overset{*O}{\|}}{P}}-O-\underset{\underset{O}{\|}}{\overset{\overset{O}{\|}}{P}}-O \Biggr\}^{\equiv} \rightleftharpoons$$

$$Ad-O-\underset{\underset{O}{\|}}{\overset{\overset{O}{\|}}{P}}-\overset{*}{O}\Biggl(\underset{\underset{*O}{\|}}{\overset{\overset{*O}{\|}}{P}}-O\Biggr\}^{\equiv} + PO_3^- \rightleftharpoons \quad (16)$$

$$Ad-O-\underset{\underset{O}{\|}}{\overset{\overset{O}{\|}}{P}}-\overset{*}{O}-\underset{\underset{O}{\|}}{\overset{\overset{O}{\|}}{P}}-\overset{*}{O}-\underset{\underset{O}{\|}}{\overset{\overset{O}{\|}}{P}}-O \Biggr\}^{\equiv}$$

(Here the symbol * stands for an atom of ^{18}O.)

On the other hand, Knowles and Blättler investigated several enzymes, including pyruvate kinase, using ATP where the gamma phosphate residue was chiral because of substitution with ^{16}O, ^{17}O and ^{18}O. Using this chiral ATP, they showed that phosphorylation proceeds with inversion about phosphorus; a fully free monomeric metaphosphate ion would presumably lead to racemization. This experiment argues against monomeric metaphosphate, or more precisely against free monomeric metaphosphate, as the intermediate in the phosphorylation. The detailed pathway of enzymic phosphorylation must still be elucidated; knowledge of the chemistry of monomeric metaphosphate is a necessary component of the solution.

The major new idea implied by the chemistry here discussed is that monomeric metaphosphate, and perhaps ATP can act to phosphorylate carbonyl groups. The chemical mechanism for the amidation of ethyl benzoate by such a pathway is reasonably secure; if the enzymic processes also proceed by way of the phosphorylation of carbonyl compounds (whether or not they proceed by way of free metaphosphate) then ATP plays a kinetic as well as a thermodynamic role. In one of the most fruitful advances in biochemistry, F. Lipmann (Ref. 21) showed how thermodynamically unfavorable processes can be carried forward by ATP. Now it is necessary to examine the possibility that ATP is important for its kinetic role as well.

Acknowledgement - The work here described was supported by the National Science Foundation under Grant No. CHE 77-05948.

REFERENCES

1. W. E. McEwen and K. D. Berlin, Organophosphorus Stereochemistry, Parts I and II, Dowden, Hutchinson and Ross, Stroudsberg, Pennsylvania (1975); J. Emsley and D. Hall, The Chemistry of Phosphorus, Halsted Press, New York (1976); A. J. Kirby and S. G. Warren, The Organic Chemistry of Phosphorus, Elsevier, Amsterdam (1967), p. 281 ff; T. C. Bruice and S. Benkovic, Bioorganic Mechanisms, Vol. 2, W. A. Benjamin, New York (1966), pp. 22-25, 157-159; S. J. Benkovic and K. J. Schray, in The Enzymes, Vol. VIII, 3rd ed., P. D. Boyer, Ed., Academic Press, New York (1973), p. 201; W. P. Jencks, Catalysis in Chemistry and Enzymology, McGraw-Hill, New York (1969), pp. 81-83, 112-115, 151, 160-161, 608.

2. F. H. Westheimer, Accts. Chem. Res. 1, 70 (1968); I. S. Sigal and F. H. Westheimer, J. Am. Chem. Soc. 101, 752 (1979).

3. G. DiSabato and W. P. Jencks, J. Am. Chem. Soc. 83, 4400 (1961); A. J. Kirby and A. G. Varvoglis, ibid., 80, 415 (1967); P. Haake and P. S. Ossip, ibid., 93, 6924 (1971); D. G. Gorenstein, ibid., 94, 2523 (1972); R. Kluger, J. Org. Chem. 38, 2721 (1973); J. Rebek and F. Gaviña, J. Am. Chem. Soc. 97, 1591, 3221 (1975); A. F. Gerrard and N. K. Hamer, J. Chem. Soc. B, 539 (1968), 369 (1969); E. Niecke and W. Flick, Angew. Chem., Int. Ed. Engl. 13, 134 (1974); O. J. Scherer and N. Kuhn, Chem. Ber. 107, 2123 (1974); N. T. Kulbach and O. J. Scherer, Tetrahedron Lett., 2297 (1975); M. Regitz, H. Scherer, W. Illger and H. Eckes, Angew. Chem. Int. Ed. Engl. 12, 1010 (1973); M. Regitz, A. Liedhegener, W. Anschutz and H. Eckes, Chem. Ber. 104, 2177 (1971); M. Regitz, H. Scherer and W. Anschutz, Tetrahedron Lett. 753 (1970); D. Brown and N. Hamer, J. Chem. Soc. (London) 1155 (1960); D. Samuel and B. Silver, ibid., 4321 (1961); D. L. Miller and T. Ukena, J. Am. Chem. Soc. 91, 3050 (1969); F. Ramirez and J. F.

Maracek, J. Am. Chem. Soc. 101, 1460 (1979).

4. M. C. Bailly, Bull. Soc. Chim. 9, 421 (1942); A. Desjobert, Compt. Rend. 224, 575 (1947); A. Desjobert, Bull. Soc. Chim. 14, 809 (1947).

5. P. W. C. Bernard, C. A. Bunton, D. R. Llewellyn, K. G. Oldham, B. L. Silver and C. A. Vernon, Chem. Ind. (London) 760 (1955); W. W. Butcher and F. H. Westheimer, J. Am. Chem. Soc. 77, 2420 (1955).

6. C. H. Clapp and F. H. Westheimer, J. Am. Chem. Soc. 96, 6710 (1974); C. H. Clapp, A. Satterthwait and F. H. Westheimer, ibid., 97, 6873 (1975).

7. J. B. Conant and A. A. Cook, J. Am. Chem. Soc. 42, 830 (1920); J. B. Conant and S. M. Pollack, ibid., 43, 1665 (1921); J. B. Conant and E. L. Jackson, ibid., 46, 1003 (1924); J. B. Conant and B. B. Coyne, ibid., 44, 2530 (1922); J. A. Maynard and J. M. Swan, Aust. J. Chem. 16, 596 (1963).

8. G. L. Kenyon and F. H. Westheimer, J. Am. Chem. Soc. 88, 3557, 3561 (1966).

9. A. C. Satterthwait and F. H. Westheimer, J. Am. Chem. Soc. 100, 3197 (1978).

10. C. M. Suter, P. B. Evans and J. M. Kiefer, J. Am. Chem. Soc. 60, 538 (1938).

11. M. Horiguchi and H. Rosenberg, Biochem. Biophys. Acta 404, 333 (1975).

12. I. A. Rose, J. Biol. Chem. 235, 1170 (1960); J. L. Robinson and I. A. Rose, ibid., 247, 1096 (1972); D. J. Kuo and I. A. Rose, J. Am. Chem. Soc. 100, 6288 (1978).

13. J. Buchanan, Adv. in Enzym. 39, 91 (1973).

14. A. Levitzki and D. E. Koshland, Jr., Biochem. 10, 3365 (1971).

15. I. Lieberman, J. Biol. Chem. 223, 327 (1956).

16. D. M. Greenberg, L. K. Wynston and A. Nagabhushanam, Biochem. 4, 1872 (1965).

17. P. van der Werf, M. Orlowski and A. Meister, Proc. Nat. Acad. Sci. USA 68, 2982 (1971).

18. G. D. Markham and G. H. Reed, J. Biol. Chem. 253, 6184 (1978).

19. G. Lowe and B. S. Sproat, Chem. Comm. 783 (1978).

20. W. A. Blättler and J. R. Knowles, J. Am. Chem. Soc. 101, 510 (1979); S. J. Abbott, S. R. Jones, S. A. Weinman and J. R. Knowles, ibid., 100, 2558 (1978).

21. F. Lipmann, Adv. in Enzym. 1, 99 (1941).

NUCLEOSIDE PHOSPHOROTHIOATES - TOOLS FOR THE INVESTIGATION OF ENZYME MECHANISMS

F. Eckstein

Max-Planck-Institut für experimentelle Medizin, Abt. Chemie, Göttingen, Federal Republic of Germany

A variety of phosphorothioate analogues of ATP and ADP have been synthesized in which a non-bridging oxygen of the phosphate group is replaced by sulfur. This substitution - when not in the terminal phosphate-leads to the generation of chiral phosphorus. These compounds therefore exist as pairs of diastereomers which can be separated by enzymatic methods. They can be used for the determination of the stereochemistry of enzymatic phosphate or nucleotidyl transfer reactions.

Although adenosine 5'-triphosphate (ATP) is a substrate for a large number of enzymes it is only recently that one begins to understand details of its interactions with enzymes. In this article I will briefly discuss certain stereochemical aspects of such interactions in as much as they are amenable to analysis by phosphorothioate analogues of ATP (1).

Looking at a model or a structural formula of ATP it is apparent that on chelation with metal ions one can obtain a pleora of diastereomers since the phosphate group to which the metal is coordinated becomes chiral in this process. Considering the most prevalent metal ion involved in forming enzyme - nucleotide substrate complexes, Mg^{2+}, and only taking into account α, β - as well as β, γ-bidentate complexes one comes up with six diastereomers. The question arises as to which of these isomers is exclusively or preferentially recognized by the active site of an enzyme. The answer to this question is not easy to obtain because one cannot separate these diastereomers since the ligands around the metal ion are exchanged fast. Two methods are at present available to stabilize these isomers: 1. By changing the metal from Mg^{2+} to Cr^{3+} or Co^{3+} the group of W. Cleland has obtained stable diastereomers which they could separate (2,3). 2. By exchanging one of the non-bridging oxygens at the α - or β-phosphorus by sulfur one fixes the chirality at this particular phosphorus and makes it independent at the metal ion.

In this lecture I will discuss this latter class of compounds which has recently been the subject of a review (1).

The phosphorothioate analogues of ATP permit one to investigate the following questions which arise from the interaction of a nucleoside triphosphate with an enzyme:
1. Does an enzyme have a preference for one of the diastereomers of the substrate and if so which?
2. To which of the phosphate groups is the metal ion of the substrate metal ion complex bound at the active site of an enzyme, if at all?
3. What is the stereochemical course of enzymatic phosphate or nucleotide transfer? Does it proceed with inversion or retention of configuration?

Phosphorothioate analogues can easily be obtained. The mixture of diastereomers of ADPαS can be synthesized chemically (4). For the separation of diastereomers enzymatic methods have been applied. Various kinases preferentially phosphorylate one or the other isomer of ADPαS to ATPαS (4). Although originally the diastereomers of ATPαS were analyzed by enzymatic methods, this can now also be done by hplc (5) or ^{31}P-nmr (6,7). The diastereomers of ATPβS can be obtained by kinase reactions from ADPβS (4). The absolute configurations of both sets of diastereomers, ATPαS as well as ATPβS, have been determined by various groups by a number of methods (5,8,9,10,11,12). Since none of these diastereomers exists in crystalline form X-ray structural analysis was not directly applicable.

Looking at the stereoselectivity of various enzymes for one or the other diastereomer of ATPαS or ATPβS one finds that most exhibit preferences although of various degrees. As is obvious from the synthesis, many kinases fall into this class of enzymes (4) as do various nucleoside di- or triphosphate polymerases (13,14,15,16). It is beyond the scope of this lecture to list all the enzymes investigated with respect to this particular problem.

What about the second question? How can it be answered with the help of phosphorothioates? It is not difficult to understand that the Co^{3+} or Cr^{3+} complexes of ATP can provide an answer since they have very long half lifes. The crucial experiments with the phosphorothioates were carried out by Jaffe and Cohn (11) who found that ATPβS, R-isomer is a substrate for hexokinase in the presence of Mg^{2+}, but in the presence of Cd^{2+} it is the S-isomer.

The basis for this swich-over is the preference of Mg^{2+} for complexation with oxygen and that of Cd^{2+} with sulfur. In this way the Mg-R isomer has the same screw sense as the Cd-S isomer and both can fit into the active site of hexokinase to be substrates. Other metal ions such as Co^{2+}, Mn^{2+} or Zn^{2+} bind to oxygen as well as sulfur and, therefore, both diastereomers become substrates in the presence of these cations. This conclusion of course demands that the metal is coordinated to the β-phosphorus even when the substrate is at the active site. Otherwise, if the metal was replaced by a positively charged side group on the enzyme, there would be no reason for this swich-over. This observation provides a handle to study various enzymes for this change of requirement of diastereomers in the presence of Mg^{2+} and Cd^{2+}.

It must suffice here to cite as examples the DNA dependent DNA as well as RNA polymerases from E. coli (10,17) where such changes in stereoselectivity with respect to the β-phosphorus of the substrate occur by an appropriate change of cation. However, both enzymes take only the diastereomers with S-configuration at α-phosphorus as substrate not only with Mg^{2+} but also Co^{2+}, Mn^{2+} or Cd (at least in the case of RNA polymerase). These results would indicate that the metal ion remains liganded to the β-phosphorus at the active site but might not to the α-phosphorus. It is gratifying to see that there is complete agreement in the results obtained by the two methods, the phosphorothioate approach as discussed here and the one using stable Co^{3+} or Cr^{3+} ATP complexes (2,3).

The stereochemical course of enzymatic phosphate or nucleotide transfer has been investigated by principally two methods. The former has become approachable by the synthesis of $^{16,17,18}O$ γ-labelled ATP and the analysis of the phosphorylated product (18,19), as well as by the availability of ^{18}O γ-labelled adenosine 5'-O-(3-thio-triphosphate) (20,21). The latter has been studied mainly using phosophorothioate analogues of nucleotides (1).

Two enzymatic phosphate transfer reactions play a central role in the elucidation of the stereochemical course taken by many of the enzymes studied. Myokinase phosphorylates adenosine 5'-phosphorothioate to the S-diastereomer of ADPαS which in turn is stereospecifically phosphorylated to ATPαS of the same configuration (6,7). Methods have been developed which can analyze whether in an ATP molecule an ^{18}O is attached to the α-phosphorus in the αP-O-βP bridging or non-bridging position (22,23,24). Since these methods are also applicable to the diastereomers of ATPαS (12,25,26) it is possible now to determine the absolute configuration of [^{18}O]adenosine 5'-phosphorothioate, the product of many hydrolytic enzymes.

The results of both approaches, the one using $^{16,17,18}O$-γ-ATP and the one using phosphorothioates, have been elegantly reviewed and discussed recently (27).

They can be summarized by saying that each nucleophilic substitution reaction on phosphorus proceeds by an in-line mechanism leading to inversion of configuration. Two such substitution reactions which occur when the reaction proceeds via a covalent enzyme intermediate result in retention of configuration. The evidence accumulated so far justifies the conclusion that retention of configuration is, indeed, an indication for such an intermediate.

In this lecture I have briefly discussed what aspects of nucleotideenzyme inter-actions can be analyzed by the use of nucleotide phosphorothioate analogues. I hope I could make a convincing case that they permit a deeper insight into such complexes than is normally possible with the unmodified substrates.

References

1. F. Eckstein, Accounts of Chem. Res. 12, 204-210 (1979).
2. R.D. Cornelius and W.W. Cleland, Biochemistry 17, 3279-3286 (1978).
3. D. Dunaway-Mariano and W.W. Cleland, Biochemistry, in press.
4. F. Eckstein and R.S. Goody, Biochemistry 15, 1685-1691 (1976).
5. P.M.J. Burgers and F. Eckstein, Proc. Natl. Acad. Sci. USA 75, 4798-4800 (1978).
6. E.K. Jaffe and M. Cohn, Biochemistry 17, 652-657 (1978).
7. K.F.R. Shen and P.A. Frey, J. Bio. Chem. 252, 4445-4448 (1977).
8. P.M.J. Burgers, B.K. Sathyanarayana, W. Saenger and F. Eckstein, Europ. J. Biochem., 1979, in press.

9. Bryant and S.J. Benkovic, Biochemistry 18, 2825-2828 (1979).
10. R.L. Jarvest and G. Lowe, J. C. S. Chem. Comm. 1979, 364-366.
11. E.K. Jaffe and M. Cohn, J. Biol. Chem. 253, 4823-4825 (1978).
12. J.P. Richard, H.T. Ho and P.A. Frey, J. Amer. Chem. Soc. 100, 7756-7757 (1978).
13. F. Eckstein, V.W. Armstrong and H. Sternbach, Proc. Natl. Acad. Sci. USA 73, 2987-2990 (1976).
14. P.M.J. Burgers and F. Eckstein, Biochemistry 18, 450-454 (1979).
15. F. Eckstein, H. Sternbach and F. v. d. Haar, Biochemistry 16, 3429-3432 (1977).
16. P.M.J. Burgers and F. Eckstein, J. Biol. Chem. 254, 6889-6893 (1979).
17. V.W. Armstrong, D. Yee and F. Eckstein, Biochemistry 18, in press (1979).
18. W.A. Blättler and J.R. Knowles, J. Amer. Chem. Soc. 101, 510-511 (1979).
19. S.J. Abbott, S.R. Jones, S.A. Weinman and J.R. Knowles, J. Amer. Chem. Soc. 100, 2558-2560 (1978).
20. J.P. Richard and P.A. Frey, J. Amer. Chem. Soc. 100, 7757-7758 (1978).
21. G.A. Orr, J. Simon, S.R. Jones, G.J. Chin and J.R. Knowles, Proc. Natl. Acad. Sci. USA 75, 2230-2233 (1978).
22. C.F. Midelfort and I.A. Rose, J. Biol. Chem. 251, 5881-5887 (1976).
23. M. Cohn and A. Hu, Proc. Natl. Acad. Sci. USA 75, 200-203 (1978).
24. G. Lowe and B.S. Sproat, J. C. S. Chem. Commun. 1978, 565-566.
25. C.F. Midelfort and I. Sarton-Miller, J. Biol. Chem. 253, 7127-7129 (1978).
26. R.L. Jarvest and G. Lowe, J. C. S. Chem. Commun. 1979, 364-366.
27. J.R. Knowles, Ann. Res. Biochemistry, in press.

THE STEREOCHEMICAL COURSE OF SEVERAL ENZYME CATALYZED REACTIONS AT THE PHOSPHODIESTER LEVEL

F. R. Bryant, J. F. Marlier and S. J. Benkovic

Department of Chemistry, The Pennsylvania State University, University Park, Pennsylvania 16802, U.S.A.

Abstract - A methodology is outlined for examining the stereochemical course at phosphorus of enzyme catalyzed reactions at the phosphodiester level utilizing phosphorothioates. Experiments with intestinal and venom phosphodiesterases that catalyze the formation of 5'-phosphonucleotides and polynucleotide phosphorylase that catalyzes the de novo polymerization of nucleotide diphosphates are reported. All of the reactions examined proceed with retention of configuration at phosphorus.

INTRODUCTION

The stereochemical course of phosphoryl transfer reactions for a number of kinases, nucleotidyl and nucleoside transferases, and polymerases recently has been established and in a majority of the cases the result is inversion of configuration (Refs. 1-4). In the discussion to be presented here, the methodology for elucidating the overall stereochemistry of the phosphoryl transfer process to water catalyzed by venom and intestinal phosphodiesterase as well as the transfer to phosphate catalyzed by polynucleotide phosphorylase from M. Luteus will be outlined. The consequences of these results for the enzyme mechanism also will be discussed.

METHODS

The approach employed is derived from the seminal work of Usher and Eckstein and features the use of a phosphorothioate chiral center. Combining the discovery that the resulting two diastereomers of ATP(αS) are distinguished by various kinases (Ref. 5) with the recent elucidation of the absolute configuration of the P_α center in each of the diastereomers (Refs. 3,6,7) a method can be devised to elaborate the overall stereochemical course of reactions that involve substituted AMPS derivatives. A typical protocol is depicted in Scheme I for a reaction proceeding with retention of configura-

SCHEME I

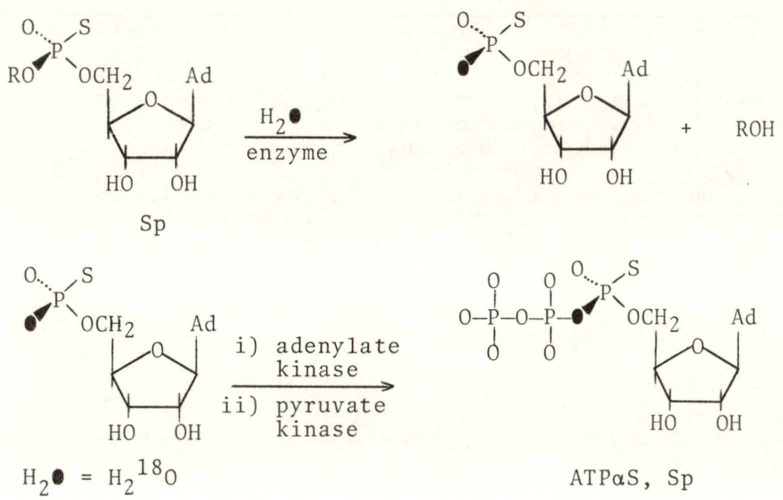

tion. Elaboration and quantitation of the location of the ^{18}O label in the product AMPS is achieved by a modification of a gc-mass spectrometric method employed by Frey et al. (Refs. 8,9) in which the enzymatically generated Sp diastereomer of ATPαS is chemically degraded to a mixture of trimethyl phosphate and trimethyl phosphorothioate that is separated by gas chromatography before molecular weight determination. A control indicates a 65:35 cleavage favoring hydrolytic attack at P_α.

The phosphodiesterases isolated from snake venom and bovine intestine evince a marked preference for the Rp diastereomer of the AMPS derivative; $k_{Rp}/k_{Sp} > 50$. Utilizing the Rp diastereomer of ATPαS and of the p-nitrophenyl derivative of AMPS the overall stereochemical course of the hydrolysis reaction catalyzed by these diesterases is retention. It is significant that the Rp diastereomer of ApSA, whose absolute configuration has been assigned on the basis of ^{31}P NMR chemical shift documented for the diastereoisomers of UpSA also is the preferred diastereomer hydrolyzed by these enzymes, inferring that our results from the chemically more reactive substrates probably can be extended to DNA degradation. A similar stereochemical result has been obtained for the venom diesterase with the Rp diastereomer of UpSA (Ref. 10).

Polynucleotide phosphorylase (M. Luteus) catalyzes the reversible phosphorolysis and polymerization of nucleotide diphosphates.

$$nNDP \xrightleftharpoons{M^{2+}} (Np)n + nPi$$

The enzyme exists in two forms: the T form that catalyzes the nonprocessive primer dependent polymerization and the nonprocessive phosphorolysis of oligonucleotides; the I form catalyzes processive de novo polymerization and the nonprocessive phosphorolysis of oligonucleotides (Ref. 11). The I form also catalyzes an exchange reaction:

$$^{32}Pi + ADP \rightleftharpoons AD^{32}P + Pi$$

which has the same kinetic characteristics of the de novo polymerization and should enable one to study only the first initiation step in the absence of competing polymerization.

Polynucleotide phosphorylase (T-form) preferentially polymerizes the Sp diastereomer of ADPαS (Ref. 12) forming in the presence of ApU primer and UDP a mixed Up(S)A .... Ap(S)A polymer. The overall stereochemical course established by enzymatic degradation of the polymer indicates an inversion of the configuration at phosphorus. In contrast the exchange process catalyzed by the I-form of the enzyme utilizes the Sp diastereomer of ADPαS and forms the ^{32}P-ADPαS (Sp) species. Thus the exchange process and possibly the initiation step in de novo polymerization proceeds with a mechanism involving retention of configuration at phosphorus. However a retention mechanism arising from partial phosphorolysis of de novo polymer must first be excluded.

The finding of a stereochemical retention process is in accord with a mechanism involving a single or odd number of phosphoryl-enzyme intermediates whose formation and decomposition occur via inversion processes, if one excludes a pentacovalent species and an associated permutational isomerization. Although there is no other evidence for the existence of an enzyme-AMPS for the venom enzyme, some data supporting this type of intermediate has been reported for the intestinal enzyme (Ref. 13). It may be that hydrolysis of a phosphodiester linkage which cannot readily proceed through an uncoupled pentacovalent transition state as opposed to a more coupled one is more readily achieved through a double displacement process. In the case of the phosphorylase catalyzed polymerization process the advantage gained by incorporation of such a species in the reaction sequence is evident; namely the ability to carry out a processive polymerization process through an achor point at the active site.

Acknowledgement - This work was supported by a grant from the National Institutes of Health (GM 13306).

REFERENCES

1. W. A. Blattler and J. R. Knowles, J. Am. Chem. Soc., in press.
2. P. M. J. Burgers and F. Eckstein, Proc. Natl. Acad. Sci. U.S.A. 75, 3798-4800 (1978).
3. F. Eckstein, H. Steinbach, and F. von der Haar, Biochemistry 16, 3429-3432 (1977).
4. J. P. Richard, D. C. Prasher, D. H. Ives, and P. A. Frey, J. Biol. Chem. 254, 4339-4341 (1979).
5. D. A. Usher, E. S. Erenrich, and F. Eckstein, Proc. Natl. Acad. Sci. U.S.A. 69, 115-118 (1972).
6. R. L. Jarvest and G. Lowe, Chem. Commun. 364 (1979).
7. F. R. Bryant and S. J. Benkovic, Biochemistry 18, 2825-2828 (1979).
8. K. F. Rex Sheu and P. A. Frey, J. Biol. Chem. 253, 3378-3380 (1978).
9. J. P. Richard, H. Ho, and P. A. Frey, J. Am. Chem. Soc. 100, 7756-7757 (1978).
10. P. M. J. Burgers and F. Eckstein, Biochemistry 18, 592-596 (1979).
11. T. Godefroy-Colbrun and M. Grunberg-Manago, The Enzymes Vol. VII, Academic Press, New York, 1972, pp. 532-574.
12. P. M. J. Burgers and F. Eckstein, Biochemistry 18, 450-454 (1979).
13. S. J. Kelley and L. G. Butler, Biochemistry 16, 1102-1104 (1977).

MODELS OF BIOPOLYMERS BY RING-OPENING POLYMERIZATION OF CYCLIC PHOSPHORUS CONTAINING COMPOUNDS

S. Penczek

Centre of Molecular and Macromolecular Studies, Polish Academy of Sciences, 90-362 Łódź, Boczna 5, Poland

Abstract - Methods are described, allowing the preparation of the high molecular weight polyesters of phosphoric acid, with a sequence of atoms in the main chain similar to that in naturally occuring polymers, teichoic acids or nucleic acids. Although the six-membered esters give only low molecular weight polymers, because of an extensive transfer involving the exo-cyclic group of these non-strained rings, the polymerization of the six-membered phosphate or phosphoroamidates readily leads to linear polymers with $\bar{M}_n > 10^5$.
This approach was further applied to bicyclic compounds with deoxyribose moiety and polyphosphate; backbones bearing deoxyribose were prepared by ionic polymerization method. The same approach was used for preparation of teichoic acids. The ionic (mostly anionic) polymerization of much more strained five-membered monomers gives high molecular weight polymers converted into the corresponding linear polyacids.
Kinetics and thermodynamics of polymerization of the model monomers described in this paper show the scope and limitation of the method.

INTRODUCTION

Models of biopolymers, i.e. polypeptides, polysaccharides, and polyphosphates (nucleic acids (NA), teichoic acids (TA)) can be prepared by various methods and are finding a number of different applications. These include studies of specific interactions, catalytic activities, matching enzymatic catalysis, biomedical uses (as, for instant, biologically active substances or as drug carriers) etc.
Method of preparing chains of polyphosphates are limited. There are practically no methods that could give the high-molecular weight polymers contaning only some desired fragments of NA, e.g. sugar moieties bound to polyphosphate backbone or bases bound to this backbone directly. These and related new polymers are now being available by applying a new method of preparation, based on the direct polymerization, by ring-opening reaction, of the cyclic phosphorus containing compounds. This concept is illustrated below :
e.g.:

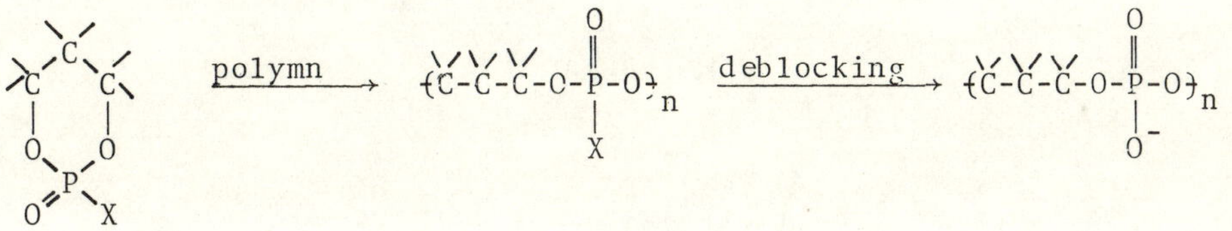

$$\text{(cyclic monomer)} \xrightarrow{\text{polymn}} \{C-C-C-O-\underset{\underset{X}{|}}{\overset{\overset{O}{\|}}{P}}-O\}_n \xrightarrow{\text{deblocking}} \{C-C-C-O-\underset{\underset{O^-}{|}}{\overset{\overset{O}{\|}}{P}}-O\}_n \quad (1)$$

Polymerization involves the ionic mechanisms[1], necessary to cleave the P-O bond. Thus, the direct polymerization of cyclic acid ($-X=-O^-(H^+)$) is not compatible with a mechanism to be used, because the ionizable group would react with the anionic or cationic active species in polymerization (chain carrier). This reaction would break the chain growth and should therefore be eliminated.

GENERAL CONDITIONS OF THE RING-OPENING POLYMERIZATION OF CYCLIC PHOSPHATES AND OF RELATED MONOMERS

Conversion of a cyclic monomeric phosphate into a linear macromolecule of high polymerization degree requires that a number of conditions is met simultaneously. In order to obtain macromolecules of desired polymerization degree ($\overline{DP}_n$) it is essential to obey the following rules :
1°. Conditions have to be found at which the side reactions are eliminated and the only processes involved are : initiation and propagation. The number average polymerization degree of the polymer is equal at these conditions : $\overline{DP}_n = (|M|_o - |M|_e)/|I|_o$, where $|M|_o$ and $|M|_e$ are the starting and equilibrium monomer concentrations and $|I|_o$ is the starting concentration of initiator.
2°. The opening of the monomeric ring has to proceed in the same way throughout the polymerization process, e.g. exclusively by head-to-tail addition, leading to the compositionaly unique structures, i.e. having only one, unique kind of mers in the chain.

On top of the conditions listed above there is a polymerizability of a given monomer, i.e. its thermodynamic ability to be converted into the linear chain. Ring strain is the major factor, governing the polymerizability.

THERMODYNAMICS OF POLYMERIZATION OF CYCLIC ESTERS OF PHOSPHORIC ACID

Data on ring strain in cyclic phosphates has been available from studies of their hydrolysis. For the five-membered methylethylenephosphate Westheimer reported[2] 7-9 kcal·mol^{-1} on the bases of comparison of the heats of hydrolysis of trimethylphosphate ($\sim$20 kcal·mol^{-1}) and methylethylene phosphate (29.5 kcal·mol^{-1}). Thus, assuming, that the change of entropy due to polymerization is lower than 10 e.u. it is not surprising that in the polymerization of methylethylene phosphate, at or close to 25°, the polymer yield was practically quantitative.

For three-and four-membered rings, that are not however of interest for our purpose, the strain energy arises from distortion of the normal bond angles. For five-and six-membered rings, discussed in this review, the strain energy is much smaller and the sign of ΔG is very sensitive to small changes in the physical conditions and chemical structure.

For a given basic structure, substitution leads to decreased polymerizability. Assuming a chair structure for a monomer (e.g. six-membered one) and a planar zig-zag structure for the polymer the changes of enthalpy of polymerization with substitution can be interpreted as arising from the release of gauche interactions[3,4]. Substitution reduces the change in the number of gauche interactions on polymerization and leads to the reduced polymerizability.

In contrast to the five-membered esters the six-membered ones are much less strained and in their polymerization the polymer-monomer equilibrium is clearly observed.

In the anionic and cationic processes polymerization comes to equilibrium; this has been detected by dilatometric or ^{31}P-NMR measurements. The most recent results involve equilibrium anionic polymerization of 2-oxo-1,3,2 λ^5-dioxaphosphorinane :

$$n \begin{array}{c} \text{CH}_2\text{-CH}_2 \\ \text{CH}_2 \quad \text{CH}_2 \\ | \quad \quad | \\ \text{O} \quad \quad \text{O} \\ \diagdown \text{P} \diagup \\ \text{O} \quad \text{H} \end{array} \underset{k_d}{\overset{k_p}{\rightleftharpoons}} \quad \begin{array}{c} \text{O} \\ \| \\ \text{+CH}_2\text{CH}_2\text{CH}_2\text{OPO+}_n \\ | \\ \text{H} \end{array} \quad (2)$$

Fragments of the ^{31}P{^{1}H}NMR spectra of equilibrates (mixtures of monomer and polymer) at various temperatures are given in Fig.1 :

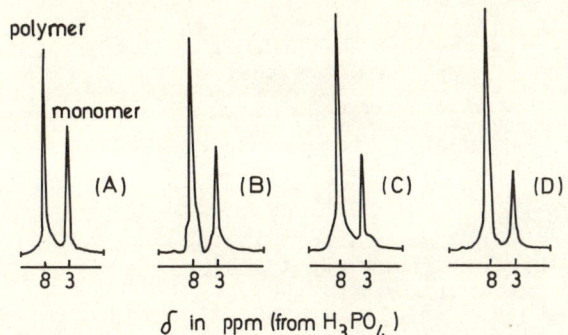

Fig.1. $^{31}P\{^1H\}$NMR spectra of the monomer-polymer equilibrium mixture at various temperatures. (A): at 40°C; (B): at 60°C; (C): at 80°C; (D): at 120°C (in bulk)[6].

Another method of determination of the equilibrium monomer concentration $|M|_e$, that can directly be measured from the spectra, as shown in Fig.1, consists of the studies of the kinetics of polymerization. Polymerization stops when $|M|_t$ achieves its equilibrium value; this can easily be estimated for instance, when dilatometry is used for studying kinetics[7].
Remembering that for ideal solution

$$\ln|M|_e = \frac{\Delta H_p}{RT} - \frac{\Delta S_p^o}{R} \qquad (3)$$

plotting of $\ln|M|_e$ as a function of $1/T$ gives ΔH_p and ΔS_p^o.
For the non-ideal solutions, when the monomer-polymer interaction has to be taken into account we have:

$$\Delta G_p^o = RT\{\ln\phi_1 + 1 - \chi(\phi_2 - \phi_1)\} \qquad (4)$$

where ϕ_1 and ϕ_2 are volume fractions of monomer and polymer respectively and χ is the monomer-polymer interaction parameter. Using eqns (3) or (4) we determined ΔH_p and ΔS_p^o values for a number of six-membered monomers. Below some of these data are listed, obtained for anionic polymerization:

monomer:	⌬P(=O)H	⌬P(=O)OCH₃	⌬P(=O)OC₂H₅	⌬P(=O)OSi(CH₃)₃
ΔH_p kcal·mol^{-1}	1.5±0.2	-0.7±0.5	1.4±0.6	3.9±1.3
ΔS_p^o e.u.	4.6±0.6	-2.8±1.5	2.6±1.4	8.7±3.3
Reference:	6)	5)	7)	7)

In spite of the large errors, the results obtained for various monomers are in reasonably good agreement. Heats of polymerization are close to zero or slightly positive and reflect the small ring strains. The bond angles in cyclic monomers and those in the open-chain esters are almost identical. There is an additional possibility of the skeletal d_π-p_π interactions, but there is no clear cut evidence, that any homologue is preferentially stabilized by these effects. Our value ΔS_p^o = -2.8 e.u. for the methylpropylenephosphate is to be compared with ΔS_p^o calculated for a hypothetical polymerization of cyclohexane (-2.5 e.u.) or a six-membered lactam (-6.6 e.u.)[8]. Conversion of a six-membered cyclic phosphate into a polymer results in a large increase in the rotational and vibrational entropy because of the enhanced flexibility of the unit in polymer chain. As a consequence, the entropy change, being negative when mostly due to aggregation processes, is

becoming positive, especially for larger exocyclic groups. Their rotational freedom, according to the studies of molecular models, can be highly restricted in cyclic monomers, provided that the exocyclic groups are at the axial position. Calculation of ΔG_p^0 indicates, that polymerization is not feasible under standard conditions, i.e. 1.0 mol·l^{-1}. Polymerization at room temperature becomes feasible in bulk, yet kinetically inhibited for all of the triesters (cf. next paragraph), because the activation energy of propagation is high enough to make the rate of polymerization impractically low.
Starting from ethylpropylenephosphate polymerization becomes endothermic and driven by a positive entropy change. The endothermicity and positive change of entropy may have their common origin in the increase of a number of degrees of freedom, higher apparently in polymer than in monomer and energetically outweighing the energy difference due to the ring strain. Thus, the polymerization of monomers with bulky substituents resembles polymerization of sulfur, selenium, and some cyclosiloxanes, for which simultaneously $\Delta H_p > 0$ and $\Delta S_p > 0$ [9].

The values determined for ΔH_p and ΔS_p indicate that the majority of the six-membered monomers with unsubstituted rings will polymerize, although some monomer will be left at equilibrium, its concentration only slightly depending on temperature (because ΔH_p are close to zero). One of the examples is shown in Fig.1, where in the temperature range from 40° to 120° the monomer equilibrium concentration changes less than two times.

The described above polymerizability of the unsubstituted six-membered monomers means that these rings build-up over the sugar rings should have higher polymerizability, because the strain increases, as evidenced from the studies of the hydrolysis. Khorana compared rates of hydrolysis of cyclic--1,2-isopropylidene-D-xylofuranose-3',5'-cyclic phosphate and 1,3-propylene phosphate and found that the former is hydrolysed in basic medium approx. 10 times faster[10].

KINETIC AND MECHANISTIC FEATURES OF POLYMERIZATION

There are following distinctive groups of monomers converted into polymers that lead eventually to polyacids-models of biopolymers (shown below schematically).

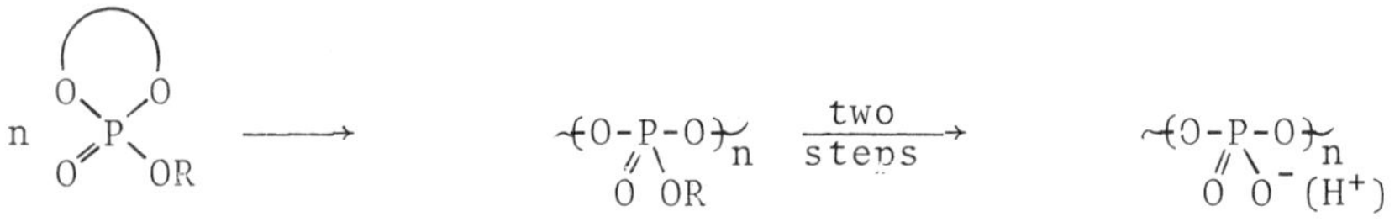

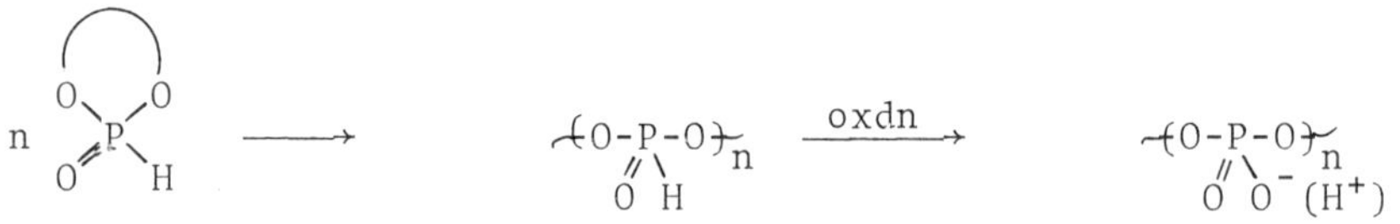

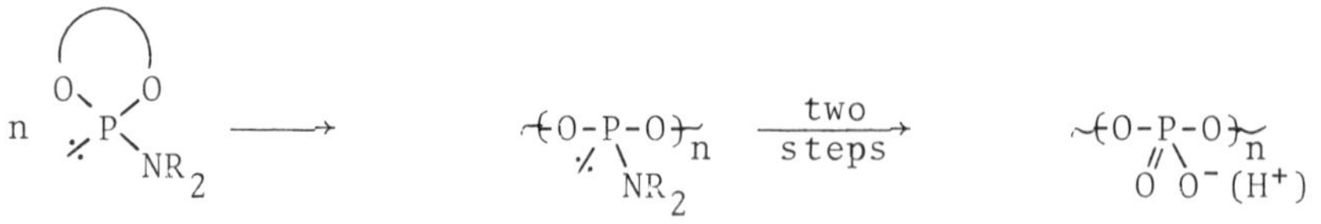

Thus, all of the three routes, namely the ester, phosphite, and phosphoroamidate methods have been used in our work and, therefore, the studies of the kinetics and mechanism of polymerization became of particular interest. These studies are at present completed only for the six-membered esters, but the understanding of their polymerization allowed us to choose conditions leading to high polymers according to eqs (5) and (6).

Structure of polyphosphates prepared by ionic polymerization of the six-membered cyclic monomers
Analyses of the ^{1}H and ^{31}P-NMR spectra of polymers prepared by both anionic and cationic mechanisms revealed that linear products with expected structures were formed. These products are of rather low polymerization degree, as evidenced by vpo measurements and contain, according to NMR, cyclic end-groups. Thus, the average oligomeric macromolecule has the following

structure (for both anionic and cationic initiation, and R=CH_3 in equ. 4):

$$CH_3-O-\underset{\underset{OCH_3}{|}}{\overset{\overset{O}{\|}}{P}}-O-CH_2CH_2CH_2 \{O-\underset{\underset{OCH_3}{|}}{\overset{\overset{O}{\|}}{P}}-OCH_2CH_2CH_2\}_n \quad O-\overset{\overset{O}{\|}}{P}\underset{OCH_2}{\overset{OCH_2}{<}}CH_2$$

$\delta^{31}P\{^1H\}$: 1.0   0.0   -7.0

$\delta^1H\{^31P\}$:   (t)4.23   (m) $\left.\begin{array}{l}4.35\\4.38\\4.45\\4.50\end{array}\right\}$ axial and equatorial

Good agreement was observed by comparison of the $\overline{DP}_n$ measured directly by vapour pressure osmometry and $\overline{DP}_n$ determined from the NMR spectra, assuming that every macromolecule contains one cyclic end-group[5]. The structure of the end-groups indicates that the transfer to monomer is responsible for the formation of polymers of low molecular weight. The mechanism of this transfer reaction is discussed in the next section.
Polymerization of 2-oxo-1,3,2 λ^5-dioxaphosphorinane[11]:

$$n \; \underset{\underset{O\overset{}{=}P-H}{}}{\underset{O\diagdown \;\; \diagup O}{\overset{CH_2\diagdown \;\; \diagdown CH_2}{CH_2 \;\; CH_2}}} \longrightarrow \{CH_2CH_2CH_2-O-\underset{\underset{O \;\; H}{\nearrow \diagdown}}{P}-O\}_n \quad (7)$$

$\delta^{31}P\{^1H\}$: 8.0   3.0

leads to the high molecular weight polymer with $\overline{M}_n$ up to 10^5. The $^{31}P\{^1H\}$-NMR spectrum of the polymer contains only one singlet, spectra of various equilibrated monomer— polymer mixtures are given in Fig.1. Anionic polymerization of the phosphoroamidates fulfils, according to the not yet completed measurements, the two requirements formulated in the first paragraph. This seems to be a living system, in which the calculated and measured $\overline{DP}_n$ equal to each other :

e.g.:
$$n \; \underset{\underset{N(C_2H_5)_2}{|}}{\underset{O\diagdown \;\; \diagup O}{\overset{CH_2\diagdown \;\; \diagdown CH_2}{CH_2 \;\; CH_2}}} \quad \text{anionic} \quad \{CH_2CH_2CH_2-O-\underset{\underset{N(C_2H_5)_2}{|}}{P}-O\}_n$$

In the equilibrated monomer-polymer system only two singlets are observed and, for instance, the following $\overline{DP}_n$ were calculated and found by osmometry:

$\overline{DP}_n$(calcd) = $(|M|_o-|M|_e)/|I|_o$ = 135   ($\overline{M}_n$ = 21300)

$\overline{DP}_n$(found) =   150   ($\overline{M}_n$ = 22900)

Both calculated and found value are close to each other indicating that polymerization proceeds without an appreciable transfer or termination. Thus, from the six-membered monomers the most promising for using as rings fused to the sugar rings are these that do not contain the exocyclic ester group. Polymerization of the five-membered monomers is very much facilitated by the ring strain. The ring strain increases the rate of the ring opening and the side reactions, for instance reactions involving the exocyclic group, are becoming even less important. Therefore, anionic polymerization of cyclic esters with ester exocyclic group gives high polymers with polymerization degrees close or identical to the calculated values.

Kinetics and mechanism of propagation
The complete polymerization process has to consist of at least two elementary reactions, namely initiation and propagation reactions. In the former one the active species, participating then in the propagation process are formed.

It has been established for the six-membered 2-alkyl-2-oxo-1,3,2-dioxaphosphorinanes that the cationic and anionic active species have the following structures:

a) Cationic initiation and formation of active species

e.g.:

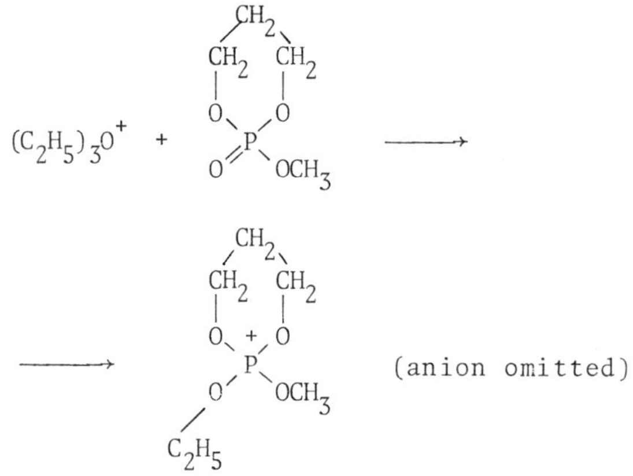

$$(C_2H_5)_3O^+ + \text{[6-membered 2-methoxy-2-oxo-1,3,2-dioxaphosphorinane]} \longrightarrow$$

$$\longrightarrow \text{[2-ethoxy-2-methoxy-1,3,2-dioxaphosphorinanium cation]} \quad \text{(anion omitted)} \tag{8}$$

The formation of the tetralkoxyphosphonium cation in reaction (8) established earlier[13] was confirmed[5]. This species regenerate at the molecular chain ends at very step of propagation:

e.g.:

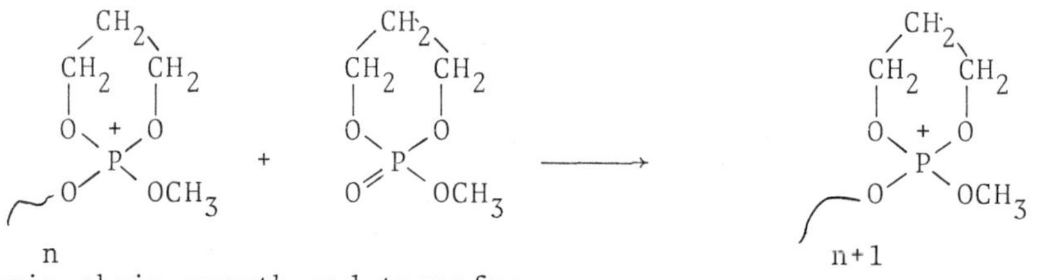

$$\text{[chain-cation]}_n + \text{[monomer]} \longrightarrow \text{[chain-cation]}_{n+1} \tag{9}$$

b) Cationic chain growth and transfer

The observed structure of polymers formed from the six-membered monomers (cf. preceding paragraph) can be explained by assuming the simultaneous propagation and chain transfer to monomer. This is shown in scheme 10 for the cationic polymerization:

e.g.:

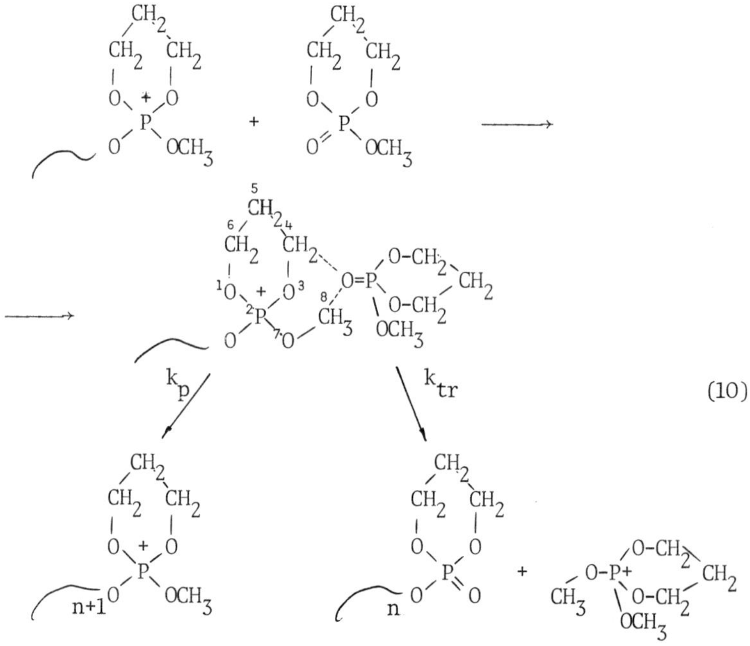

$$\text{[chain-cation]} + \text{[monomer]} \longrightarrow \text{[intermediate complex with labeled atoms }^1O,^2P,^3O,^4CH_2,^5CH_2,^6CH_2,^7O,^8C\text{]}$$

$$\xrightarrow{k_p} \text{[chain-cation]}_{n+1} \qquad \xrightarrow{k_{tr}} \text{[chain]}_n + \text{[new cation]} \tag{10}$$

$O^3\text{-}C^4$ (or $O^1\text{-}C^6$) breaking: propagation

$O^7\text{-}C^8$ breaking: transfer of kinetic chain and termination of molecular chain

The tetraalkoxyphosphonium ions formed during initiation were observed throughout the propagation process by $^{31}P\{^1H\}NMR$ at -0.25δ [7]; their concentration was constant. These species are regenerated in two ways: by propagation and by chain transfer as shown in scheme (10). In tetraalkoxyphosphonium ions (cf. scheme 10) a partial positive charge is localised on carbon atoms C_4 and C_6, being a part of the ring, and on the exocyclic C_8 atom. Therefore, when an O_3-C_4 bond (or equivalent O_1-C_5 bond) is broken chain propagation takes place if however the O_7-C_8 bond is broken then the chain transfer proceeds. Because of the low ring strain in the six-membered cyclic triphosphates both reactions proceed with comparable rate constants. Thus, chain transfer, as shown in scheme 10, leaves an oligomer with a cyclic end-group and leads to reinitiation by forming a new tetraalkoxyphosphonium ion. Its reactivity is close or identical to that of the parent macromolecular tetraalkoxyphosphonium ion and, therefore, chain transfer has no kinetic effect (reinitiation is fast, there is no slow-down of the polymerization due to transfer).

c) Anionic initiation
Anionic polymerization was initiated in various ways; aluminum alkyls (e.g. $(i-C_4H_9)_3Al$), $n-C_4H_9Li$, Na, K, Cs, and metal alkoxides were used. In case of simple initiators, like metal alkoxides, initiation proceeds by a direct nucleophilic attack of an alcoholate anion on phosphorous atom, leading to the trigonal bipyramid structure, as shown below (equ.11) for 2-methyl-2-oxo--1,3,2-dioxaphosphorinane :

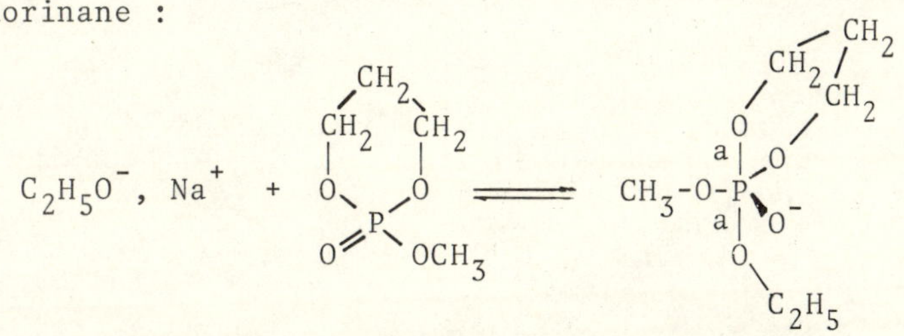

$$C_2H_5O^-, Na^+ + \text{[cyclic phosphate]} \rightleftharpoons \text{[trigonal bipyramid]} \qquad (11)$$

In the formed anion the negative charge is distributed among the oxygen atoms and, in principle, the P-O bond can be broken at any ligand, but the apical bonds are preferred because of their enhanced weakness. The six-membered ring is shown in scheme 11 to occupy the axial-equatorial position: this is the preferred structure according to the X-ray studies of related compounds[14]. Therefore, the apical (marked with a) bond is broken and the new alcoholate anion is produced; this one is already a growing center of propagation :

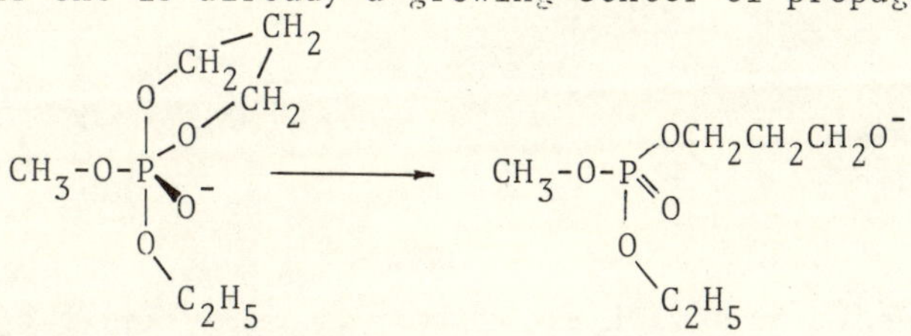

$$\text{[structure]} \longrightarrow \text{[structure]} \qquad (12)$$

d) Anionic propagation and chain transfer [1]
Thus formed growing anionic centre participates in the chain growth by attacking the phosphorus atom in the next monomer molecule and reproducing again an alcoholate anion.
Thus, in short, the growing active species have the structure of the alcoholate anion and the chain propagation consists of an attack by these species on the electrophilic phosphorus atom in the monomer molecule. This attack is directed along the apical position and leads to the formation of the transition state (or a high energy intermediate product), having the structure of a trigonal bipyramid.
Chain growth is reversible and ring opening — ring closure equilibrium should use the same reaction pathway. Both involve an attack along the apical position, as shown according equ. 13 :

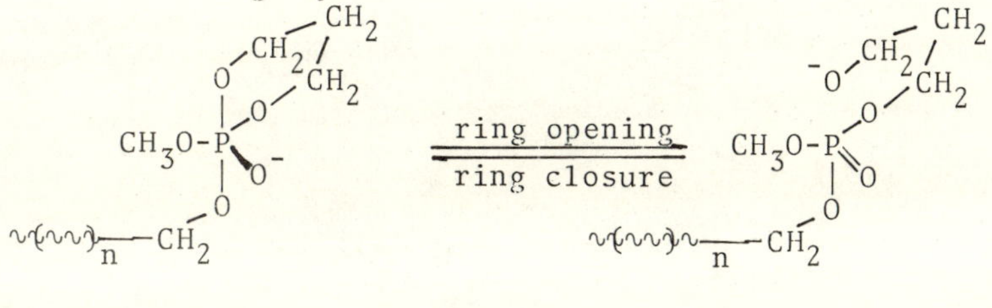

$$\text{[structure]} \underset{\text{ring closure}}{\overset{\text{ring opening}}{\rightleftharpoons}} \text{[structure]} \qquad (13)$$

Polymers prepared by anionic polymerization of 2-alkyl-2-oxo-1,3,2-dioxaphos-phorinanes have, as it was already mentioned, rather low molecular weights and every macromolecule has one cyclic end-group, as shown in scheme 15 This could easily be explained if we assume that in the intermediate state, having a structure of bipyramid, the ring remains intact and that the other axial P-O bond breaks, giving a CH_3O^- anion and a neutral macromolecule with a ring at its end. This would, however, require the exocyclic group to leave from a more stable equatorial position. A similar problem has been faced in the hydrolysis of the cyclic (five-membered) phosphates and it was proposed that within a bipyramid substituents can change their positions by a pseudo-rotation. Application of this approach to the propagation/transfer competition requires a change of the equatorial position of the exocyclic group to the apical one. Thus, the corresponding pseudorotation, along the P-O$^-$ bond taken as a pivot, changes the positions of the methoxy and polymeric chain substituents :

$$\text{(14)}$$

Thus, the CH_3-O-P bond, being now in the apical position, breaks, providing a cyclic end-group, and an alcoholate anion CH_3O^-, becoming through reinitiation the second end-group:

$$\text{(15)}$$

Anionic polymerization with eliminated transfer and termination

Formation of the low molecular weight polymers in the polymerization of six--membered phosphates stems from two interconnected factors : the low ring strain and the similarity of reactivities of the exocyclic ester groups and the endocyclic ester functions within a ring. Studies of the thermodynamics of polymerization (cf. the corresponding section of this paper) indicate that the change of the exocyclic group does not appreciably influence the values of ΔH_p^0 and ΔS_p^0 (standard enthalpy and entropy of the conversion of monomer into its polymer). Thus, the only way to obtain higher molecular weight products has been the change of the structure of the exocyclic group. The structure of these groups should allow their conversion (deblocking) into the acidic groups. Conditions of this conversion should be mild enough in order to leave the main chain intact. As it was already discussed two six--membered monomers fulfilled these conditions :

$$\text{and} \qquad \text{(16)}$$

2-oxo-1,3,2 λ^5-dioxaphosphori-nane

2-dialkylamino-1,3,2-dioxa-phosphorinane

Although ΔH_p for both of these monomers are close to zero, but the exocyclic groups do not react with growing species (alcoholate anions) and, thus, the propagation proceeds in the uninterrupted way. Polymers of both of these monomers were prepared with $\overline{M}_n$ above 10^5 and with expected structures of the backbones.

Preparation of simple macromolecular models of polyacids
Oxidation of poly (2-oxo-1,3,2 λ^5-dioxaphosphorinane) leads directly to the corresponding polyphosphate. Its structure is similar to the backbones of nucleic acids and some teichoic acids :

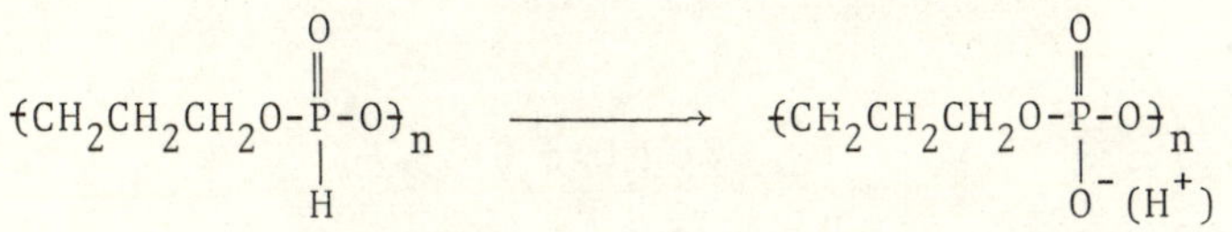

$$\{CH_2CH_2CH_2O-\underset{\underset{H}{|}}{\overset{\overset{O}{\|}}{P}}-O\}_n \longrightarrow \{CH_2CH_2CH_2O-\underset{\underset{O^-(H^+)}{|}}{\overset{\overset{O}{\|}}{P}}-O\}_n \qquad (17)$$

Oxidation proceeds smoothly in CH_2Cl_2 solution at room temperature by using N_2O_4 as the oxidating agent; the polyacid precipitates gradually out from solution with a progress of oxidation[15]. Molecular weights determined for the starting and final polymer do not differ appreciably indicating, that oxidation proceeds without the main chain scission. Molecular weight of the polyacid was measured by light scattering in 0.1 N NaCl water solution. Remembering the high instability of phosphites towards hydrolysis the oxidation can be considered as a quantitative process. The isolated polyacid is a highly hygroscopic, colourless, elastic material. Another simple polymer, which can be prepared of high molecular weight by anionic polymerization, is the poly (2-diethylamino-1,3,2-dioxaphosphorinane)[12]. Its acidolysis with anhydrous acetic acid in CH_2Cl_2 solution leads to the corresponding poly (2-oxo-1,3,2λ^5-dioxaphosphorinane), which can be converted into a polyester of phosphoric acid as described in equ.17.

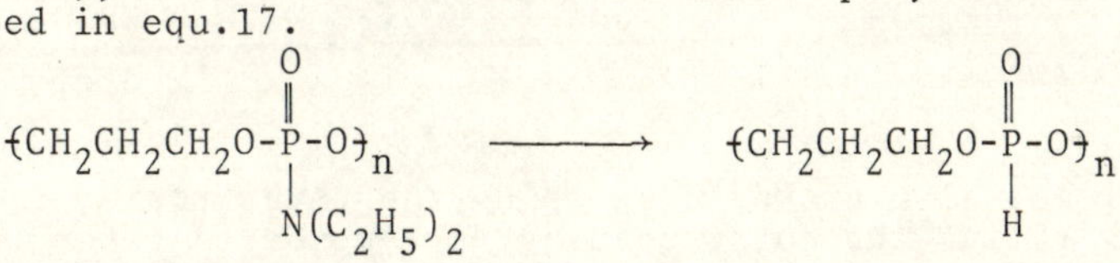

$$\{CH_2CH_2CH_2O-\underset{\underset{N(C_2H_5)_2}{|}}{\overset{\overset{O}{\|}}{P}}-O\}_n \longrightarrow \{CH_2CH_2CH_2O-\underset{\underset{H}{|}}{\overset{\overset{O}{\|}}{P}}-O\}_n \qquad (18)$$

Some of the teichoic acids differ in the structure of the backbones from nucleic acids; in place of the repeating unit (mer) consisting of six atoms, they contain only five, namely a sequence

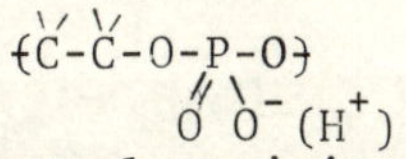

$$\{\overset{\vee}{C}-\overset{\vee}{C}-O-\underset{\underset{O^-(H^+)}{\diagdown}}{\overset{\overset{O}{\diagup\!\!\!\diagup}}{P}}-O\}$$

This sequence can be approached by polymerizing the corresponding five-membered monomer. Much higher ring strain (when compared with the six-membered rings) eliminates chain transfer involving the exocyclic group and allowed us to prepare the high molecular weight polyesters :

e.g.:

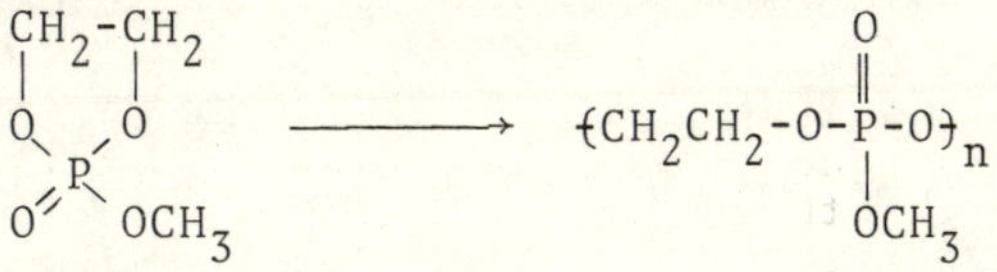

$$\underset{O\diagdown\,\,\,\diagup O}{\overset{CH_2-CH_2}{|\quad\,\,\,|}}\underset{O^{\diagdown\!\diagup}\overset{|}{P}\diagdown OCH_3}{} \longrightarrow \{CH_2CH_2-\underset{\underset{OCH_3}{|}}{\overset{\overset{O}{\|}}{P}}-O\}_n$$

Poly(methylethylenephosphate) (poly(2-methyl-1,3,2-dioxaphospholane)) was converted into the corresponding polyacid in two consecutive reactions: dealkylation and ion-exchange, as shown below in scheme 19 [15]:

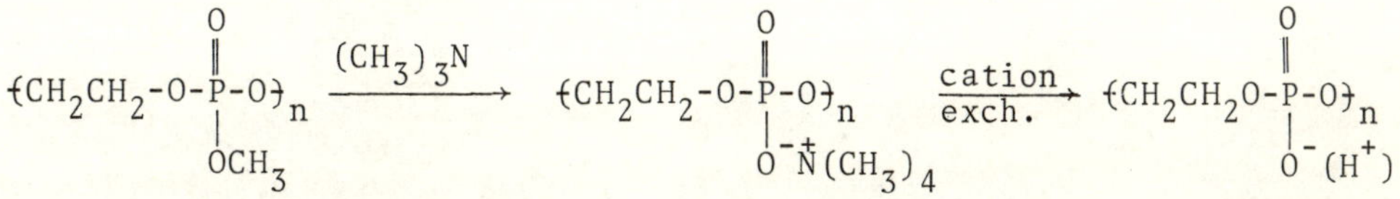

$$\{CH_2CH_2-O-\underset{\underset{OCH_3}{|}}{\overset{\overset{O}{\|}}{P}}-O\}_n \xrightarrow{(CH_3)_3N} \{CH_2CH_2-O-\underset{\underset{O^-\overset{+}{N}(CH_3)_4}{|}}{\overset{\overset{O}{\|}}{P}}-O\}_n \xrightarrow[\text{exch.}]{\text{cation}} \{CH_2CH_2O-\underset{\underset{O^-(H^+)}{|}}{\overset{\overset{O}{\|}}{P}}-O\}_n$$

(19)

Here, however, the conditions of conversion have not yet been optimized and the resulting polyacid still contains a few percent of ester bonds in the side chains.

Macromolecular models bearing elements of teichoic or nucleic acids as the side groups

Recently, the first model of teichoic acid has been prepared by ring-opening polymerization[17]:

$$\begin{array}{c} CH_2OAc \\ | \\ CHOH \\ | \\ CH_2OH \end{array} + PCl_3 \longrightarrow \begin{array}{c} CH_2OAc \\ | \\ CHO \\ | \\ CH_2O \end{array}\!\!\diagdown\!\!P\!-\!Cl \longrightarrow \begin{array}{c} CH_2OAc \\ | \\ CHO \\ | \\ CH_2O \end{array}\!\!\diagdown\!\!P\!\!\diagup\!\!\begin{array}{c} O \\ H \end{array} \quad (20a)$$

$$n \begin{array}{c} CH_2OAc \\ | \\ CH-CH_2 \\ | \quad | \\ O \quad O \\ \diagdown P \diagup \\ O \quad H \end{array} \longrightarrow \{CHCH_2\text{-O-}\overset{O}{\underset{H}{\overset{\|}{P}}}\text{-O}\}_n \xrightarrow{oxdn} \{CHCH_2\text{-O-}\overset{O}{\underset{O^-(H^+)}{\overset{\|}{P}}}\text{-O}\}_n \quad (20b)$$

where CH$_2$OAc is a substituent on the CH.

^{31}P and ^{31}P{^{1}H}-NMR spectra (in H$_2$O) show that essentially there are two P atoms in the chain of equal concentration (0.8δ and 1.5δ). Tentatively we explain the presence of two signals on the basis of two different chemical environments of P atoms (iso- (ll or dd) and syndio-(ld or dl) placements) along the chain :

$$\ldots\text{-O-}\overset{H\ CH_2OAc}{\underset{}{C}}\text{-CH}_2\text{-O-}\overset{}{\underset{O\ O^-(H^+)}{\overset{\diagup\diagdown}{P}}}\text{-O-}\overset{H\ CH_2OAc}{\underset{}{C}}\text{-CH}_2\text{-O-}\ldots \quad \text{and} \quad \ldots\text{-O-}\overset{H\ CH_2OAc}{\underset{}{C}}\text{-CH}_2\text{-O-}\overset{}{\underset{O\ O^-(H^+)}{\overset{\diagup\diagdown}{P}}}\text{-O-}\overset{CH_2OAc\ H}{\underset{}{C}}\text{-CH}_2\text{-O-}\ldots \quad (21)$$

iso-   syndio-

The same polymer was also prepared by using the ester route, i.e. by synthesizing and polymerizing the corresponding five-membered ester :

$$\begin{array}{c} CH_2OAc \\ | \\ CH-CH \\ | \quad | \\ O \quad O \\ \diagdown P \diagup \\ O \quad OCH_3 \end{array} \longrightarrow \{CH_2\text{-CH-O-}\overset{O}{\underset{OCH_3}{\overset{\|}{P}}}\text{-O}\}_n \quad (22)$$

with CH$_2$OAc on the CH.

with further dealkilation and conversion of the polysalt into the free polyacid.

Another recently prepared model of a biopolymer contains a moiety of deoxyribose :

$$\text{HO-}\overset{}{\underset{OH}{\bigcirc}}\text{-OCH}_3 + P[N(C_2H_5)_2]_3 \longrightarrow \quad (23)$$

(deoxyribose with cyclic phosphoramidite, $(C_2H_5)_2N$ on P)

$$\overset{}{\underset{N(C_2H_5)_2}{P\diagdown O}}\text{-deoxyribose-OCH}_3 \xrightarrow{polymn} \{O\text{-deoxyribose-OCH}_3\text{-O-P}\}_n \underset{N(C_2H_5)_2}{} \xrightarrow{two\ steps} \{O\text{-deoxyribose-OCH}_3\text{-O-P}\}_n \underset{O^-(H^+)}{\overset{O}{\diagdown}} \quad (24)$$

Although, according to the ^{31}P and ^{1}H-NMR spectra, the polyacid is formed in the sequence of reactions (23) and (24)^{12} but the purity of the polymer is not yet established. Another route to the same structure, by using the cyclic phosphite, does not seem at present to give the advantageous results.

REFERENCES

1. S.Penczek, Pure & Appl.Chem. 48, 363 (1976).
2. J.R.Cox, Jr, R.E.Wall and F.W.Westheimer, Chem.Ind. 929 (1959); F.W.Westheimer, Accounts Chem.Res. 1, 70 (1968).
3. K.J.Ivin, "Thermodynamics of Addition Polymerization Processes" in "Reactivity, Mechanism and Structure in Polymer Chemistry", A.D.Jenkins and A.Ledwith, Eds. J.Wiley, London-New York, 1974.
4. R.C.B.Cubbon, Makromol.Chem. 80, 44 (1964).
5. G.Łapienis and S.Penczek, Macromolecules 7, 167 (1974).
6. K.Kałużyński and S.Penczek, Makromol.Chem. 180, 2289 (1979).
7. G.Łapienis and S.Penczek, J.Polym.Sci., Polym.Chem.Ed. 15, 371 (1977).
8. F.S.Dainton and K.J.Ivin, Quart.Rev. (London) 12, 61 (1958); K.J.Ivin, Angew.Chem. 85, 533 (1973).
9. H.Sawada, J.Macromol.Sci., Revs.Macromol.Chem. 5, 151 (1970).
10. H.G.Khorana, G.M.Tener, R.S.Wright and J.G.Moffat, J.Amer.Chem.Soc. 79, 430 (1957).
11. K.Kałużyński, J.Libiszowski and S.Penczek, Makromol.Chem. 178, 2943 (1977)
12. K.Kałużyński, in preparation.
13. J.H.Finkley, D.Z.Denney and D.B.Denney, J.Am.Chem.Soc. 91, 5826 (1969).
14. G.Aksnes and K.Bergesen, Acta Chem.Scand. 20, 2508 (1966).
15. K.Kałużyński, J.Libiszowski and S.Penczek, Macromolecules 9, 365 (1976).
16. J.Libiszowski and S.Penczek, J.Polymer Sci., Polymer Chem.Ed.16,1275 (1978).
17. P.Kłosiński, M.S.Thesis, Polytechnic Inst.of Łódź, 1978.

SOME ASPECTS OF OLIGONUCLEOTIDE SYNTHESIS

C. B. Reese

Department of Chemistry, King's College, Strand, London WC2R 2LS, U.K.

Abstract - Some recent developments crucial to the establishment of the phosphotriester approach to the synthesis of oligonucleotides are discussed. First, the use of the conjugate base of syn-p-nitrobenzaldoxime (8) in the unblocking of aryl (especially 2-chlorophenyl)-protected internucleotide linkages is described. Secondly, the use of o-chlorophenyl p-nitrophenyl phosphorochloridate (15b) and, more generally, of o-chlorophenyl phosphorodi-(1,2,4-triazolide) (35) in the first phosphorylation step and 1-mesitylenesulphonyl-3-nitro-1,2,4-triazole (MSNT, 18) in the second phosphorylation step of the phosphotriester approach are described. Thirdly, a method, involving the use of O-2,4-dichlorophenyl S-methyl phosphorochloridothioate (19), for the introduction of terminal mono-, di- and tri-phosphate residues into oligonucleotides is described. Fourthly, the use of the o-dibromomethylbenzoyl (DBMB, as in 37) group for the protection of alcoholic hydroxy functions is described. Finally, the synthesis of 5'-O-triphosphoryladenylyl-(2'→5')-adenylyl-(2'→5')-adenosine (2-5A, 25) and the 3'-terminal decanucleotide sequence (UpCpGpUpCpCpApCpCpA, 39) of yeast alanine tRNA is described.

It now seems clear that it is desirable that internucleotide linkages should be protected during the course of the chemical synthesis of oligo- and poly-nucleotides. The advantages of adopting this so-called phosphotriester approach rather than the alternative phosphodiester approach have been discussed elsewhere (Ref. 1). If the internucleotide linkages are to be protected, the choice of a protecting group is clearly a decision of crucial importance. In 1968, we suggested (Ref. 2) the use of the phenyl group for this purpose and subsequently investigated the possibility of using other aryl protecting groups including o-fluorophenyl, o-chlorophenyl and p-chlorophenyl (Refs. 3 and 4). Three other protecting groups, namely benzyl (Ref. 5), 2-cyanoethyl (Ref. 6) and 2,2,2-trichloroethyl (Ref. 7), had earlier been suggested by other workers.

Scheme 1

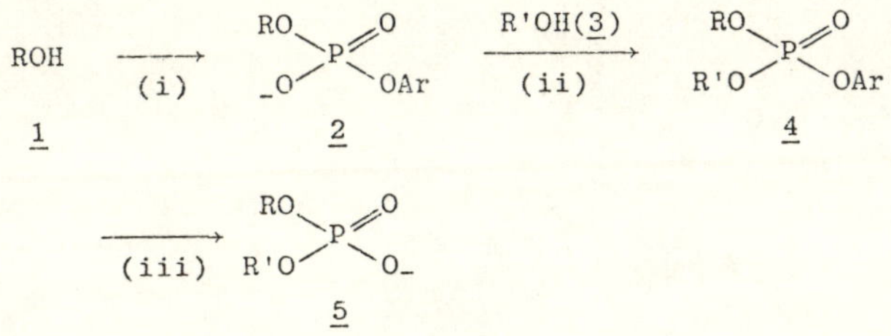

$$ROH \xrightarrow{(i)} \underset{2}{\overset{RO}{\underset{}{P}}}\!\!\!\overset{O}{\underset{OAr}{\diagdown}}\!\!\!\!\overset{}{} \xrightarrow[(ii)]{R'OH(3)} \underset{4}{\overset{RO}{\underset{R'O}{P}}}\!\!\!\overset{O}{\underset{OAr}{\diagdown}}$$

$$\xrightarrow{(iii)} \underset{5}{\overset{RO}{\underset{R'O}{P}}}\!\!\!\overset{O}{\underset{O^-}{\diagdown}}$$

Ar = Ph, 2-ClC₆H₄

Two separate phosphorylation steps [(i) and (ii), Scheme 1] are required in the phosphotriester approach. In addition, a third step involving the unblocking of the internucleotide linkages [step (iii)] is necessary at the end of the synthesis. It is appropriate to consider step (iii) first as any protecting group (Ar) will be unsuitable if it cannot be removed in a satisfactory manner. We found that alkaline hydrolysis of oligonucleotides [Scheme 2] with phenyl-protected internucleotide linkages (represented by 4; Ar = Ph) did not yield exclusively the desired products (represented by 5) but that significant quantities of cleavage products (represented by 6a and 6b) were also obtained (Ref. 8). Indeed, even when o-chlorophenyl-protected oligonucleotides (represented by 4; Ar = 2-ClC₆H₄) were treated with sodium hydroxide in aqueous dioxan, ca. 2% internucleotide cleavage (to give products represented by 6a and 6b; Ar = 2-ClC$_6$H$_4$) per phosphotriester group was observed (Ref. 9).

Scheme 2

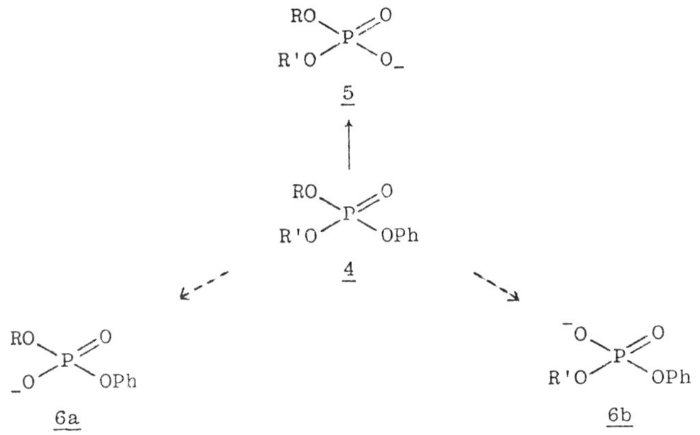

If an aryl protecting group derived from a phenol more acidic than o-chlorophenol is used, the extent of internucleotide cleavage during unblocking is likely to be decreased. However, the phosphotriester intermediates then tend to be very susceptible to hydrolysis even under mildly alkaline conditions and extremely difficult to handle (Ref. 10). It therefore seemed that, if aryl protecting groups were to be used successfully in the synthesis of relatively high molecular weight oligo- and poly-nucleotides, it would be necessary to unblock the internucleotide linkages with a nucleophile other than hydroxide ion. Furthermore, if as would seem to be desirable, the unblocking process is to involve only one nucleophilic substitution at phosphorus, it is essential that an oxygen nucleophile [XO$^-$, Scheme 3] should be used and that the second (and preferably not rate-determining) step of the unblocking process should proceed by O-X cleavage as indicated in Scheme 3. We have recently found (Ref. 9) that the conjugate bases (pK$_a$'s ~ 10) of syn-p-nitrobenzaldoxime (8) and syn-pyridine-2-carboxaldoxime (9) are excellent nucleophiles for the present purpose and that they meet all of the above requirements. Indeed, when they are used in conjunction with o-chlorophenyl-protected oligonucleotides, the unblocking process occurs at a satisfactory rate and the extent of internucleotide cleavage appears to be negligible (Ref. 9). We therefore believe that the crucial problem regarding the protection of the internucleotide linkages in the phosphotriester approach has been solved.

Scheme 3

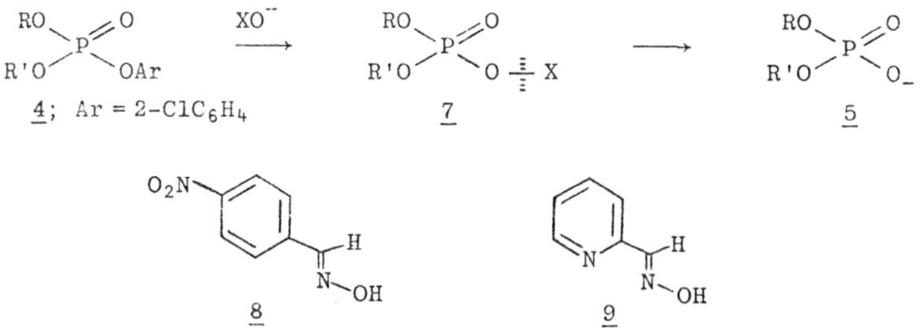

Let us now return to the first phosphorylation step [(i), Scheme 1] of the phosphotriester approach. One procedure which has been used with some success involves (Ref. 1) the reaction [Scheme 4] between the 3'-hydroxy function of a protected nucleoside or oligonucleotide (ROH, 1) with the anion of phenyl or o-chlorophenyl phosphate (10) in the presence of 2,4,6-tri-isopropylbenzenesulphonyl chloride (11) in pyridine solution to give the desired intermediate (2). Although this procedure has proved to be useful, it is unsatisfactory in at least two respects. First, its use can lead to the formation of unwanted symmetrical products (12) especially in the synthesis of oligodeoxyribonucleotides (Ref. 11) and secondly, yields are generally poor when guanine residues (protected by acylation on N-2) are present (Ref. 12).

It seemed likely that the first of the latter problems, that is the formation of unwanted symmetrical products (12) could be prevented if a monofunctional phosphorylating agent [13, Scheme 5] were used. It was envisaged that such a phosphorylating agent would have to fulfil at least three requirements: (i) it would need to be reactive enough to

Scheme 4

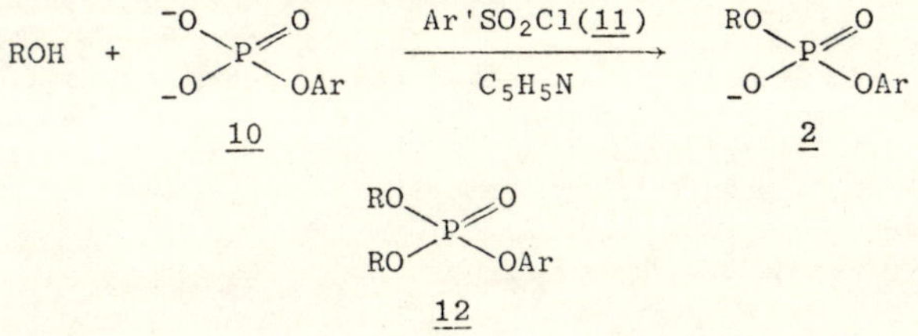

Ar = Ph or 2-ClC$_6$H$_4$; Ar' = 2,4,6-(Me$_2$CH)$_3$C$_6$H$_2$

phosphorylate the relatively hindered 3'-hydroxy functions of 2'-protected ribonucleoside building blocks, (ii) the products (<u>14</u>) obtained would need to be such that their conversion into the desired phosphodiester intermediates (<u>2</u>) would proceed under conditions which were mild enough to prevent any removal whatsoever of the acid- or base-labile protecting groups present and (iii) it must be possible for any by-products formed in the conversion of <u>14</u> into <u>2</u> to be removed easily so that the latter phosphodiester intermediates (<u>2</u>) can be isolated in a pure state.

Scheme 5

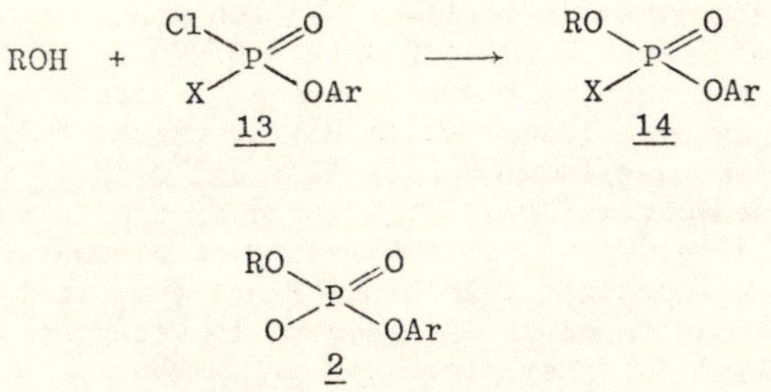

p-Nitrophenyl phenyl (Ref. 13) and o-chlorophenyl p-nitrophenyl (Ref. 9) phosphorochloridates (<u>15a</u> and <u>15b</u>, respectively) are readily obtainable reagents which appeared to meet all of the above requirements for a monofunctional phosphorylating agent. These reagents (<u>15a</u> and <u>15b</u>) react rapidly [Scheme 6] even with the 3'-hydroxy functions of ribonucleoside building blocks to give the corresponding phosphotriesters (<u>16</u>). Treatment of the latter intermediates (<u>16</u>) with the triethylammonium salt of p-thiocresol in acetonitrile solution at room temperature leads (Ref. 13) to nucleophilic attack at the position indicated (arrow) to give the triethylammonium salts of the desired phosphodiesters (<u>2</u>), which may be isolated in a pure state by precipitation. This procedure is suitable for the stepwise synthesis but unfortunately (see below) not for the block synthesis of oligonucleotides.

Scheme 6

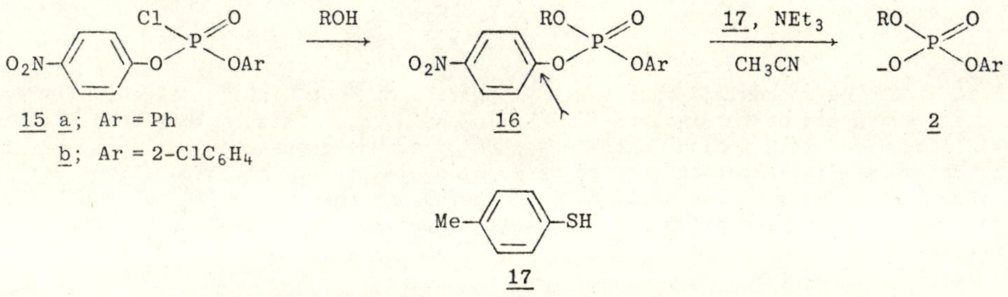

<u>15</u> <u>a</u>; Ar = Ph
    <u>b</u>; Ar = 2-ClC$_6$H$_4$

The second phosphorylation step [(ii), Scheme 1] of the phosphotriester approach involves the reaction between a phosphodiester intermediate (<u>2</u>) and the 5'-hydroxy function of a protected nucleoside or oligonucleotide (R'OH, <u>3</u>) in the presence of an activating agent. 2,4,6-Tri-isopropylbenzenesulphonyl chloride (<u>11</u>) (Ref. 14) has been used successfully as the activating agent but the condensation reactions then tend to be rather slow. Furthermore, the reactions are often accompanied by O-sulphonation (of R'OH) and darkening of the reaction medium. As noted above, yields tend to be low if guanine residues are present and 11 is used as the activating agent. In 1973, Russian workers reported (Ref. 15) that O-sulphonation and darkening did not occur when 1-arenesulphonylimidazole derivatives were used as activating agents instead of arenesulphonyl chlorides. The same advantages obtain but reaction rates are very much faster if arenesulphonyl derivatives of tetrazole (Ref. 16) and 3-nitro-1,2,4-triazole (Ref. 17) are used. We especially favour the use of the mesitylenesulphonyl derivative of 3-nitro-1,2,4-triazole (MSNT, 18) (Refs. 9 and 17) as the activating agent in the second phosphorylation step [(ii), Scheme 1].

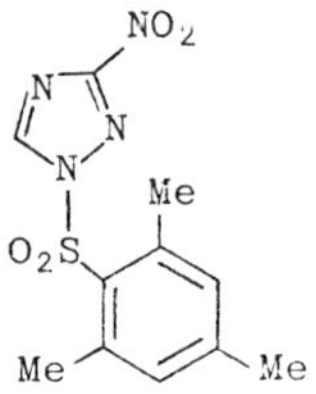

18

Before ending the discussion on phosphorylation, it is relevant to mention a method which we have recently developed (Ref. 18) for the introduction of terminal phosphate residues in synthetic oligonucleotides. The availability of such a method is of much importance as a number of the synthetic procedures used in the phosphotriester approach lead to oligonucleotides containing one more nucleoside residue than phosphate group. We believe that O-2,4-dichlorophenyl S-methyl phosphorochloridothioate (19) (Ref. 18) which can easily be prepared from 2,4-dichlorophenyl phosphorodichloridite, is a particularly useful reagent for the introduction of terminal phosphate residues. Alcoholic hydroxy functions of partially-protected nucleosides and oligonucleotides [ROH, Scheme 7] react with 19 in pyridine solution to give fully-protected phosphorothioates (20). Treatment of the latter with the conjugate base of syn-p-nitrobenzaldoxime (8) gives the S-methyl phosphorothioate anion (21) (Ref. 18). It should be noted that the unblocking of 2-chlorophenyl-protected internucleotide linkages will occur concomitantly. Oxidation of 21 with iodine in wet pyridine solution leads to the corresponding monophosphate (22) (Ref. 18).

Scheme 7

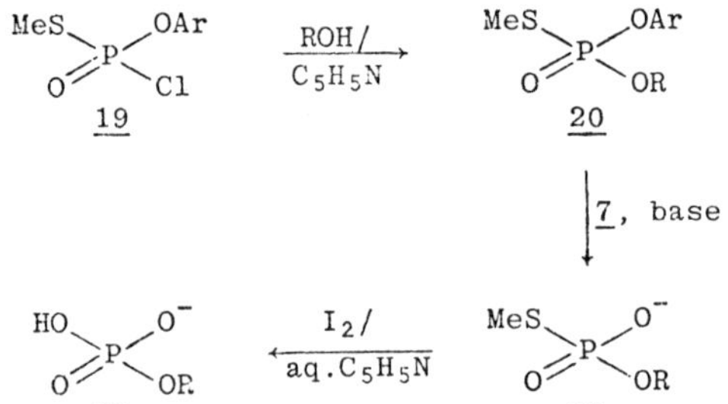

Ar = $2,4-Cl_2C_6H_3$

It should be emphasized that S-methyl phosphorothioate intermediates (21) may also be used in the synthesis of terminal di- and tri-phosphates. This is indicated in Scheme 8. Thus if 21 is oxidized with iodine in the presence of orthophosphate or pyrophosphate ion in rigorously dried pyridine solution, the corresponding di- or tri-phosphate (23 or 24) is obtained (Ref. 19). An interesting example of the use of this method in the synthesis of a naturally-occurring occurring oligonucleotide with a terminal triphosphate residue is discussed below.

It has recently been reported (Ref. 20) that 5'-O-triphosphoryladenylyl-(2'→5')-adenylyl-(2'→5')-adenosine (25, 2-5A) is formed on incubation of extracts from interferon-treated cells or rabbit reticulocytes with double-stranded ribonucleic acids and adenosine 5'-triphosphate. Furthermore this unusual oligoribonucleotide derivative with unnatural 2'→5'-internucleotide linkages has been found to be an extremely powerful inhibitor of cell-free protein synthesis, even at sub-nanomolar concentrations (Ref. 20). We undertook the synthesis of 2-5A (25) in order to confirm the structure assigned to the natural product and

Scheme 8

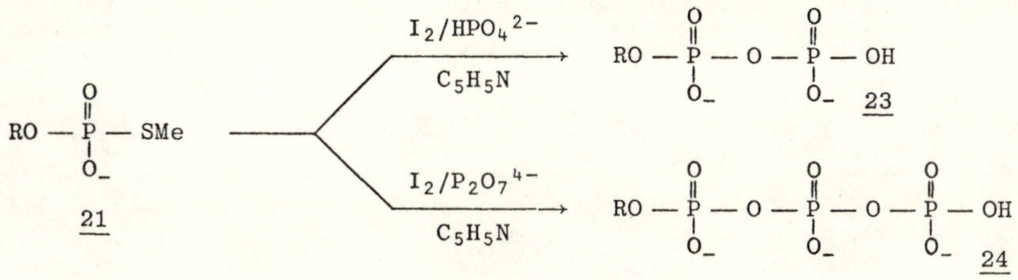

$$RO-\overset{O}{\underset{O_-}{P}}-SMe \quad \underset{21}{} \quad \begin{array}{c} \xrightarrow{I_2/HPO_4{}^{2-}} \\ C_5H_5N \\ \\ \xrightarrow{I_2/P_2O_7{}^{4-}} \\ C_5H_5N \end{array} \quad \begin{array}{c} RO-\overset{O}{\underset{O_-}{P}}-O-\overset{O}{\underset{O_-}{P}}-OH \quad \underline{23} \\ \\ RO-\overset{O}{\underset{O_-}{P}}-O-\overset{O}{\underset{O_-}{P}}-O-\overset{O}{\underset{O_-}{P}}-OH \quad \underline{24} \end{array}$$

to examine the general applicability of the synthetic methods described above.

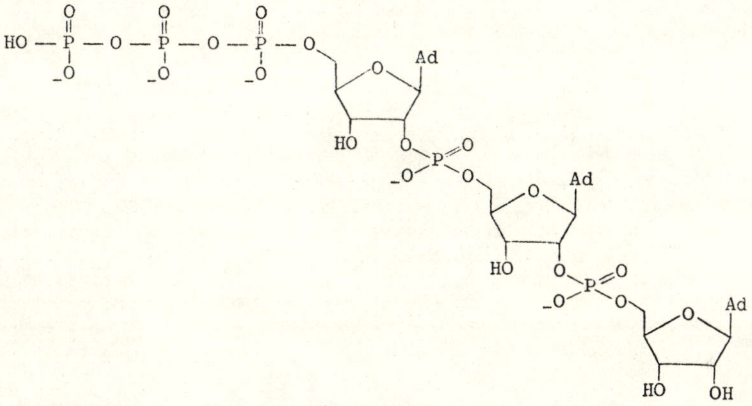

25; Ad = adenin-9-yl

The key nucleoside building block used in the synthesis of 2-5A (25) was 5'-O-p-chlorophenoxyacetyl-3'-O-methoxytetrahydropyranyl-6-N-benzoyladenosine (26). This compound, which was obtained as a pure crystalline solid (Ref. 21) was converted [Scheme 9] into the triethylammonium salt of its 2'-o-chlorophenyl phosphate (27) in 86% overall yield by the phosphorylation procedure (Ref. 13) indicated in Scheme 6 [i.e. phosphorylation with o-chlorophenyl p-nitrophenyl phosphorochloridate (15b) followed by reaction with p-thiocresol (17) and triethylamine in acetonitrile]. The phosphodiester intermediate (27) so obtained, was then allowed to react with 2',3'-O-methoxymethylene-6-N-benzoyladenosine (28) in the presence of an excess of MSNT (18) to give the expected fully-protected dinucleoside phosphate which, on brief treatment with sodium hydroxide gave the partially-protected dinucleoside phosphate (29) in 64% overall yield. Phosphorylation of 29 with 27 in the presence of an excess of 18, followed again by a brief treatment with sodium hydroxide gave the partially-protected trinucleoside diphosphate (30) in 75% yield, based on 29. The required intermediate trinucleotide derivative (31) was obtained by phosphorylation of 30 with O-2,4-dichlorophenyl S-methyl phosphorochloridothioate (19) (Ref. 18), followed by a four step unblocking process: (i) treatment with the conjugate base of syn-p-nitrobenzaldoxime (8) to unblock the phosphotriester groups, (ii) treatment with concentrated aqueous ammonia to remove the N-benzoyl groups, (iii) treatment with 0.01 M-hydrochloric acid to hydrolyze the methoxytetrahydropyranyl and methoxymethylene groups, and (iv) brief treatment with dilute aqueous ammonia (pH 9) to deformylate the terminal 2',3'-diol system.

The trinucleotide phosphorothioate (31) was converted into 2-5A (25) by treating it with tetra-(tri-n-butylammonium) pyrophosphate (5 molecular equivalents) and iodine (50 molecular equivalents) in anhydrous pyridine solution according to the procedure indicated in Scheme 8. The products were purified by chromatography on DEAE-Sephadex A25 and pure 2-5A (25) was isolated in ca. 40% yield. The biological activity of synthetic 2-5A was identical to that of the naturally-occurring material isolated by Kerr et al (Ref. 20). The h.p.l.c. retention times of the synthetic and natural materials were also identical.

The synthesis of 2-5A (25) was carried out in a satisfactory manner by means of the synthetic procedures described above. However, from these and other studies it became clear that it would be necessary to make at least two modifications to these procedures before they

Scheme 9

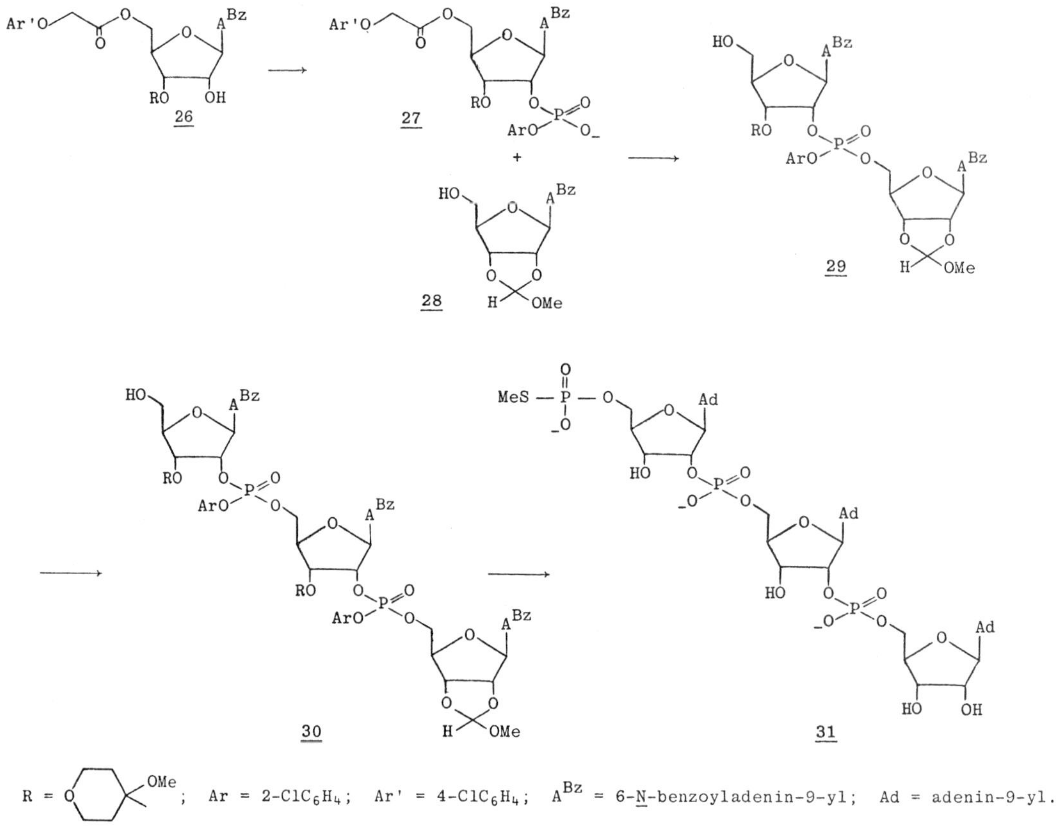

[Scheme 9 structural diagrams showing compounds 26, 27, 28, 29, 30, 31]

R = (tetrahydropyranyl-OMe group); Ar = 2-ClC$_6$H$_4$; Ar' = 4-ClC$_6$H$_4$; A^{Bz} = 6-$\underline{N}$-benzoyladenin-9-yl; Ad = adenin-9-yl.

would be suitable for the synthesis of high molecular weight oligonucleotides. In the first place, the procedure used for the first phosphorylation step [(i), Scheme 1] involving o-chlorophenyl p-nitrophenyl phosphorochloridate [15b, Scheme 6] is suitable only for the phosphorylation of nucleoside building blocks [e.g. the conversion of 26 into 27, Scheme 9]. Unfortunately, the conjugate base of p-thiocresol (17) can attack at C-5' adjacent to a phosphotriester group as well as at a position para to an aromatic nitro group [Scheme 6, 16, arrow]. Thus when the fully-protected dinucleoside phosphate [32, Scheme 10] was treated (Ref. 22) with 17 (13 molecular equivalents) and triethylamine (12 molecular equivalents) at 20°C, it was ca. 50% converted into 33 after 8 hr. This result suggests that ca. 5-10% thiolate ion promoted cleavage would occur adjacent to each protected internucleotide linkage under the conditions required for 16 to be completely converted into 2 [Scheme 6].

Scheme 10

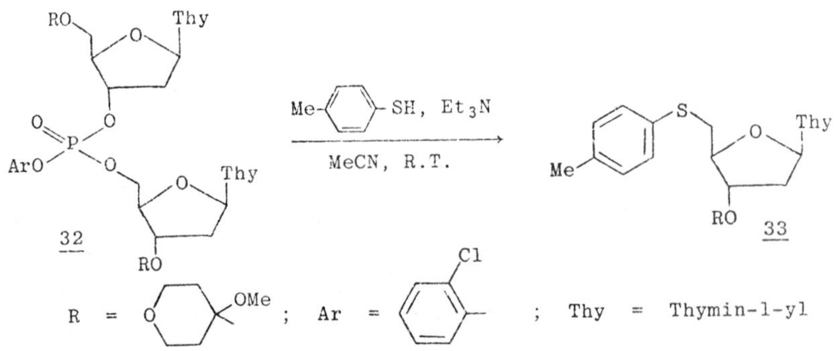

[Scheme 10 structural diagram showing conversion of 32 to 33 with Me-C$_6$H$_4$-SH, Et$_3$N in MeCN, R.T.]

R = (tetrahydropyranyl-OMe group); Ar = 2-chlorophenyl; Thy = Thymin-1-yl

Fortunately, it was possible to find (Ref. 23) an alternative phosphorylation procedure which could be used both in the stepwise and block synthesis of oligonucleotides. p-Chlorophenyl phosphorodi-(1,2,4-triazolide) has been used (Ref. 24) as a bifunctional phosphorylating agent in the phosphotriester approach. However, if, for example, 2',5'-protected ribonucleoside derivatives (34) are treated [Scheme 11] with an excess of o-chlorophenyl phosphorodi-(1,2,4-triazolide) (35) in acetonitrile-pyridine solution and the products are then subjected to hydrolysis, the desired 3'-o-chlorophenyl phosphates are obtained in high yields (Ref. 25). The latter may readily be isolated as their pure triethylammonium salts (36), free from o-chlorophenyl phosphate. Thus an apparently bifunctional phosphorylating agent can behave as though it were monofunctional.

Scheme 11

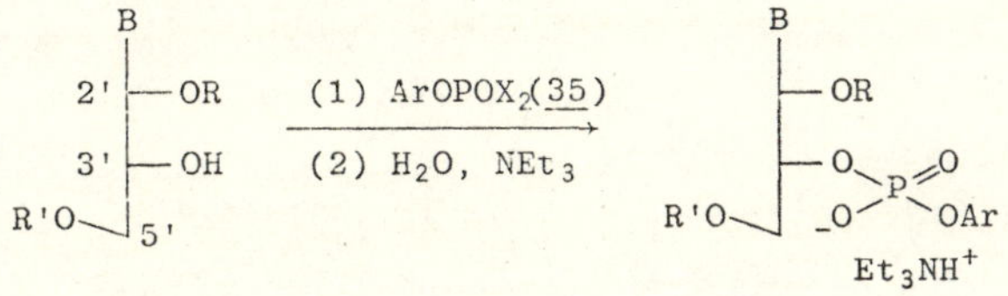

$R = $ tetrahydropyranyl-OMe ; $Ar = 2\text{-}ClC_6H_4$; $X = $ 1,2,4-triazolyl

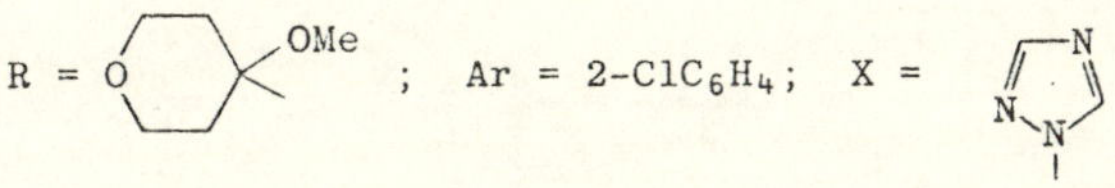

In the preparation (Ref. 21) of the partially-protected trinucleoside diphosphate (30) which was required as an intermediate in the synthesis of 2-5A (25), it was necessary to remove a 5'-O-p-chlorophenoxyacetyl protecting group by alkaline hydrolysis. Inevitably some concomitant unblocking of the two o-chlorophenyl-protected internucleotide linkages must have occurred, thereby lowering the yield of 30. Clearly this problem becomes more serious as the molecular weight of the oligonucleotide and consequently the number of phosphotriester groups increases. Therefore, if the synthesis of really high molecular weight oligonucleotides is to be undertaken, the p-chlorophenoxyacetyl group must be replaced by an acyl protecting group which is removable under much milder conditions.

We now believe that this problem has been solved. Indeed, we have recently introduced the use of the o-dibromomethylbenzoyl (DBMB) protecting group (as in 37) (Ref. 26) which is, in our opinion, an ideal acyl protecting group for the present purpose. When o-dibromomethylbenzoate esters (37), which are easily prepared (Ref. 26) by treating the corresponding hydroxy compounds (ROH) with o-dibromomethylbenzoyl chloride in pyridine solution [Scheme 12], are treated with an excess of silver perchlorate and 2,4,6-collidine in acetone-water (98:2 v/v), they are quantitatively converted into the corresponding o-formylbenzoates (38). When the latter intermediates (38) are treated with an excess of morpholine (Ref. 27), deacylation occurs rapidly and the unprotected hydroxy functions are liberated.

Scheme 12

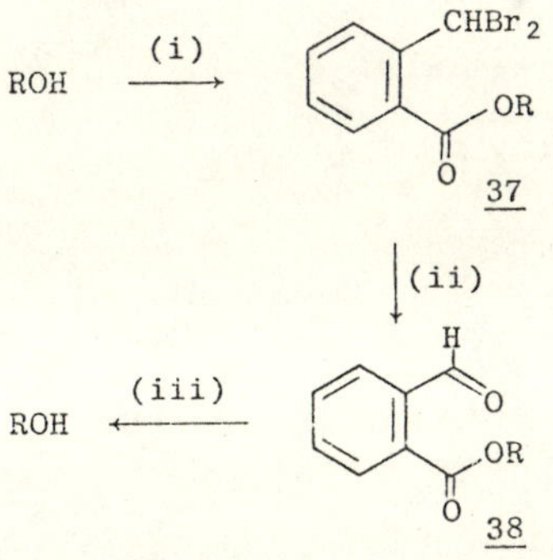

Reagents: (i) $2\text{-}Br_2CHC_6H_4COCl/C_5H_5N$; (ii) $AgClO_4$, 2,4,6-collidine/$Me_2CO\text{-}H_2O$ (98:2 v/v); (iii) morpholine.

Now that a satisfactory procedure for the first phosphorylation step [(i), Scheme 1] had been developed and an acyl protecting group removable under very mild conditions was available, we believed that we were in a position to undertake the synthesis of high molecular weight oligo- and indeed poly-nucleotides in both the ribose and the 2'-deoxyribose series. I should like to end this lecture with a description of the synthesis (Ref. 25) of the 3'-terminal decanucleotide (39) of yeast alanine tRNA.

$$UpCpGpUpCpCpApCpCpA$$

$$\underline{39}$$

Five nucleoside building blocks were used in the synthesis of the decamer (39). It is not proposed to discuss the details of the nucleoside chemistry here except to note that both of the 5'-O-DBMB derivatives (41a and 41b) were obtained as pure crystalline solids and that 41b was obtained in 71% isolated yield by reacting 40c directly with o-dibromomethyl-benzoyl chloride (Ref. 26).

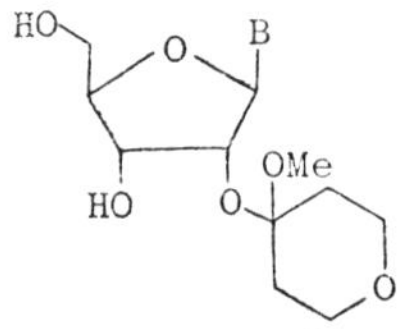

40 a; B = C^{Bz}
  b; B = G^{Bz}
  c; B = U

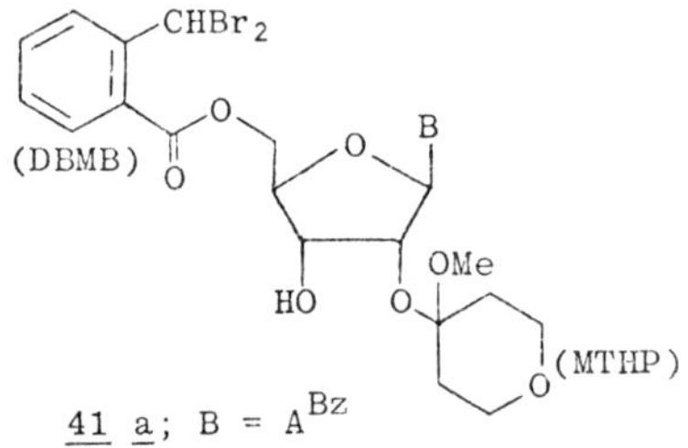

41 a; B = A^{Bz}
  b; B = U (71%)

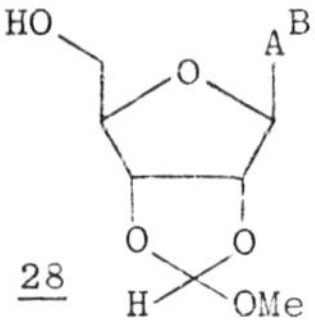

28

A^{Bz} = 6-N-benzoyladenin-9-yl; C^{Bz} = 4-N-benzoylcytosin-1-yl;
G^{Bz} = 2-N-benzoylguanin-9-yl;   U = uracil-1-yl

Scheme 13

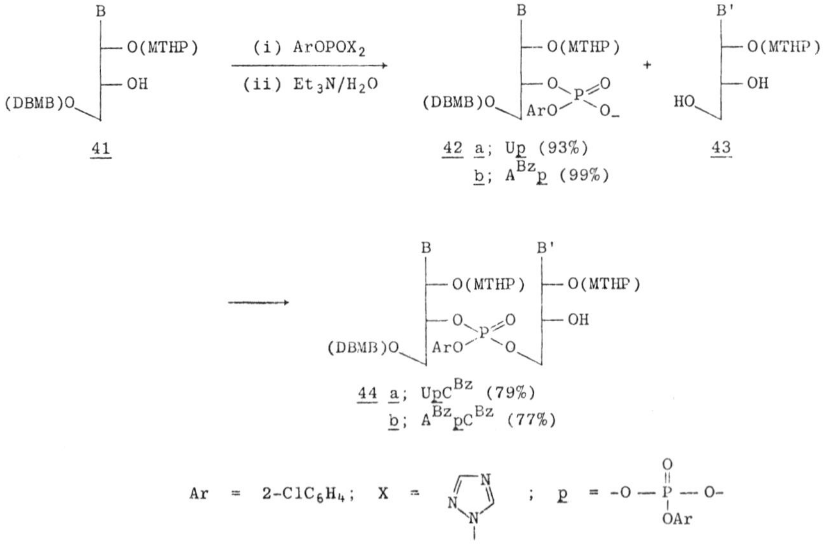

41 $\xrightarrow{\text{(i) ArOPOX}_2}_{\text{(ii) Et}_3\text{N/H}_2\text{O}}$ 42 a; Up (93%)
  b; A^{Bz}p (99%)

+ 43

⟶ 44 a; UpC^{Bz} (79%)
  b; A^{Bz}pC^{Bz} (77%)

Ar = 2-ClC$_6$H$_4$;   X = (imidazol-1-yl);   p = $-O-\overset{\overset{O}{\|}}{\underset{OAr}{P}}-O-$

The procedure outlined in Scheme 11 (Ref. 23) was used to convert the 5'-O-(o-dibromomethylbenzoyl)-2'-O-methoxytetrahydropyranyl derivatives (41) of uridine and 6-N-benzoyladenosine into the triethylammonium salts of their 3'-o-chlorophenyl phosphates (42a and 42b, respectively), which were isolated as colourless solids in 93 and 99% yields [Scheme 13]. The latter were then allowed to react with the appropriate 2'-O-methoxytetrahydropyranyl derivatives (43) (Ref. 28) in the presence of an excess of MSNT (18) to give the desired partially-protected dinucleoside phosphates (44a and 44b) in the very satisfactory isolated yields indicated [Scheme 13]. As far as can be judged, phosphorylation occurs on the 5'-hydroxy functions of the 2'-O-methoxytetrahydropyranyl derivatives (43) with a very high degree of regioselectivity. No dinucleoside phosphates with 3'→3'-internucleotide linkages could be detected in the products.

Phosphorylation of 44a and 44b with o-chlorophenyl phosphorodi-(1,2,4-triazolide) [35, Scheme 11] gave the corresponding protected dinucleotide derivatives (45a and 45b) in very high yields [Scheme 14]. The latter were allowed to react with the appropriate 2'-O-methoxytetrahydropyranyl derivatives (46) in the presence of MSNT (18) to give the partially-protected trinucleoside diphosphates (47a-c) in satisfactory to good yields. Phosphorylation of the latter with o-chlorophenyl phosphorodi-(1,2,4-triazolide) (35) gave the three protected trinucleotides (48a-c), required for the synthesis of the decamer (39), in very high yields.

Scheme 14

44; Ar = 2-ClC$_6$H$_4$

45 a; UpC^{Bz}p (98%)
   b; A^{Bz}pC^{Bz}p (98%)

46

47 a; UpC^{Bz}pG^{Bz} (78%)   →   48 a; UpC^{Bz}pG^{Bz}p (96%)
   b; UpC^{Bz}pC^{Bz} (82%)   b; UpC^{Bz}pC^{Bz}p (96%)
   c; A^{Bz}pC^{Bz}pC^{Bz} (72%) c; A^{Bz}pC^{Bz}pC^{Bz}p (98%)

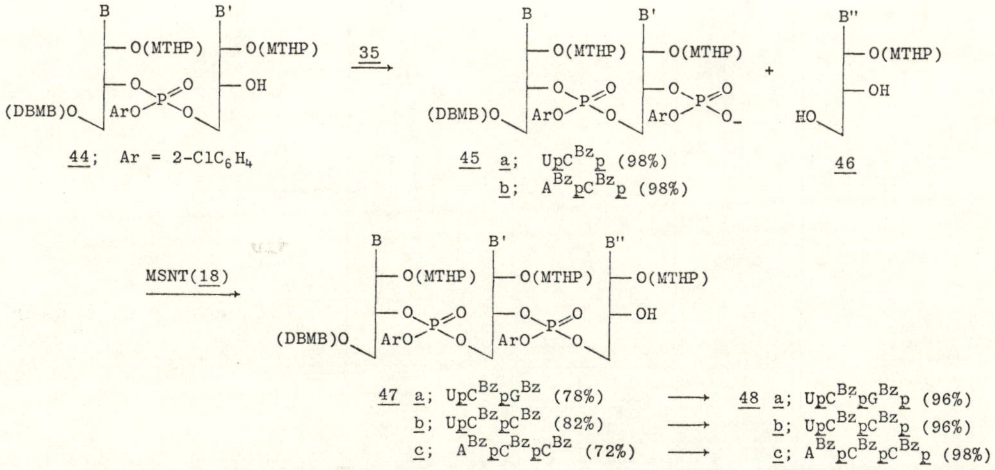

The strategy used for the synthesis of the fully-protected decamer (51) from the three protected trinucleotides (48a-c) is outlined in Scheme 15. In the first place [sequence (a), Scheme 15], 48c was allowed to react with 2',3'-O-methoxymethylene-6-N-benzoyladenosine (28; ca. 1.4 molecular equivalents) and MSNT (18, 5.8 molecular equivalents) in pyridine solution at room temperature for 30 min to give the fully-protected tetramer in 75% isolated yield. The DBMB group was removed (Ref. 26) from the latter material by stirring it with silver perchlorate (0.33 M, ca. 14 molecular equivalents) and 2,4,6-collidine (7 molecular equivalents) in acetone-water (98:2 v/v) for 1 hr at room temperature, then removing the excess of silver ion (by the addition of lithium bromide) and treating the products with morpholine (ca. 20 molecular equivalents) for 5 min. In this way, the partially-protected tetramer (49) was obtained in 94% isolated yield, based on its fully-protected precursor. The second reaction sequence [(b), Scheme 15], i.e. the preparation of the partially-protected heptamer (50) [obtained in ca. 56% overall yield, based on 49], was carried out in the same way. In the final step [reaction sequence (c)], a twofold excess of the protected trinucleotide (48a) was used. However, the isolated yield of the fully-protected decamer (51) was only ca. 50%.

The decamer (51) can be unblocked at its 5'-end and then extended to give a longer sequence of the yeast alanine tRNA molecule. Indeed, studies directed towards this end are now in progress at King's College. However, in order to characterize the fully-protected decamer (51), a small quantity of it was unblocked by the three step procedure indicated in Scheme 16. The fully-unblocked material was then chromatographed on DEAE-Sephadex A25. It can be seen from Figure 1 that one single component (eluted with ca. 1.0 M-triethylammonium bicarbonate) accounts for nearly all of the nucleotide material eluted from the column. The latter main component, i.e. the putative decamer (39), was completely degraded by treatment with aqueous sodium hydroxide, ribonuclease A, calf spleen phosphodiesterase and crotalus adamanteus snake venom phosphodiesterase. Satisfactory qualitative and quantitative

Scheme 15

(a) (DBMB)$A^{Bz}\underline{p}C^{Bz}\underline{p}C^{Bz}\underline{p}$ + (HO)A^{Bz}(MM) $\xrightarrow[\text{(ii) [94%]}]{\text{(i) [75%]}}$ (HO)$A^{Bz}\underline{p}C^{Bz}\underline{p}C^{Bz}\underline{p}A^{Bz}$(MM)
       <u>48c</u>   <u>28</u>   <u>49</u>

(b) (DBMB)$U\underline{p}C^{Bz}\underline{p}C^{Bz}\underline{p}$ + <u>49</u> $\xrightarrow[\text{(ii) [78%]}]{\text{(i) [71.5%]}}$ (HO)$U\underline{p}C^{Bz}\underline{p}C^{Bz}\underline{p}A^{Bz}\underline{p}C^{Bz}\underline{p}C^{Bz}\underline{p}A^{Bz}$(MM)
       <u>48b</u>   <u>50</u>

(c) (DBMB)$U\underline{p}C^{Bz}\underline{p}G^{Bz}\underline{p}$ + <u>50</u> $\xrightarrow{\text{(i) [ca. 50%]}}$ (DBMB)$U\underline{p}C^{Bz}\underline{p}G^{Bz}\underline{p}U\underline{p}C^{Bz}\underline{p}C^{Bz}\underline{p}A^{Bz}\underline{p}C^{Bz}\underline{p}C^{Bz}\underline{p}A^{Bz}$(MM)
       <u>48a</u>   <u>51</u>

(i) MSNT(<u>18</u>)/C$_5$H$_5$N; (ii) (a) AgClO$_4$- 2,4,6-collidine/Me$_2$CO-H$_2$O (98:2 v/v), (b) morpholine.

data were obtained by analysis of the hydrolysates.

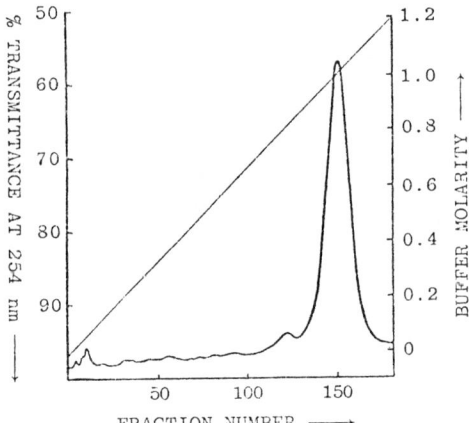

Figure 1. DEAE-Sephadex chromatography of the products obtained after the complete unblocking [by the procedure indicated in Scheme 16] of the fully-protected UpCpGpUpCpCpApCpCpA. A linear gradient of triethylammonium bicarbonate buffer (pH 7.5) was used.

Scheme 16

(DBMB)$U\underline{p}C^{Bz}\underline{p}G^{Bz}\underline{p}U\underline{p}C^{Bz}\underline{p}C^{Bz}\underline{p}A^{Bz}\underline{p}C^{Bz}\underline{p}C^{Bz}\underline{p}A^{Bz}$(MM) (<u>51</u>)

$\Big\downarrow$ (i), (ii), (iii)

UpCpGpUpCpCpApCpCpA (<u>39</u>)

Reagents: (i) $\underline{N}^1,\underline{N}^1,\underline{N}^3,\underline{N}^3$-Tetramethylguanidinium <u>syn</u>-4-nitro-
           benzaldoximate in dioxan-water (1:1 v/v), 20°C,
           20 hr; (ii) aqueous ammonia (<u>d</u> 0.88), 50°C, 24 hr;
           (iii) 0.01 <u>M</u>-hydrochloric acid (pH 2), 20°C, 6 hr.

In conclusion, we believe that the main problems which have been encountered in the chemical synthesis of oligonucleotides have been virtually solved. First, it now seems clear that the phosphotriester approach with aryl (especially o-chlorophenyl) protecting groups is the method of choice. Secondly, at the present time, the procedure illustrated in Scheme 11 (Ref. 23) appears to be the most suitable for the first phosphorylation step [(i), Scheme 1] and MSNT (<u>18</u>) (Refs. 9 and 17) appears to be a particularly useful activating agent for the second phosphorylation step [(ii), Scheme 1]. Thirdly, the o-chlorophenyl protecting groups

are best removed from the internucleotide linkages by treatment of the fully-protected oligonucleotides with the conjugate base of syn-p-nitrobenzaldoxime (8) or syn-pyridine-2-carbox-
aldoxime (9) (Ref. 9).   Finally, the o-dibromomethylbenzoyl (DBMB) group (Ref. 26) (as in 37)
appears to be the most suitable acyl group yet devised for the protection of hydroxy-functions
in oligonucleotide synthesis, especially in the ribose series.

> Acknowledgements - I should like to express my extreme gratitude to a number
> of excellent coworkers without whose skill and hard work our studies in the
> chemical synthesis of oligonucleotides would have been impossible.   In con-
> nection with our recent work described above, I should especially like to
> acknowledge the contributions of Dr. Jyoti Chattopadhyaya, Mr. Simon Jones,
> Dr. Bernard Rayner, Mr. Richard Titmas, Dr. Masaru Ubasawa, Mrs. Aiko Ubasawa
> and Dr. Laval Yau.   I should also like to thank the Science Research Council
> for generous financial support.

REFERENCES

1. C. B. Reese, Tetrahedron Report No. 56, Tetrahedron 34, 3143-3179 (1978).
2. C. B. Reese and R. Saffhill, Chem. Comm., 767-768 (1968).
3. J. C. M. Stewart, Ph.D. Thesis, Cambridge University, 39-45 (1969).
4. C. B. Reese, Colloques Internationaux du C.N.R.S. 182, 319-328 (1970).
5. A. M. Michelson and A. R. Todd, J. Chem. Soc., 2632-2638 (1955).
6. R. L. Letsinger and V. Mahadevan, J. Am. Chem. Soc. 87, 3526-3527 (1965).
7. F. Eckstein and I. Rizk, Angew. Chem. internat. Edit. 6, 695-696 (1967).
8. N. J. Cusack, C. B. Reese and J. H. van Boom, Tetrahedron Letters, 2209-2212 (1973).
9. C. B. Reese, R. C. Titmas and L. Yau, Tetrahedron Letters, 2727-2730 (1978).
10. R. W. Adamiak, R. Arentzen and C. B. Reese, Tetrahedron Letters, 1431-1434 (1977).
11. R. Arentzen and C. B. Reese, J.C.S. Perkin I, 445-460 (1977).
12. T. E. England and T. Neilson, Canad. J. Chem. 54, 1714-1721 (1976).
13. C. B. Reese and Y. T. Yan Kui, J.C.S. Chem. Comm., 802-804 (1977).
14. R. Lohrmann and H. G. Khorana, J. Am. Chem. Soc. 88, 829-833 (1966).
15. Y. A. Berlin, O. G. Chakhomakhcheva, V. A. Efimov, M. N. Kolosov and V. G. Korobko,
    Tetrahedron Letters, 1353-1354 (1973).
16. J. Stawinski, T. Hozumi and S. A. Narang, Can. J. Chem. 54, 670-672 (1976).
17. Y. T. Yan Kui, Ph.D. Thesis, London University, 146-155 (1977).
18. C. B. Reese and L. Yau, J.C.S. Chem. Comm., 1050-1052 (1978).
19. S. S. Jones and C. B. Reese, unpublished observations.
20. A. G. Hovanessian, R. E. Brown and I. M. Kerr, Nature 268, 537-540 (1977); I. M. Kerr,
    R. E. Brown and A. G. Hovanessian, ibid. 268, 540-542 (1977); I. M. Kerr and
    R. E. Brown, Proc. Natl. Acad. Sci. USA 75, 256-260 (1978).
21. S. S. Jones and C. B. Reese, in the press.
22. C. B. Reese and R. C. Titmas, unpublished observations.
23. J. B. Chattopadhyaya and C. B. Reese, unpublished observations.
24. J. Stawinski, T. Hozumi, S. A. Narang, C. D. Bahl and R. Wu, Nucleic Acids Res. 4,
    353-371 (1977).
25. B. Rayner, C. B. Reese, A. Ubasawa and M. Ubasawa, unpublished observations.
26. J. B. Chattopadhyaya, C. B. Reese and A. H. Todd, J.C.S. Chem. Comm., in the press
    (1979).
27. M. L. Bender and M. S. Silver, J. Am. Chem. Soc. 84, 4589-4590 (1962).
28. C. B. Reese, R. Saffhill and J. E. Sulston, J. Am. Chem. Soc. 89, 3366-3368 (1967);
    Tetrahedron 26, 1023-1030 (1970).

APPLICATIONS OF 31P NMR TO BIOLOGICAL SYSTEMS WITH EMPHASIS ON INTACT TISSUE DETERMINATIONS

T. Glonek

Nuclear Magnetic Resonance Laboratory, Chicago College of Osteopathic Medicine, Chicago, IL 60615, U.S.A.

Abstract - Phosphorus-31 nuclear magnetic resonance (NMR) has been applied to the study of intact tissues and whole biological fluids and their corresponding perchloric acid extracts. The spectroscopic phosphorus profiles obtained reveal resonance signals from the low molecular weight biophosphates which are present at concentrations greater than about 1 mmolar, and the spectra can be integrated to yield quantitative information. For example, P-31 spectra recorded from intact muscles showed resonances from adenosine triphosphate, phosphocreatine, inorganic orthophosphate, and the sugar phosphates, and quantitation of these metabolites by P-31 spectroscopy was in good agreement with values obtained by chemical analysis. Moreover, spectra obtained from various muscles showed considerable variation in their phosphorus profiles; differences could be detected between normal and diseased muscle, cardiac and fast muscle, and similar muscle types from different animals.

Other tissues which have been examined include: endocrine glands (dog adrenal and rat testes), kidney, liver, erythrocytes, reticulocytes, platelets, neuroblastoma clones, human amniotic fluid, and human saliva. The studies which have been conducted include: determinations of disease states, determinations of nutritional effects, and the effects of chronic feeding of environmental levels of heavy metal (Cd and Pb) toxins.

Because no fractionation procedures are involved, the P-31 spectrometer detects all of the soluble phosphorus-containing molecules present in a tissue, and this has led to some unexpected findings. For example, in muscle, a number of signals have been detected in the phosphodiester region of the spectrum which could not be correlated with any of the known common phosphates of muscle tissue. One of these signals has been identified as that from glycerol 3-phosphorylcholine. Another, which has yet to be characterized, is an N-derivative of glycerol 3-phosphorylethanolamine, exhibiting a rate of hydrolysis at pH 7 at 31° of 3%/min. The resonance signal from tetrametaphosphate has also been detected, as have resonances from phosphonic acids and resonances from other phosphorus-containing molecules yet to be identified.

INTRODUCTION

Discussed in the following paper is a very practical application of nuclear magnetic resonance (NMR) spectroscopy, namely, the use of the NMR spectrometer to obtain a distribution of tissue metabolites for the description, diagnosis, and treatment of human diseases. When considering the biochemistry of living systems, intact tissues, and their extracts, NMR of the phosphorus atom has, perhaps, the greatest utility, since within such complex chemical systems phosphorus has the status of a heteroatom, and there are relatively few compounds which contain phosphorus. These compounds, however, represent some of the more interesting and abundant biological chemicals. Phosphorus-31 NMR spectroscopic tissue profiles, as a rule, are quite simple. Most of the resonance signals detected represent single spectroscopic transitions arising from a single phosphorus atom within a molecule (see Note a.). The only exceptions appear to be the nucleoside di- and triphosphates, and, even in these cases, the multiplets are simple doublets and triplets, which are readily resolved by the spectrometer and which serve a useful purpose in contributing to the identification of the source phosphate.

The technique had not been extensively applied at physiological or clinical levels until the decade of the seventies principally because of sensitivity limitations. Whereas signal-to-noise ratios are almost never considered in applications of other forms of spectroscopy to biological research, in NMR spectroscopy, they dominate experimental design. The steady advance of scientific technology, however, has largely overcome the sensitivity limitation, so that, in 1979, it is a relatively simple matter to obtain highly resolved resonance data from the intracellular metabolites of intact, metabolizing tissues.

Note a. This statement is true only when the spectra are broad band, proton decoupled to eliminate H-1--P-31 spin-spin couplings. In most P-31 NMR applications, however, as in C-13 NMR applications, the spectra are routinely proton-decoupled.

BACKGROUND: CONDENSING AGENT-MEDIATED PHOSPHORYLATIONS

In 1963, John R. Van Wazer and myself were conducting experiments investigating condensing agent-mediated polymerizations of inorganic orthophosphate. We knew the progress of the reactions would involve complicated pathways and the production of numerous intermediates and, further, that attempts to analyze the reaction mixtures would, in all probability, lead to the disruption of some of the more interesting intermediates. We, therefore, needed an analytical method which was capable of the simultaneous analysis of a number of phosphatic compounds, and we needed one which could examine the reaction mixtures without disrupting the reaction processes. Phosphorus NMR was ideally suited for this type of analytical work. Unfortunately, it was not a very sensitive method in 1963. In order to obtain phosphorus spectra, it was usually necessary to prepare a solution of the phosphate of interest in the concentration range 0.5 to 3.0 molar. Fig. 1 shows typical spectra obtained in 1963 on a carbodiimide-mediated condensation of phosphates (Ref. 1, 2, & 3). The data show reasonably resolved resonance signals, with a signal-to-noise ratio for the largest peaks approaching 40 to 1. Such spectra could be effectively analyzed to obtain concentrations of a number of inorganic orthophosphates; however, one was forced to work with extremely concentrated solutions. The reaction mixture from which these spectra were obtained was 1.5 molar in phosphorus.

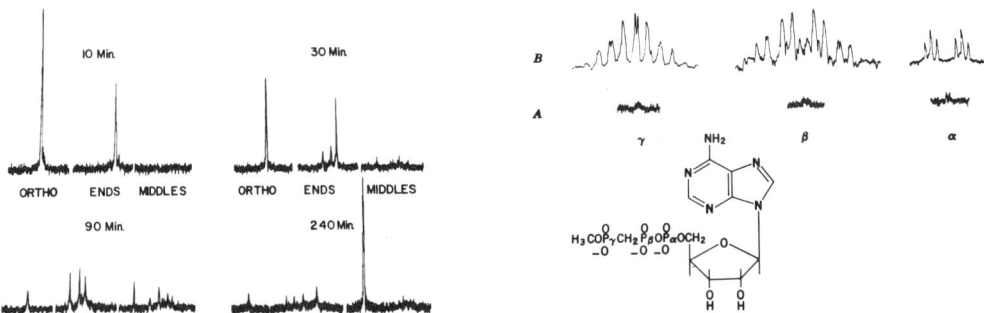

Figure 1. Phosphorus NMR spectra of an inorganic orthophosphoric acid condensation obtained in the middle 1960's on a Varian HR-100 spectrometer using 5 mm spinning sample tubes. Each spectrum is a single slow passage scan taken without field-frequency stabilization. The starting reaction mixture contained 1.5 mmoles orthophosphoric acid, 1.5 mmoles tri-n-butylamine, and 1.5 mmoles trichloroacetonitrile in tetramethylurea to a volume of one ml. Only the ortho, end, and middle phosphate regions of the spectra are shown for each time. Note the low signal-to-noise ratios of the 90 min spectrum signals.

Figure 2. 40.5 MHz P-31 NMR spectra of an analog of adenosine triphosphate taken on Monsanto Corporation's (St. Louis, Missouri) Varian HR-100 spectrometer system shortly after its conversion to a field-frequency stabilized system. The resonance patterns of spectrum A were obtained from a single sweep without field-frequency stabilization; the spectra of B were obtained with field-frequency stabilization, using signal-averaging. The peaks of B represent a theoretical enhancement of 7.0 obtained from a time-average of 50 scans taken at 30 scans/hr. (Note: The alpha multiplet was recorded at a signal level which differed from the gain setting used in recording multiplets beta and gamma. The relative peak areas and intensities, therefore, are not comparable.)

The phosphate condensation work, exemplified by Fig. 1, was carried out at Monsanto Corporation in St. Louis, Missouri, with a Varian HR-100 NMR spectrometer, a device which employed five millimeter spinning sample tubes, requiring about 1/2 milliliter of sample. While Van Wazer and I were at Monsanto, the first commercial field-frequency stabilization units became available. These permitted signal-averaging processes, and we successfully applied the combined technology to phosphorus magnetic resonance. The spectrum from the very first sample which was analyzed is shown in Fig. 2 (Ref. 4). The compound of interest was an analog of a biological phosphate, a methylene methylester derivative of adenosine triphosphate (ATP). You will note from Spectrum A, which was obtained from a single slow passage scan, that the signal-to-noise ratio from this particular molecule in as concentrated a solution as was possible to obtain was such that the phosphorus resonance signals could just barely be detected. After signal-averaging for about a half-day (Spectrum B), the resonance bands were clearly defined. These could now be expanded and analyzed in the manner that one ordinarily applies to proton magnetic resonance spectra. Although such data represented biomedical applications of phosphorus magnetic resonance, phosphorus spectroscopy still lacked the sensitivity needed for the examination of intact tissues.

Approximately ten years later, practical Fourier transform NMR spectroscopy became a reality. For phosphorus magnetic resonance, this meant that it was now possible to obtain good P-31 data at phosphorus concentrations only 1/100th that which was necessary in 1963. The method was immediately exploited by Moon and Richards in 1973 (Ref. 5) in their first published spectrum of the 2,3-diphosphoglycerate signals and other phosphorus resonances from intact whole blood. At about this same time, Michael Bárány came to the University of Illinois from the Muscle Institute in New York, and we began to discuss the possibility of studying muscle components by phosphorus magnetic resonance. The effort was designed to obtain the possible location of binding sites and other specifics involved with muscle action and myosin filaments.

One day, Michael casually asked me if it was possible to put a piece of muscle in the NMR tube and examine it with the spectometer. I informed him that this had never been done to my knowledge, but that one could certainly analyze red cells with phosphorus resonance, and so I suggested we try the experiment. Fig. 3 shows the first NMR spectrum we subsequently obtained from an intact muscle (Ref. 6).

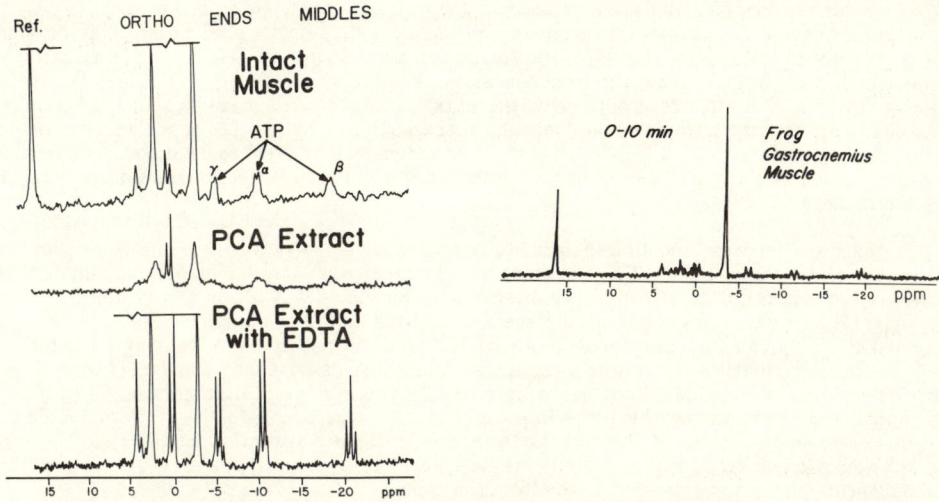

Figure 3. The first intact tissue phosphate profile obtained in Chicago and the corresponding perchloric acid extract profiles. The tissue examined was a fresh, intact frog gastrocnemius muscle. Ref. 8, is the signal from the reference compound, methylenediphosphonic acid (Na countercation, pD 9.5 in D_2O), contained in a sealed 1 mm capillary tube coaxially mounted in the NMR tube and inserted through the muscle. The spectra were gathered through the use of Fourier-transform technology on a Bruker HFX-5 spectrometer operating at 36.43 MHz for phosphorus: sweep width, 2,500 Hz; cycling time, 832 msec; 4,096 data points per free induction decay; temperature, 28°C. The accumulation time for the gathering of the muscle spectrum was 90 min; the spectra from the relatively concentrated extracts were gathered in less than 10 min. In these spectra and all subsequent phosphate spectra presented, the chemical shift scale is given in ppm units relative to the resonance position of 85% inorganic orthophosphoric acid with increasing values corresponding to increasing radio frequency as recommended by IUPAC. The prominent resonances have been truncated for the purpose of the illustration.

Figure 4. Phosphorus NMR spectrum of a frog muscle, initially at 0°C, acquired in the first 10 min after the muscle was removed from the animal. The prominent signal arises from phosphocreatine; the signal from inorganic orthophosphate, at 1.7 ppm, is virtually absent.

At this point, it is necessary to explain some of the details of Fig. 3; all of the remaining figures of this paper are set up in a similar manner. The abscissa is the chemical shift scale, i.e., the relative position in parts-per-million (ppm) of phosphorus resonances from different chemical groupings. The scale is arranged so that the inorganic orthophosphoric acid resonance is at the zero position, with lower relative field energies to the left, and higher relative field energies to the right. The ordinate represents the intensity of the signal. Signal strengths, i.e., the areas under the resonance curves, are proportional to the amount of material giving rise to the signal, so that integrations of the spectra lead to quantitation of the data. The spectra may also show multiplicity in some of the resonance bands. For instance, in Fig. 3, multiplets can be seen in the gamma, alpha, and beta resonances from adenosine triphosphate. These arise from spin-coupling between the interacting phosphorus atoms of the ATP triphosphate side chain. Interactions between phosphorus and protons which also occur are eliminated in the spectra presented here by a process called proton broad band decoupling. Because phosphorus NMR has low sensitivity, spectral base lines are frequently noisy.

The experimental design leading to Fig. 3 was quite simple. A frog was cooled to about zero degrees where it became inactive. The gastrocnemius muscle from the leg was then removed intact, and was placed in an NMR tube - the fit to a 10 mm NMR tube is just about right. A small capillary containing a reference substance was then inserted through the middle of this muscle, and the entire assembly was placed in the NMR spectrometer, whereupon the acquisition of data commenced. After 90 minutes, the top spectrum of Fig. 3 was obtained. The resonances are as follows, from left to right: the signal from methylenediphosphonic acid, which is our reference in this case (16.3 ppm); resonances from orthophosphates, the trioses of the Embden-Meyerhof pathway (4.5-3.5 ppm); inorganic orthophosphate (1.7 ppm); phosphodiesters (1 - -1 ppm); creatine phosphate, which is the principal energy source in this particular muscle (-3.2 ppm); and the gamma (-5 ppm), alpha (-11 ppm), and beta (-17.5 ppm) resonances from ATP.

A classic experiment which may be performed with any tissue is extraction with cold 60% perchloric acid. The extract is then analyzed for its contents which, for the most part, are the low molecular weight water-soluble molecules. When a neutralized extract from muscle is examined by phosphorus magnetic resonance, one obtains a phosphorus profile similar to that of the middle spectrum of Fig. 3. Essentially the same resonance signals are present in the extract spectrum as are present in that from the intact muscle; however, except for two diester resonances, the signals are considerably broadened. Integration of the signal areas indicates that, insofar as can be determined, this profile is essentially the same as that which was obtained from the intact muscle. The broadness of the NMR lines is attributed to the interaction of the phosphate groups with the extract calcium, magnesium, and transition metal ions. If the sample is treated with either a chelating resin, such as Chelex-100, or a chelating molecule, such as ethylenediaminetetracetate (EDTA), these interactions are prevented, and the signals sharpen (Fig. 3, bottom). Most of the separate resonances detected from chelating agent-treated extracts can be accurately quantitated.

Despite limitations arising from the time required for analysis, our early intact tissue work provided us with considerable useful information. For example, the chemical shifts and coupling constants exhibited by adenosine triphosphate told us that in intact muscles, the molecule was present essentially completely as its monomagnesium salt. Integration of the spectra showed that there was no detectable adenosine diphosphate (ADP), although its resonances could be easily detected in corresponding perchloric acid extracts. Thus we know that, in the intact muscle, the adenosine diphosphate was bound in such a way that its NMR resonance signals could not be detected under the high resolution conditions employed in our experiments. The most logical interpretation was that ADP was bound to the actin molecule. This interpretation corresponded to one of the current theories of muscle contraction and, in fact, the amount of ADP bound, which could be determined from the perchloric acid extract, correlated with the number of actin binding sites known to be present in the intact muscle.

The time required to obtain the intact muscle spectrum of Fig. 3 was long relative to the metabolic rate of the muscle at 31°C, the temperature at which the profile was obtained. Under the best of conditions, however, spectra could be obtained in much less time. Fig. 4 shows the analysis of a similar muscle, obtained during the first 10 minutes after excision from the animal. A spectrum of reasonable quality was obtained, from which metabolites, such as creatine phosphate and inorganic orthophosphate, could be precisely quantitated. The Fig. 4 profile showed that the amount of creatine phosphate in the intact frog muscle, at 38 millimolar, was very high, and that the inorganic orthophosphate, which could not be detected, was extremely low. These amounts corresponded to the best values which have been reported by other investigators employing freeze clamp procedures and wet-chemical analysis. The phosphorus profile thus provided supporting evidence for yet another theory of muscle biochemistry.

PHOSPHATE PROFILES OF NORMAL AND DISEASED MUSCLES

Shortly after we began our research on intact muscles, we initiated a program to compare muscle profiles from normal tissues and tissue afflicted with various metabolic diseases. Fig. 5 compares phosphate profiles obtained from the normal chicken pectoralis muscle and a similar muscle from a chicken afflicted with hereditary chicken dystrophy (see Note b.) These results created considerable excitement. In examining the normal chicken pectoralis profile, we observed the usual signals from the sugar phosphates, inorganic orthophosphates, creatine phosphate, and adenosine triphosphate; however, the spectrum obtained from an equivalent amount of the dystrophic chicken pectoralis differed in two respects. The first, and most exciting, was that a new signal, which was not in the profile of the normal chicken pectoralis, appeared at about zero ppm. The second was that the overall concentration of the phosphates in the tissue was reduced (note the greater noise level in the dystrophic spectrum).

The new resonance appeared in a region of the spectrum characteristic of phosphodiesters, such as diethylphosphate; however, at that time, this resonance could not be ascribed to any of the known muscle phosphates of dystrophic chicken (Ref. 6). We proceeded to isolate (Ref. 7) the substance, using the magnetic resonance spectrometer to assay the steps in the isolation procedure (Fig. 6). Except for the assay method, conventional biochemical procedures were used during the isolation: an extraction procedure followed by barium precipitation; fractionations, first with an organic solvent, and then by column chromatography, eventually leading to a chemically pure preparation of the compound.

The chemical shift of the compound's resonance signal had already indicated that the phosphate functional group was a phosphodiester. The spectrum also provided one other piece of important information concerning this molecule, however, and this is shown in Fig. 7.

<u>Phosphorus NMR-pH Titration of Serine Ethanolamine Phosphodiester.</u> Phosphorus NMR chemical shifts, especially those from the ionized (or protonated) phosphates or phosphonates, undergo transitions

---

Note b. The chicken is one of the few animals for which there is a true breeding dystrophic strain.

associated with protonation-deprotonation reactions (Ref. 4, 5, 7, 9, 10); hence, one can use NMR to obtain pH titration curves. Fig. 7 shows the NMR pH titration curve obtained from the new diester phosphate. In addition to the strong acid ionizations, we observed a single shift inflection with a pK of about 9 corresponding to two equivalents of acid per phosphorus atoms, leading to the conclusion that two amino groups were also present in the molecule. With the purified molecule in hand, we called upon a number of our colleagues in the Chicago Metropolitan Area who had proton and carbon magnetic resonance facilities, and were able to obtain the corresponding resonance spectra. This data, in combination with more classic procedures, led to the identification of the molecule as serine enthanolamine phosphodiester (Fig. 8) (Ref. 8). This rather simple metabolite can be found in dystrophic tissues to concentrations approaching 10 millimolar. It does not appear to be in any way involved with phospholipid metabolism, and at this time, we have no idea as to its biological function, except to say that it is a marker for the hereditary dystrophy of chicken.

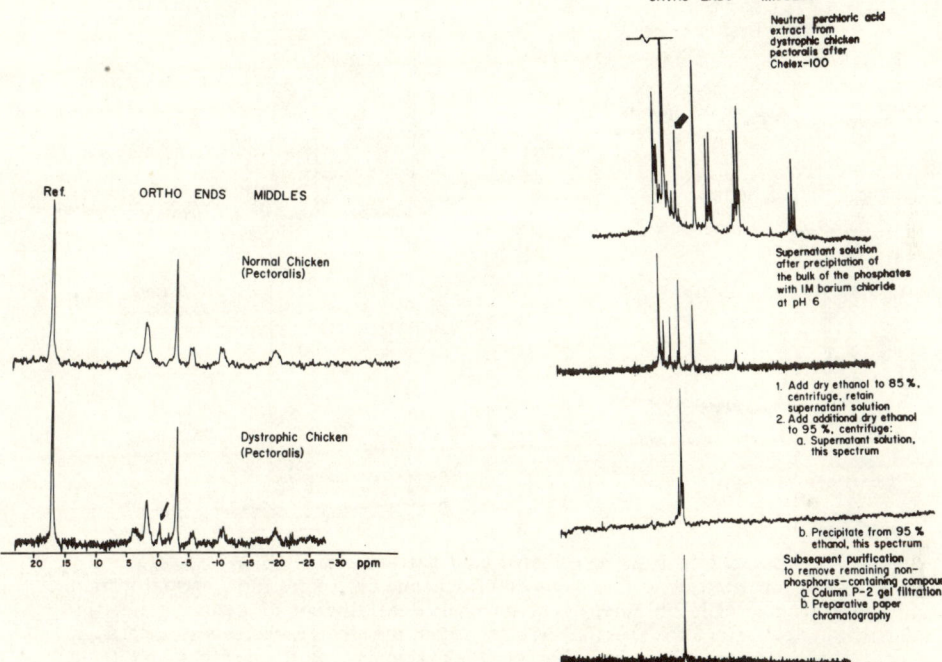

Figure 5. Comparison of phosphate spectral profiles of intact normal and dystrophic chicken breast muscles (Department of Avian Sciences, University of California, Davis; lines 200 and 304, respectively). The arrow at -0.4 ppm indicates a resonance arising from the dystrophic muscle but not from the normal control.

Figure 6. Phosphorus spectra taken at various stages during the purification of serine ethanolamine phosphodiester from dystrophic chicken breast muscle.

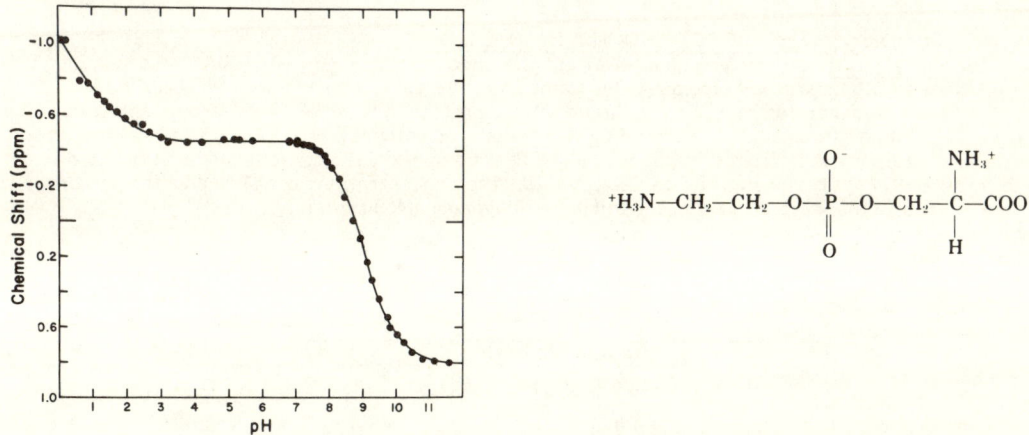

Figure 7. Phosphorus NMR-pH titration curve of 60 mmolar serine ethanolamine phosphodiester in pure water at 32°C with 1.0 and 6.0 normal HCl or KOH.

Figure 8. Serine ethanolamine phosphodiester (SEP).

<u>Duchenne Dystrophy</u>. The results from the chicken tissue samples encouraged us to pursue similar lines of research with human muscle biopsy samples. Fig. 9 shows the very first data we obtained from human muscle samples; the spectra were obtained from perchloric acid extracts of approximately 75 milligrams of tissue (Ref. 11). The data showed pronounced differences between the spectrum obtained from the Duchenne dystrophy and that obtained from a corresponding healthy muscle sample. In this Duchenne muscle, the high energy phosphates were essentially absent. To compensate, the muscle relied upon glycolysis for energy production; the sugar phosphate levels were elevated, and there was a marked elevation in the level of nicotine adenine dinucleotide, the key reducing element cofactor.

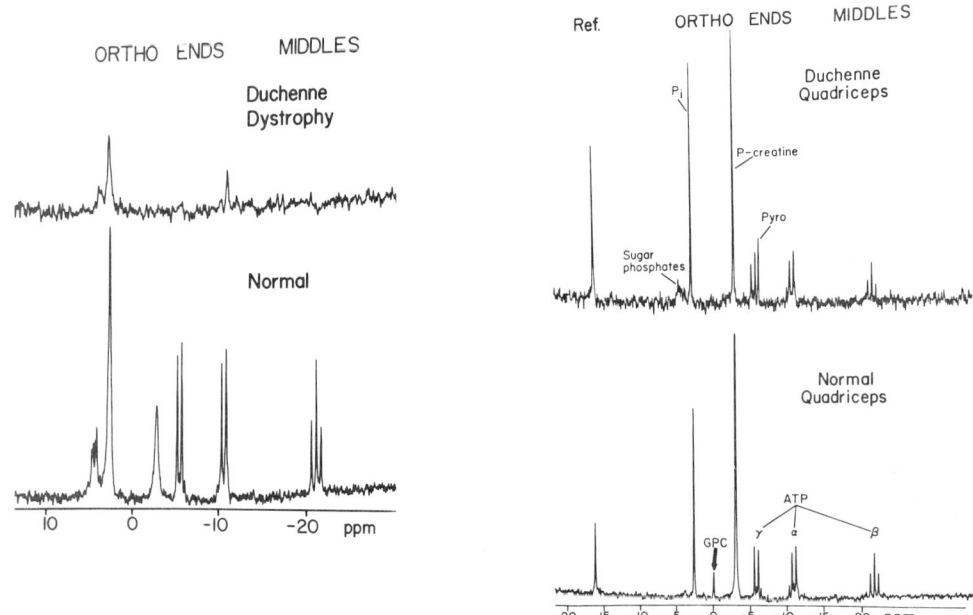

Figure 9. A phosphorus NMR perchloric acid extract profile of a vastus lateralis muscle from a patient diagnosed as having Duchenne dystrophy and a normal human muscle extract profile obtained from an equivalent amount of a similar healthy human muscle. Before the spectra were recorded, the samples were passed through Chelex-100 columns to further sharpen their signals. Both extracts and their phosphorus profiles were obtained by exactly the same procedures.

Figure 10. Comparison of phosphorus NMR spectroscopic profiles from perchloric acid extracts of human quadriceps muscle.

As it has developed, the Duchenne profile of Fig. 9 proved to be idiosyncratic. A Duchenne spectrum that is much more typical is shown in Fig. 10, which compares a human Duchenne quadriceps muscle profile to that from the healthy muscle. The common phosphate metabolites are present at similar concentrations in both types of tissues; however, the resonance of glycerol 3-phosphorylcholine (GPC) is unobservable in the profile from the diseased muscle (Ref. 12).

<u>Glycerol 3-Phosphorylcholine (GPC), $HOCH_2CH(OH)CH_2OP(O_2^-)OCH_2CH_2N^+(CH_3)_3$, as a Possible Marker for Human Dystrophies</u>. The characteristic phosphodiester of mammalian slow muscles, including those of man, is GPC. We have isolated this compound from beef heart and have observed it in frog, toad, tortoise, and human leg muscles, in rabbit soleus muscles, and in hearts of frog, rabbit, cat, dog, and rat. In contrast, we have not detected GPC in rabbit extensor digitorum longus or skeletal muscle. Tables 1 and 2 present data from phosphate profiles describing the GPC contents of healthy and diseased human muscle (Ref. 12). In addition to GPC, we list the concentrations of ATP, phosphocreatine plus inorganic orthophosphate, total phosphate, and non-collagenous (NC) protein.

TABLE 1. Phosphate Profiles of Normal Human Muscles

Muscle	GPC*	ATP*	PCr + Pi*	Total Phosphate*	NC Protein* (mg/g)	n
	(micromoles/g)					
Adult Biceps	ND**	4.9 ± 0.7	31.8 ± 10.5	51.3 ± 9.5	168	3
Adult Gastrocnemius	1.7 ± 0.7	4.9 ± 0.8	31.9 ± 4.1	53.3 ± 8.0	160	4
Adult quadriceps	0.8 ± 0.1	4.1 ± 0.5	30.4 ± 5.1	49.7 ± 7.4	156 ± 8	4
Child quadriceps	0.6 ± 0.2	5.0 ± 0.7	25.6 ± 2.1	47.3 ± 3.5	170 ± 13	10

* Results are means ± SE, where applicable.
**Not detectable.

Table 1 shows that GPC levels vary among muscle types under conditions where the adenosine triphosphate, phosphocreatine plus inorganic phosphate, total phosphate, and non-collagenous protein contents of these muscles are rather constant. Glycerol 3-phosphorylcholine is absent from biceps muscles, and exhibits its highest concentration in the gastrocnemius. Adult quadriceps muscle contains somewhat more GPC than do the quadriceps of children 1-16 years of age. In one case, a gastrocnemius muscle from a 9-year old boy contained 1.7 millimolar GPC.

Table 2. Phosphate Profiles of Diseased Human Muscles

Disease & Muscle	GPC*	ATP*	PCr + Pi*	Total Phosphate*	NC Protein*	n
	(micromoles/g)				(mg/g)	
Duchenne dystrophy, quadriceps and gastrocnemius	0.1 ± 0.2	1.7 ± 1.3	12.3 ± 6.5	22.9 ± 12.1	122 ± 11	30
Peripheral neuropathy, gastrocnemius	1.9 ± 0.8	3.4 ± 0.4	21.5 ± 1.2	37.5 ± 2.9	134 ± 6	4
Nonspecific muscle disorders, biceps	0.9 ± 0.2	3.7	18.9 ± 3.9	34.0 ± 9.0	209 ± 9	4

* Results are means ± SE, where applicable.

Table 2 summarizes results on the GPC contents of various diseased muscles (Ref. 12). Of primary interest to us is Duchenne human muscular dystrophy. We have now studied over 30 examples of Duchenne dystrophy, all verified by histologic examination. The data from Table 2 show that Duchenne dystrophic muscles do not contain GPC, that is, their GPC contents remain below the detection limits of our determinations. It should also be noted that the total phosphate contents of Duchenne muscles are decreased to about half normal values, and that their non-collagenous protein contents are decreased to 72% of control values in the case of children. To raise the sensitivity of our determinations, we have combined several Duchenne dystrophic muscle samples, concentrated the extracts, and examined them by NMR spectroscopy with protracted signal-averaging. Even under these conditions, it is not possible to detect the GPC resonance.

There are other human muscle diseases which may affect GPC levels. Fig. 11 shows comparative spectra obtained from intact human quadriceps muscles: a healthy control and a muscle afflicted with nemaline rod disease. The profile of the diseased muscle shows the absence of the GPC signal. Fig. 12 shows the phosphate profile of a muscle afflicted with Werdnig-Hoffmann disease. The Werdnig-Hoffmann disease is an example of a malady which causes a rather pronounced enhancement of the GPC level. In this particular example, virtually all the muscle's extractable phosphorus is in the form of GPC, the phosphocreatine and adenosine triphosphate levels being barely sufficient to sustain cellular metabolism (Ref. 12).

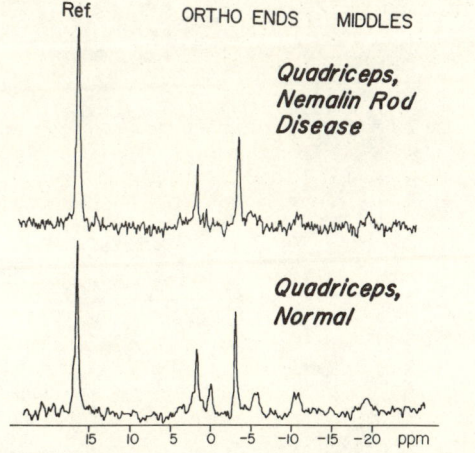

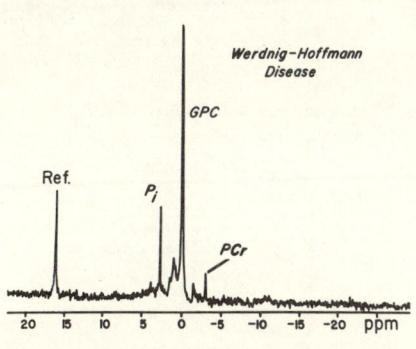

Figure 11. Intact human quadriceps muscle phosphate NMR profiles. The muscle sample for the top spectrum was obtained from a patient with nemaline rod myopathy; the bottom spectrum was obtained from a healthy human quadriceps muscle sample.

Figure 12. A phosphorus NMR perchloric acid extract profile of a gastrocnemius muscle from a patient afflicted with Werdnig-Hoffmann disease.

Other Phosphate Markers for Human Muscle Diseases. Fig. 13 shows a spectrum obtained from a human quadriceps muscle afflicted with an unknown myopathy. This particular disorder is characterized by the appearance of the resonance signal from pyrophosphate. Fig. 14 shows another example of a disease leading to the production of pyrophosphate. This spectrum was obtained from a muscle afflicted with limb girdle dystrophy, which is a myogenic disorder. The pyrophosphate resonance in this example amounts to approximately 2% of the total extractable muscle phosphorus. Fig. 15 shows the spectrum obtained from a Type II atrophy of undetermined origin. The two prominent resonances at 3.9 and 4 ppm arise from the sugar phosphate, fructose-1,6-diphosphate. In this example, the compound accounts for 66% of the total extractable phosphorus of the muscle. Other features which are observed are the absence of the resonance from phosphocreatine, at -3 ppm, and the elevated signal from dinucleotides, at -11.3 ppm. In this example, the total amount of extractable phosphorus was in the normal range.

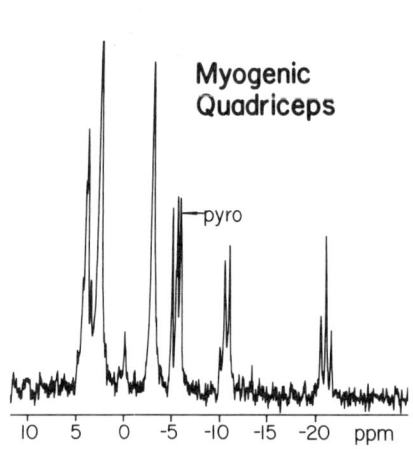
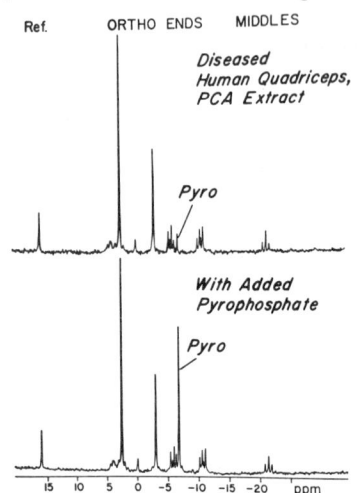

Figure 13. A phosphorus NMR perchloric acid extract profile of a quadriceps muscle from a patient afflicted with an unknown myopathy.

Figure 14. A phosphorus NMR perchloric acid extract profile of a limb girdle dystrophy. The top spectrum was obtained from the extract of the diseased muscle. The resonance from pyrophosphate occurs at -7.0 ppm. The bottom spectrum shows the spectroscopic identification of pyrophosphate. The addition of known pyrophosphate to the sample caused the resonance at -7.0 ppm to be uniformly enhanced.

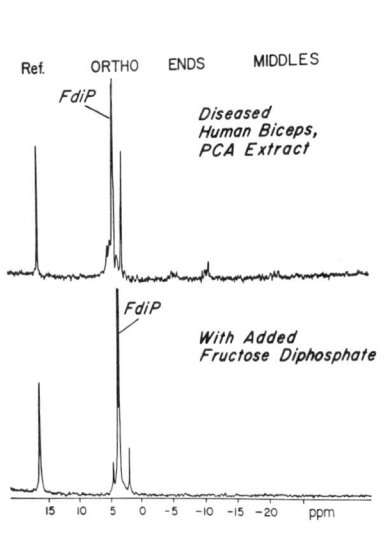
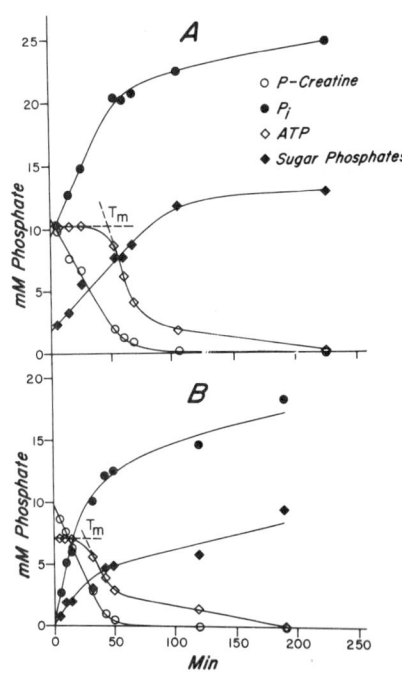

Figure 15. A phosphorus NMR perchloric acid extract profile of a Type II atrophy of undetermined origin. The top spectrum was obtained from the extract of the diseased human muscle. The pronounced signals at 3.9 and 4.0 ppm arise from fructose-1,6-diphosphate. The bottom spectrum shows the identification of the F-1,6-diP resonances through the addition of known F-1,6-diP.

Figure 16. Time-dependence of intact human muscle phosphates as determined by phosphorus NMR under anaerobic conditions at 31°C. (A) Healthy human quadriceps muscle. The maintenance time, Tm, is the point at which adenosine triphosphate begins to decrease. (B) Quadriceps muscle from a patient afflicted with nemaline rod myopathy.

The series of muscle spectra which have been presented above give some indication as to how P-31 phosphate profiles might be used in the classification of human disease types.

TIME-COURSE DATA

By measuring the phosphorus NMR spectrum of intact muscles at various times through the course of rigor, the changes in the metabolic state of the muscle can be observed (Ref. 17). Fig. 16A shows the time-dependence of phosphate metabolite levels of a healthy human quadriceps femorus muscle under the anaerobic conditions which prevail in our NMR experiments. For about 50 min., phosphocreatine, which is initially present at a concentration of 10 millimolar, decreases linearly to a concentration of about 2.3 millimolar, which is approximately 30% of its initial value; thereafter, its rate of decrease diminishes. Inorganic orthophosphate shows a corresponding increase from its initial value, also about 10 millimolar. The ATP concentration is maintained at a near constant level until it begins to decline in a sigmoidal fashion with time. The rate of decrease of phosphocreatine, which equals the rate of ATPase activity as long as ATP is constant, is roughly equal to the rate of inorganic orthophosphate formation. The parameter Tm, or maintenance time, which may be defined as the time corresponding to the intersection of the projected flat portion of the ATP curve with the projected slope of the rapidly decreasing portion, is a useful parameter characteristic of muscle metabolic decay rates under anaerobic conditions.

Fig. 16B corresponds to a time-course of a muscle taken from a patient with nemaline rod disease. The broad features of this time-course are the same as those determined on healthy muscle; however, by directly comparing the healthy muscle time-course to that from the diseased muscle, it can be seen that in the case of the diseased muscle, the phosphocreatine is lower initially and decreases at a greater rate. The inorganic orthophosphate is also at a lower initial concentration, but its rate of increase is higher than normal. The concentration of ATP declines more quickly, and its measured maintenance time is 24 minutes, or about one-half the control value. The phosphocreatine level at the Tm value in this case is only 53% of its initial concentration, indicating that in this diseased muscle, relatively higher phosphocreatine levels are required to maintain the ATP concentration.

In Fig. 17, the log of the percentage phosphocreatine concentration is plotted against time for isometrically-mounted unstimulated and caffeine-contracted frog gastrocnemius muscles. The data show that the utilization of phosphocreatine is essentially a first-order process. Although the time-course data presented represent an index of metabolic activity, the more striking changes involved in these caffeine contractures show even more clearly that the metabolism of chemical compounds, as they are occurring in intact cells and tissues, can actually be observed by phosphorus magnetic resonance.

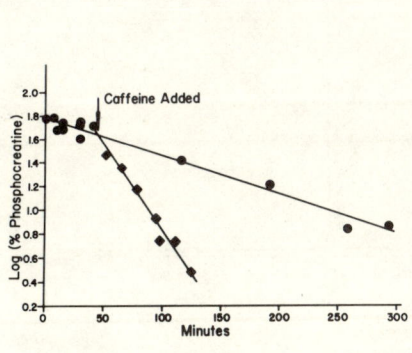
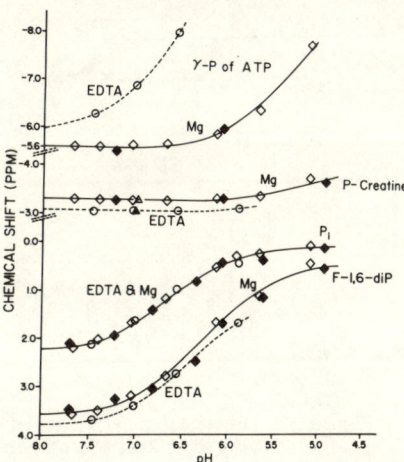

Figure 17. Phosphorus NMR time-course data for the utilization of phosphocreatine by isometrically mounted caffeine-contracted frog gastrocnemius muscles. The ordinate represents the log of the percent phosphorus present as phosphocreatine. The abscissa is time in minutes. The data from two sets of muscles are plotted in the figure: ●, not treated with caffeine; ◆, treated with caffeine at the point indicated.

Figure 18. Phosphorus-31 chemical shifts of various biological phosphates at $31^{\circ}C$ as a function of pH. ◇, a mixture made to approximate the concentration of metabolites in frog muscle: 3 mM ATP, 20 mM P-creatine, 2 mM F-1,6-diP, 0.1 M KCl, and 12 mM Mg ion. ◆, the above mixture with the omission of KCl. ○, the first mixture (with KCl) and with added 1mM EDTA and omitted 12 mM Mg ion. △, 30 mM P-creatine and 12 mM Mg ion. ▲, only 30 mM P-creatine. The phosphate anion of choice for intracellular pH measurements by phosphorus NMR is inorganic orthophosphate, P_i.

DETERMINATION OF INTRACELLULAR pH

At physiologic pH values, the chemical shifts of the weak acid phosphates, such as ATP inorganic phosphate, and the sugar phosphates, vary with the hydrogen ion concentration. Fig. 18 presents phosphorus magnetic resonance-pH titration data from an artificial mixture of biochemically relevant organic phosphates and inorganic ions present in the plasma of frog muscle (Ref. 7). At each pH value, there is a unique chemical shift for each phosphate metabolite (functional group); this is the reason for the several curves of the figure. In the physiologic pH range and in the presence of magnesium ions, there is no shift change in the resonance positions of either the gamma phosphate of ATP or phosphocreatine, whereas inorganic orthophosphate and fructose-1,6-diphosphate exhibit upfield shifts as acidity is increased from pH 7.5 to 5.0. This titration behavior makes the resonance signals from these phosphates suitable for determining intracellular pH values. Of the numerous phosphorus resonance signals which can be used for this purpose, that from inorganic orthophosphate is usually the best choice. First, this phosphate is detectable in all cellular preparations. Second, its chemical shift value is unaffected by the presence of the usual metabolic cations and anions. Thus, the chemical shift of inorganic orthophosphate is not altered by the presence of magnesium ions or calcium ions in the physiologic pH range. Also, the shift is not dependent on the presence of sodium or potassium ions. In fact, it appears that none of the solutes of the cellular sap have any effect on the inorganic phosphate chemical shift, although this hypothesis has not been rigorously tested.

Using such data, we have been able to assign an intracellular pH of 7.2 to frog gastrocnemius muscle and 7.2 to human gastrocnemius muscle, while Dawson, Gadian, and Wilkie (Ref. 13) assigned a pH value of 7.5 to the frog sartorius muscle. This same group also determined that the pH difference between two pools of inorganic phosphate in rabbit hind leg white muscle was 0.4 pH unit (Ref. 14). Similar work done at the Bell Laboratories has shown that cultured human cell lines assume the same intracellular pH as that of the medium in which they are growing (Ref. 15), whereas microorganisms maintain their intracellular pH against a gradient (Ref. 16). Fig. 19 shows phosphorus profiles of a human pectoralis major muscle where the separate P_i resonances correspond to two pH pools of inorganic orthophosphate (Ref. 17). Other work by the Bell Laboratories Group on isolated rat liver cells, involving metabolic inhibitors, has shown that the low-field resonance signal, corresponding to pH 7.5, arises from the inorganic phosphate of the mitochondria, whereas the high-field resonance, corresponding to pH 7.2, arises from the extra mitochondrial pool (Ref. 18).

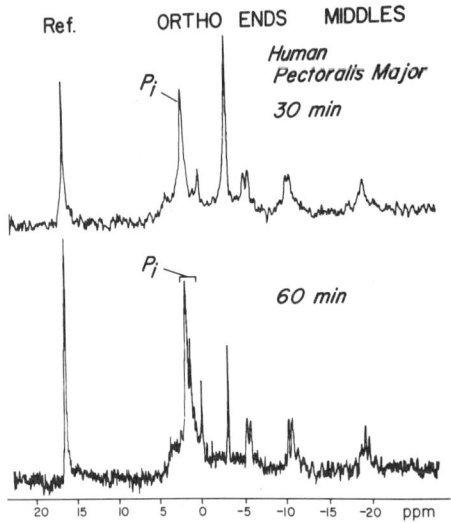
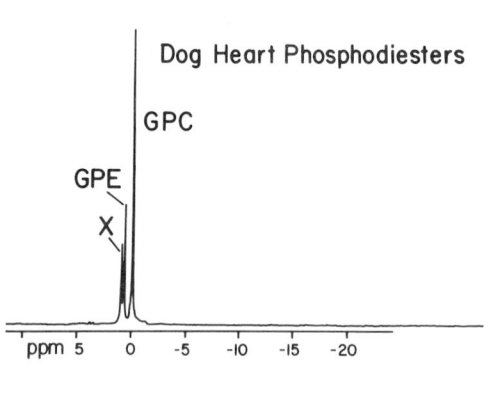

Figure 19. Phosphorus NMR profile from a portion of a human pectoralis major muscle showing the multiple orthophosphate resonances which can be resolved by the spectrometer after a period of time. The 30 min spectrum was accumulated in the period of time between 5 and 30 min after biopsy; the P_i resonance band is a single broad peak with a line width at half-height of 20 Hz. The 60 min spectrum was accumulated in the period of time between 30 and 60 min; the P_i resonance band is split into 2 signals, indicating at least two differential intracellular pH environments in the muscle. The width of the P_i resonance in both spectra, however, suggests the presence of yet other phosphate pH environments (Ref. 17).

Figure 20. Phosphorus NMR profile from a purified preparation of extracted dog heart phosphodiesters. For this spectrum, the pH of the sample was adjusted to 9.5 using perchloric acid and KOH as the titrants. (X) denotes the unknown N-derivative of glycerol 3-phosphorylethanolamine.

METABOLITES OF UNKNOWN CHEMICAL STRUCTURE

Phosphorus NMR spectra may uncover phosphorus-containing metabolites which have previously gone undetected by wet biochemical procedures. For example, the phosphodiesters (resonance position about 0 ppm) represent one such family of metabolites. In a number of muscles and other tissue sources there may be several phosphate species associated with this family of resonances (Ref. 13). Other phosphodiesters besides GPC and serine ethanolamine phosphodiester, which we have identified, are: glycerol 3-phosphorylserine, glycerol 3-phosphorylethanolamine (GPE), and a nitrogen-substituted derivative of GPE which is extremely sensitive to hydrolysis and which yields GPE upon hydrolysis (Ref. 19). The sources of yet other phosphodiester resonances, particularly those from neoplastic tissues (Ref. 20), remain unknown.

Fig. 20 shows the phosphorus profile of a purified preparation of phosphodiesters extracted from the dog heart. It should be noted that a virtually identical phosphate profile can be obtained from adrenal glands (Ref. 19). The resonance of an unknown phosphodiester is labeled "X". This compound exhibits a rate of hydrolysis at pH 7.2 at 31° of 3% per minute, which means that in the usual preparative procedures employed to detect metabolites, it would hydrolize completely to GPE. The phosphate's presence in intact tissues, therefore, would go undetected. If one examines an intact portion of brain tissue which has been chilled to 0° previous to its excision from the animal, the chemical shift value observed in the phosphodiester region is 0.87 ppm, which corresponds to the shift position of the GPE derivative. Upon perchloric acid extraction, one always observes both the resonance of GPE and its derivative. How much derivative is obtained depends on the care with which the extract is prepared and the pH to which the extract's acid is neutralized. Alkaline extracts, that is, extraction under conditions where the pH is greater than 9.5, of either brain or heart show virtually no GPE but only its derivative. The compound is stable in aqueous solution if the pH is kept above 9.5.

Fig. 21 shows a phosphate profile from a partially purified extract of muscle. Two unknown resonances are present which do not correspond to any of the usual biochemical phosphates. The resonance at -23.4 ppm has been identified as that from inorganic cyclic tetrametaphosphate (Fig. 22). The resonance at 10.1 ppm does not correspond to any ordinary biological phosphorus-containing metabolite that we are aware of, and its nature is essentially unknown at this time. Fig. 23 shows phosphate profiles obtained from perchloric acid extracts of perfused rat hearts where the experimental design involved manipulation of the cation content of the perfusate. If the perfusate contains either 20 millimolar calcium or 6 micromolar cadmium, the resonance at 11 ppm appears. In hearts perfused with the control buffer, the signal cannot be detected. Fig. 24 shows a phosphate profile of rat testes (Ref. 21). Again, the signal at 10 ppm is detectable, as well as a group of nucleoside diphosphosugars at -12.5 ppm. We are fairly certain that the resonances detected at 11 ppm, regardless of the tissue source, arise from the same chemical compounds. Fig. 25 shows an expanded scale spectrum of the region at 11 ppm determined on an extract of cadmium-treated liver. At this time, all that can be said concerning these signals is that there appear to be at least two separate compounds present. The group of signals at 18 ppm, which were discovered through this experiment, are characteristic of signals obtained from phosphonic acid residues in peptidoglycans (Ref. 22).

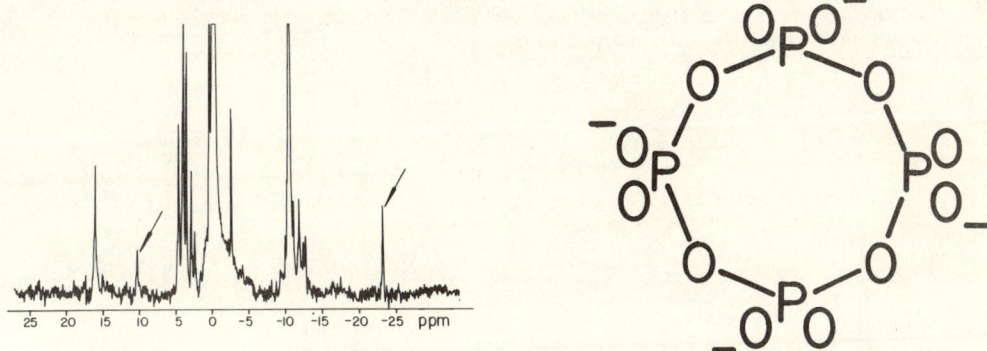

Figure 21. Phosphorus NMR spectral profile of perchloric acid extracted chicken pectoralis muscle phosphates which are not precipitated by barium acetate at pH 6. The arrows denote resonances from previously unrecognized muscle phosphorus-containing molecules. The signal at 10.1 ppm has not been characterized; the resonance at -23.4 ppm is that from inorganic tetrametaphosphate. The large peaks are not on-scale so that the minor resonances may be clearly shown.

Figure 22. The cyclic tetrametaphosphate anion.

PHOSPHATE PROFILES FROM OTHER MAMMALIAN TISSUES AND INTACT CELLULAR SYSTEMS

Endocrine Tissues. Figs. 26 and 27 show phosphorus spectra from intact rat testis and intact dog adrenal tissues, respectively. These profiles exhibit two features which, thus far, appear to be unique to endocrine tissue spectra. The first, which is easily observed in the spectrum from the adrenal gland, is that the gamma and beta phosphate resonances of ATP are extremely difficult to detect. They are, however, readily observed in the corresponding perchloric acid extract. The loss of the resonance signal from these phosphates in the intact tissue spectrum indicates that there exists a specific interaction involving some component in the cell which is unique to endocrine tissue and which results in a quenching of the phosphate resonance. The second feature is that both types of endocrine tissues give rise to a resonance band at 3.2 ppm, which we believe arises from phosphorylated proteins. The resonance position is characteristic of serine phosphates, and the signal cannot be detected in the corresponding perchloric acid extract profiles. Stimulation of the adrenal gland enhances the resonance, and this change is in accord with the known biochemistry of the adrenal gland, where stimulation is known to promote an active phosphorylation of proteins.

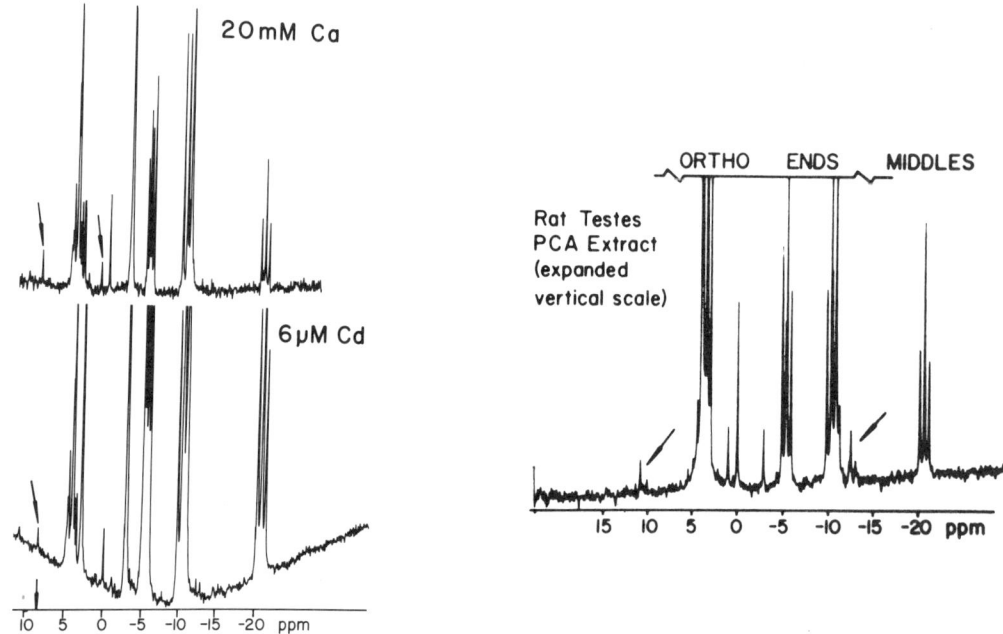

Figure 23. Phosphorus NMR profiles of perchloric acid extracted perfused rat hearts. The indicated cation chlorides were added to the Hartmann's modified Ringer's solution in the concentrations indicated. The arrows indicate unidentified phosphorus resonances. The major resonances of the spectra are off-scale for the purposes of the illustration.

Figure 24. Phosphorus NMR profile of perchloric acid extracted rat testes. The arrows indicate unidentified phosphorus resonances.

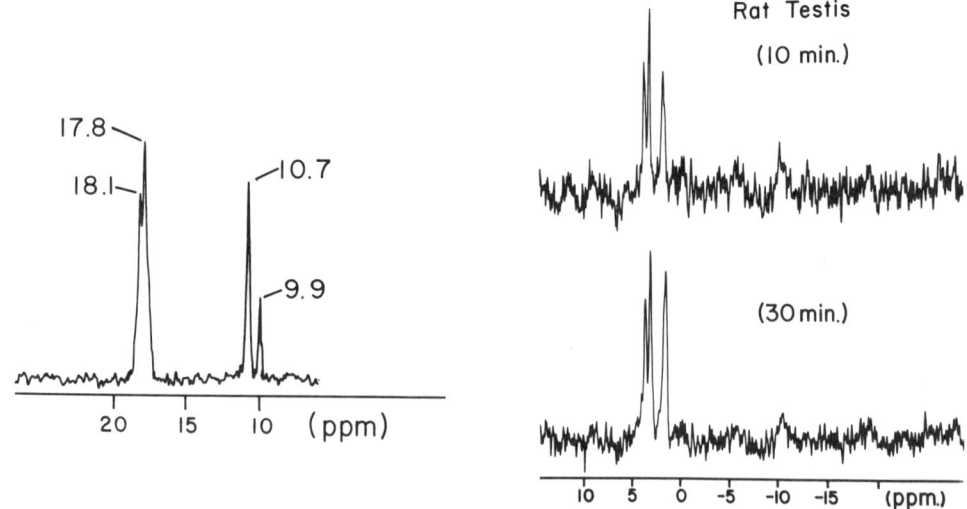

Figure 25. A 5 to 25 ppm portion of a liver perchloric acid phosphorus NMR profile obtained by extensive signal averaging. These resonances are unidentified; however, the resonances at 18 ppm are characteristic of signals obtained from natural phosphonic acid residues in peptidoglycans. Such glycans are known to contain 2-aminoethylphosphonic acid, and phosphonoalanine, $(OH)_2P(O)CH_2CH(NH_2)COOH$.

Figure 26. Phosphorus NMR profiles of an intact rat testis taken in the period of time from 2-10 min and 10-30 min after sacrifice of the animal. Of the three prominent resonance signals between 0 and 5 ppm, the center signal arises from phosphoproteins.

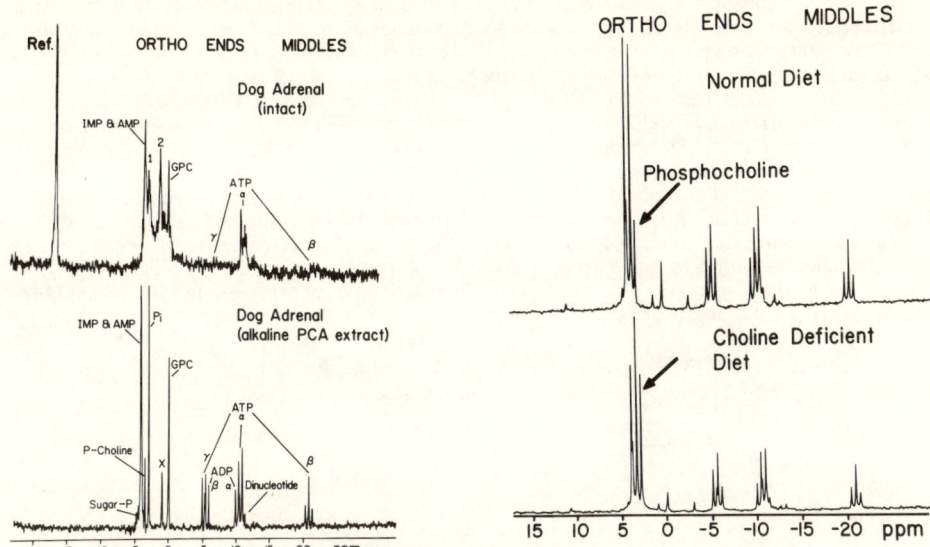

Figure 27. Phosphorus NMR profiles obtained from the intact dog adrenal gland and its corresponding perchloric acid extract. Signals 1 and 2 arise from perchloric acid insoluble material; signal 1 is at the characteristic resonance position of protein-bound phosphoserine residues. IMP and AMP indicate, respectively, inosine and adenosine monophosphates.

Figure 28. Phosphorus NMR profiles of extracted rat testes showing the effect of a choline deficient diet.

Fig. 28 shows the effect of a dietary study on the phosphate profiles of extracted rat testes. In this study, one group of rats was fed a diet deficient in choline. When rats are denied an adequate choline intake, the level of GPC, which is relatively high in the rat testes, is diminished, as one would anticipate. The level of phosphorylcholine, however, is enhanced about 40% over the control value. The other phosphate metabolites are essentially unaffected by the treatment.

Blood Cells and Sea Urchin Eggs. Fig. 29 shows phosphorus spectra obtained from intact rabbit erythrocytes, rabbit reticulocytes, and sea urchin eggs, and an extract from human platelets. A number of the resonances observed in these spectra are similar to those seen in the spectra from intact muscles. For example, the sugar phosphate resonance band at 3.7 ppm is observed in each of the spectra, as well as the signal from inorganic orthophosphate; however, the amounts of these, relative to the total phosphate profile, vary considerably. In erythrocytes and reticulocytes, for example, the inorganic orthophosphate signal is relatively weak, and appears as a small shoulder on the upfield side of the high-field signal from 2,3-diphosphoglycerate, which is a specialized phosphate associated with these systems. In the spectra from the sea urchin eggs and platelets, however, the inorganic orthophosphate signal is considerably enhanced and readily distinguished from the other phosphates in the spectrum.

ATP cannot be detected in the spectrum from intact sea urchin eggs, but it is readily observed in the profiles from the blood components. The sea urchin egg spectrum shows the resonance from phosphocreatine (-3.2 ppm), which could not be detected in the spectrum from the three blood components. This spectrum also shows a large signal due to the stored phospholipids of the microcrystalline lipovitelin-phosvitin complex. The principal resonance in this band arises from the lecithin phosphates; the downfield shoulder arises principally from, the phospholipid sphingomyelin. The upfield group of resonances, at about -11 ppm, arises from the P,P-diesterified pyrophosphate groups of dinucleotide and related cofactors.

The spectroscopic profiles from both the reticulocytes and platelets show the resonance signals from the phosphodiesters in the region of 0 ppm. The available spectroscopic evidence, along with some thin-layer chromatographic data, suggest that the compounds giving rise to these resonances are the same as those observed in the profiles obtained from muscle, i.e., GPC, GPE, and related phosphodiesters.

Human Neuroblastoma Clones. Fig. 30 shows phosphorus NMR spectra obtained from two strains of neuroblastoma clones, with and without the addition of cyclic adenosine monophosphate (AMP) to the culture medium. The top two profiles were obtained from the N-18 clonal line, which is a cellular type shown to differentiate through neurite formation when grown in the presence of cyclic AMP and observed in the confluent stage. The profile from the top spectrum was obtained from cells grown in the absence of cyclic AMP. Spectrum B was obtained from cells grown in the presence of 1 millimolar cyclic AMP. The bottom two spectra were obtained from the C-46 clonal line which does not morphologically change to any significant degree under these conditions.

The profiles of these cells resemble those obtained from other mammalian tissue types in that they show the sugar phosphate resonance band, inorganic orthophosphate, phosphodiester resonance bands, and condensed phosphate resonance signals from ATP. There are, however, numerous differences both between the spectra from the two cell types and those obtained from other mammalian cells, and among the spectra of this particular study. For example, in Profile A, there is a strong resonance at -0.65 ppm which arises from phosphodiesters of unknown nature. These phosphodiesters, of which there are two principle components, are not related to GPC or its relatives. Our best guess at this time is that this particular signal arises from adenosine monophosphorylneuraminic acid and, perhaps, one of its close analogs. In Profiles B, C, and D, there is an additional resonance band in the esterified end-group phosphate region around -13 ppm. These resonances arise from various nucleoside diphosphosugars, at least one of which is uridine diphosphoglucose. Note how the phosphate profile reflects the effect of cyclic AMP on the N-18 clonal line. The cyclic nucleotide markedly diminishes a number of the sugar phosphates while enhancing the amount of inorganic orthophosphate. It depresses the concentration of the -0.6 ppm phosphodiester while generating a considerable amount of creatine phosphate. The large signal in the unesterified end-group region at -5 ppm arises from pyrophosphate, and pyrophosphate is not a component of the N-18 profile cultured in the absence of cyclic AMP.

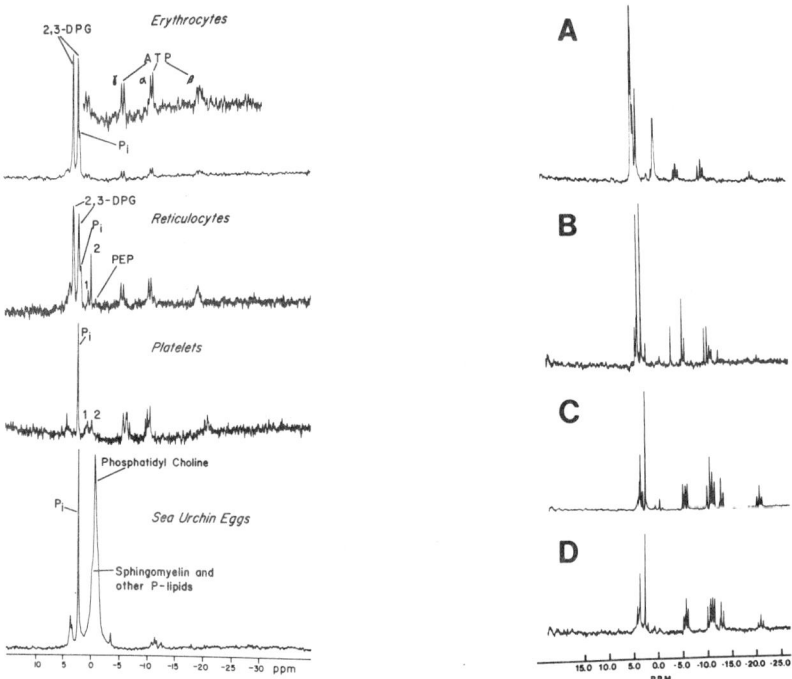

Figure 29. Phosphorus NMR profiles of intact blood components and sea urchin eggs: 2,3-DPG, 2,3-diphosphoglyceric acid; PEP, phosphoenolpyruvic acid; signal 1, uncharacterized phosphodiester resonance; signal 2, GPC. The platelet profile shows the high ADP/ATP ratio characteristic of these cells; the sea urchin egg profile shows the phospholipid resonance band, which is ordinarily unobservable in high-resolution P-31 spectra obtained from intact cellular systems.

Figure 30. Phosphorus NMR perchloric acid extract profiles of human neuroblastoma clonal lines: A, N-18 clone cultured without added cyclic AMP and harvested in the log phase; B, N-18 clone cultured with 1.0 mmolar cyclic AMP and harvested in the confluent phase; C, C-46 clone cultured without cyclic AMP and harvested in the log phase; D, C-46 clone cultured with 1.0 mmolar cyclic AMP and harvested in the confluent phase. Note the pronounced phosphodiester resonance, at 1 ppm, in profile A and the unsymetrical pyrophosphate resonance multiplets, at -12.5 ppm, in profiles C and D.

While we do not know the precise signal identities of all of the new phosphates which appear in these profiles, we feel that they represent the nucleoside sugar intermediates used in the pathway for glycoprotein synthesis, and that at least one manifestation of the neoplastic transformation is an alteration in glycoprotein synthesis. If this is the case, the profiles indicate that at least one biochemical difference between the various neuroblastoma clonal lines is found in the inhibition of the glycoprotein pathway at specific points along the biosynthetic route.

Rat Kidney and Liver. Fig. 31 shows a phosphorus profile obtained from the rat kidney. The kidney profile is quite similar to that of the muscle in that it shows the usual signals from the sugar phosphates, inorganic orthophosphate, and the nucleoside polyphosphates. In addition, it exhibits relatively elevated levels of GPC and dinucleotides. The data of this slide reflect the effect of low level chronic feeding of cadmium, a heavy metal toxin. The spectra show the diminuition of the GPC resonance in response to cadmium feeding.

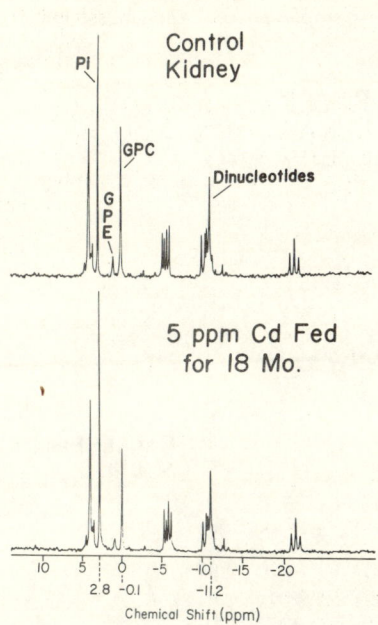
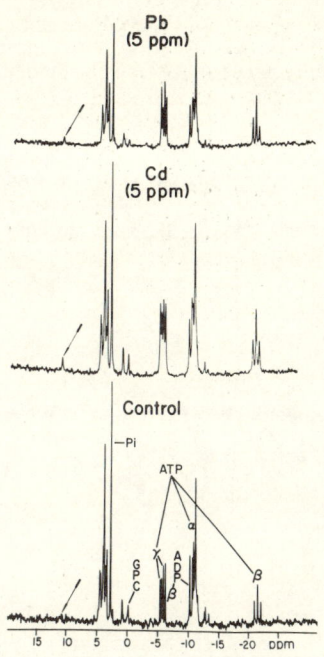

Figure 31. Phosphorus NMR perchloric acid extract profiles of freeze-clamped rat kidney. The bottom spectral profile shows the effect on kidney phosphates of chronic low level cadmium feeding; note the depressed GPC signal.

Figure 32. Phosphorus NMR perchloric acid extract profiles of freeze-clamped rat liver. The top two spectral profiles illustrate the effect on liver phosphates of chronic heavy metal feeding. Note the enhanced resonance at 11 ppm, the enhanced dinucleotide resonance at -11.5 ppm, and the increased ATP/ADP ratios in the heavy metal-fed liver profiles over that of the control.

Fig. 32 shows liver profiles obtained from another chronic heavy metal toxin feeding study. The liver profile is rich in sugar phosphates and inorganic phosphate, and relatively poor in the phosphodiesters. The signal at 0.8 ppm arises from GPE which is relatively enhanced in the liver. The ATP/ADP ratio is nearly unity in the liver, and this is reflected in the ionized end-group phosphate resonance band where the doublets of ATP and ADP are nearly identical. Liver is also rich in nucleoside diphosphocofactors, and this is reflected in the esterified end-group phosphate band where there is a strong resonance due to the cofactors of liver.

MAMMALIAN FLUIDS OTHER THAN BLOOD: AMNIOTIC FLUID AND SALIVA

Fig. 33 shows comparative phosphate profiles of human amniotic fluid taken during the third trimester. The control spectrum shows little more than the inorganic orthophosphate signal. The profile from the diabetic mother, however, shows additional resonances at 1.3 and 8.9 ppm. The 1.3 ppm resonance is characteristic of an anomeric sugar phosphate. The 8.9 ppm signal, which is unidentified as to its source, probably arises from monoesterified sugar phosphates of polymeric sugar molecules. This signal is not the same as those seen at 11 ppm in other tissues.

Fig. 34 shows phosphate profiles obtained from pelleted human salivary proteins which were resuspended in a dilute aqueous EDTA solution. The top spectrum was obtained from a middle-aged male with a history of dental caries and relatively good oral hygienic habits. The principal signal arises from salivary inorganic orthophosphate. The other resonances arise from various phosphomono- and diesters. The monoester resonance band appears at 3.95 ppm and corresponds to approximately 43% of the esterified phosphate component. The bands at 1.1 ppm and 0.6 ppm correspond to two different sets of orthodiester functional groups. These spectral patterns are similar to those observed from yeast glycans where the phosphate serves as a linkage between chains of polymerized sugar residues. The quantity present in the sample, however, precludes any possibility that these signals arise from sloughed cell wall polysaccharides of the oral flora.

The bottom spectrum was obtained from the saliva of an adolescent female exhibiting a caries-free oral cavity, but who also had a history of poor oral hygiene as well as a well-documented diet rich in carbohydrates. In this case, the phosphorus profile shows all of the resonances observed in the profile from the adult male and, in addition, resonances to quite low field at 9.44 and 8.20 ppm and an additional resonance in the diester region at 0.67 ppm. At this stage of the investigation, it cannot be said what the nature of the molecules are that give rise to these additional resonances, except to say that because they appear in pelleted material, they most probably arise from relatively high molecular weight components. We do not have enough cases at this time to state whether there is any correlation between resistance to caries infection and the appearance of these phosphodiester resonances, but the reader must certainly agree that this is a subject worth pursuing.

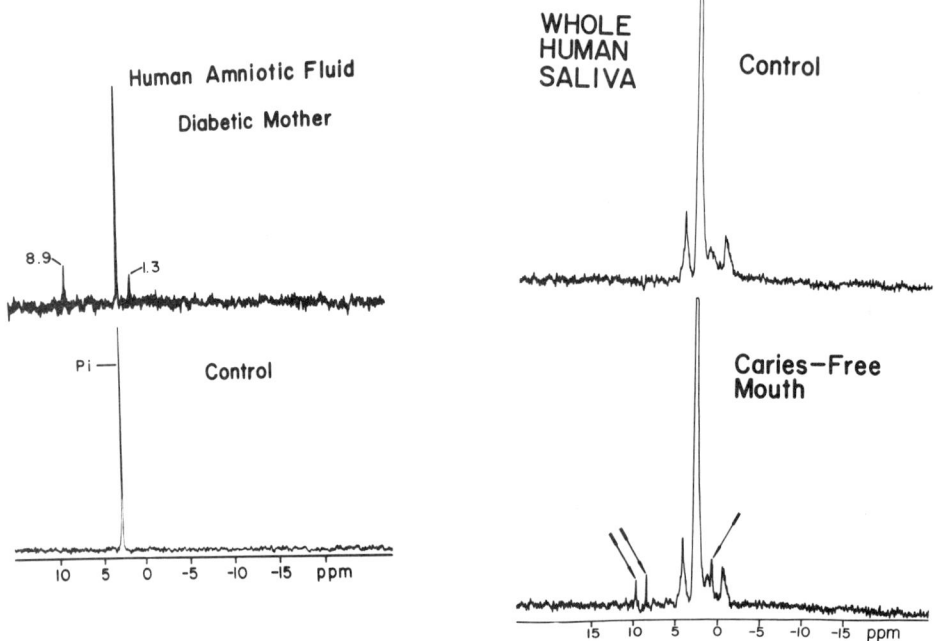

Figure 33. Phosphorus NMR spectral profile of whole human amniotic fluid obtained during the third trimester. Amniotic fluid obtained from the diabetic mother gives rise to additional phosphorus resonances at 8.9 and 1.3 ppm; only inorganic orthophosphate can be detected in the amniotic fluid from the healthy mother.

Figure 34. Phosphorus NMR spectral profiles of whole human saliva after it has been concentrated by lyophilization. The control spectrum was obtained from a middle-aged male with a common history of dental caries. The bottom spectrum was obtained from a patient with a caries-free oral cavity. Several phosphates, indicated by the arrows, are present in this profile which are not detectable in the control profile. The major resonance, which is off-scale in both profiles, arises from inorganic orthophosphate, the principal salivary phosphate.

PROSPECTS FOR THE FUTURE

The data which have been presented generalize the analysis of intact tissues and their extracts by phosphorus-31 NMR and suggest that this method will contribute significantly to the furtherance of biomedical knowledge. Work is already in progress in a number of laboratories around the world: there are muscle studies being conducted in England at Oxford (Ref. 13 and 23), at the Johns Hopkins Medical School in Baltimore (Ref. 24), and at the University of Illinois in Chicago (Ref. 11); tumor studies at the State University of New York in Brooklyn (Ref. 25), at the Bell Laboratories in Murray Hill (Ref. 15), at the University of Texas Health Science Center in Dallas (Ref. 20), and at the University of California in La Jolla (Ref. 26); nerve studies at the University of Pennsylvania; endocrine gland studies at the University of Illinois in Chicago (Ref. 19); yeast (Ref. 27) and bacterial (Ref. 16) studies at the Bell Laboratories in Murray Hill; and studies on blood and its components at a large and growing number of research institutions.

Magnetic resonance research of tissues will continue to expand for several reasons:

First, there are a large number of phosphate resonance signals which have been observed in intact tissues and their extracts which cannot be identified with any of the common phosphate metabolites of biochemistry, and it will be some time before all of these resonances are assigned to specific molecular structures. Along these lines, phosphate profiles from intact tissues have shown resonances which present us with even more perplexing problems as to their nature, since these resonances disappear when the cellular integrity is disrupted.

Second, magnetic resonance research provides a method for the in vivo verification of metabolic processes worked out through in vitro procedures. In some instances, for example, the myokinase catalyzed equilibrium $2ADP \rightleftarrows AMP + ATP$ in the ascites tumor cell, there is good agreement between in vitro and phosphorus-31 spectroscopic in vivo measurements (Ref. 15). Similarly, studies on the breakdown of phosphocreatine in aging muscle indicate clearly that the equilibrium $ADP + PCr \rightleftarrows ATP + Cr$

is shifted far to the right, as expected from the known equilibrium constant of the Lohmann reaction in vitro.

Third, phosphorus-31 spectroscopy permits the analysis of in vivo biochemical rate processes at the molecular level. Thus, the utilization of creatine phosphate during muscle contraction and relaxation has been determined by an elegant set of experiments by Professor Wilkie and collaborators (Ref. 15). The decay of phosphocreatine has also been followed during the development of aging in healthy and diseased muscle, which leads to estimates of the resting ATPase activity (Ref. 11).

Finally, phosphorus-31 spectroscopic profiles can be used in the diagnosis of diseases or to test tissue responses to drug treatment. The application to the diagnosis of diseases has considerable potential, in that the phosphate profiles of diseased tissues are frequently dramatically different from those of healthy tissues, often showing resonances from unusual phosphates which may serve as markers for the disease.

Technological Advances: Cryogenic Spectrometers. There have been many technological advances in the field of NMR spectroscopy which have made possible its eventual application to in vivo analysis. When it is considered that the field of magnetic resonance spectroscopy is only a little over twenty years old, its rate of development is seen to be impressive. Yet, today, the rate of NMR development, far from slowing down, is accelerating.

The most impressive recent gains have been made in the field of high-resolution cryogenic magnet technology. It is now possible to attain magnetic fields as high as 84 KGauss (360 MHz for H-1 resonance) without loss of field homogeneity, or work with large volume, >30 mm spinning sample tubes, again without significant loss of magnetic field homogeneity.

The use of higher magnetic fields has two effects. One, resolution is enhanced, since the chemical shift, expressed in Hz, is proportional to the magnetic field strength. This means that if signal widths remain the same as the magnetic field is increased, there will be greater separation between signals at higher field strengths. Two, the signal-to-noise ratio, which is often the limiting factor in NMR analysis, also increases with the magnetic field, and so, with all other factors being equal, the use of more powerful magnets results in enhanced spectrometer sensitivity.

NMR signals are also proportional to the number of nuclei undergoing resonance, and so it follows that larger samples will yield higher signal-to-noise ratios.

The results of applying such technological advances to the study of intact tissues is illustrated in Fig. 35. The figure shows the phosphorus-31 spectrum of an intact chicken pectoralis muscle determined at 60.75 MHz (150 MHz for H-1) in a spectrometer using 20 mm spinning sample tubes. The analysis required 5 min and should be compared to the spectrum of Fig. 3. Note that the triplet structure of the beta phosphate of ATP can now be discerned, and small peaks are observable in the bases of the bands at -3 and -12 ppm. The sugar phosphate band shows a well-defined resonance, and two nearly-equal pools of inorganic orthophosphate (1.7 ppm) are visible. The additional detail seen in this spectrum relative to that of Fig. 3 results from the much improved signal-to-noise ratio and greater resolving power of the high-field instrument. Note in Fig. 35 that the non-equivalence of the beta-gamma and alpha-beta couplings of ATP in the intact muscle are readily observed: in comparing the splitting in the gamma and alpha group doublets, it can be seen that the beta-gamma splitting is greater than the alpha-beta splitting.

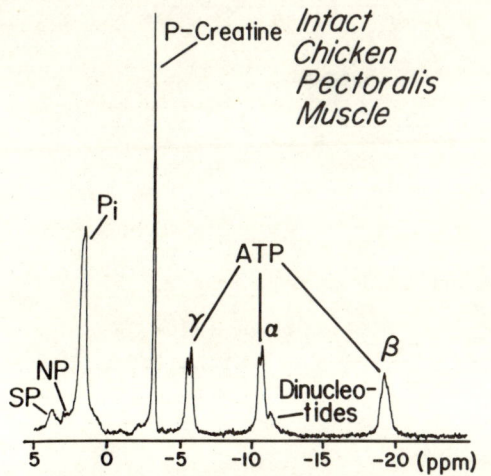
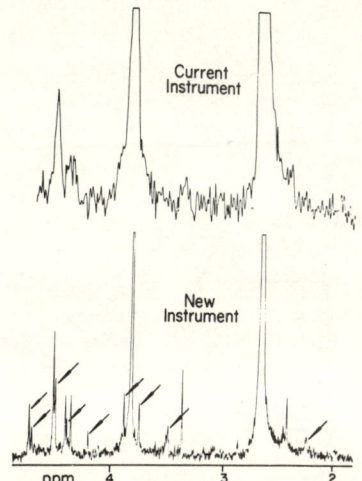

Figure 35. Phosphorus NMR spectral profile of an intact chicken pectoralis muscle taken with a cryogenic spectrometer system at 60.75 MHz for P-31 using 20 mm spinning sample tubes and equipped for quadrature detection (Ref. 28 & 29). The sample, at 24°C, was spinning during the analysis; the spectrum was obtained in 5 min. Note the well resolved J couplings in the ATP alpha and gamma group doublets and the presence of two discernible inorganic orthophosphate signals. (Thanks to L. F. Johnson and the Nicolet Technology Corporation.)

Figure 36. An expanded portion of a phosphorus NMR spectral profile showing the sugar phosphate region of a fractionated frog muscle perchloric acid extract. The top spectrum was obtained on a Bruker HFX-5 spectrometer (90 MHz for H-1). The bottom spectrum was obtained from the same sealed sample (10 mm spinning sample tube) using a Bruker cryogenic spectrometer system (270 MHz for H-1). Both spectra were taken under identical sweep conditions, at the same temperature, and for the same length of time; they are displayed on the same ppm scale. The arrows denote resonance signals which are hidden in the spectral noise of the current instrument spectrum.

Fig. 36 shows, on an expanded scale, the sugar phosphate region from a frog muscle perchloric acid extract phosphate profile. The figure compares the performance of two spectrometers in analyzing the same sample under identical sweep conditions, at the same temperature (31°), and for the same length of time. The top spectrum was run on an eleven-year old 90 MHz spectrometer; only three signals can be determined with certainty. The bottom spectrum was run on a state-of-the-art superconducting 270 MHz spectrometer. Note the much-improved signal-to-noise ratio and resolving power of the new instrument. Without doubt, the imaginative application of such technology to the study of living systems will ensure a better understanding of life processes and the infirmities to which life is heir.

Acknowledgment. The author wishes to acknowledge the support of his collaborators, without whom it would have been impossible to develop applications of phosphorus NMR to so wide a field of subjects. Special thanks to M. J. Hustad.

REFERENCES

1. J. R. Van Wazer, Bull. Soc. Chim. Fr. 1968, 1732-1739 (1968).
2. T. Glonek, J. R. Van Wazer, and T. C. Myers, Bioinorg. Chem. 1, 1-22 (1971).
3. T. Glonek, T. C. Myers, and J. R. Van Wazer, Bioinorg. Chem. 1, 23-34 (1971).
4. M. M. Crutchfield, C. H. Dungan, J. H. Letcher, V. Mark, and J. R. Van Wazer, Top Phos. Chem. 5, 1-74 (1967).
5. R. B. Moon and J. H. Richards, J. Biol. Chem. 248, 7276-7278 (1973).
6. M. Bárány, K. Bárány, C. T. Burt, T. Glonek, and T. C. Myers, J. Supramolecular Struct. 3, 125-140 (1975).
7. C. T. Burt, T. Glonek, and M. Bárány, J. Biol. Chem. 251, 2584-2591 (1976).
8. J. M. Chalovich, C. T. Burt, S. M. Cohen, T. Glonek, and M. Bárány, Arch. Biochem. Biophys. 182, 683-689 (1977).
9. M. M. Crutchfield, C. F. Callis, R. R. Irani, and G. C. Roth, Inorg. Chem. 1, 813-817 (1962).
10. T. Glonek, P. J. Wang, and J. R. Van Wazer, J. Amer. Chem. Soc. 98, 7968-7973.
11. C. T. Burt, T. Glonek, and M. Bárány, Science 195, 145-149 (1977).
12. J. M. Chalovich, C. T. Burt, M. J. Danon, T. Glonek, and M. Barany, Ann. N.Y. Acad. Sci. 317, 649-669 (1979).
13. J. Dawson, D. G. Gadian, and D. R. Wilkie, J. Physiol. 267, 703-735 (1977).
14. P. J. Seeley, S. J. W. Busby, D. G. Gadian, G. K. Radda, and R. E. Richards, Biochem. Soc. Trans. 4, 62-64 (1976).
15. G. Navon, S. Ogawa, R. G. Shulman, and T. Yamane, Proc. Natl. Acad. Sci. USA 74, 87-91 (1977).
16. G. Navon, S. Ogawa, R. G. Shulman, and T. Yamane, Proc. Natl. Acad. Sci. USA 74, 888-891.
17. S. J. W. Busby, D. G. Gadian, G. K. Radda, R. E. Richards, and P. J. Seeley, Biochem. J. 170, 103-114 (1978).
18. R. G. Shulman, T. R. Brown, K. Ugurbil, S. Ogawa, S. M. Cohen, and J. A. den Hollander, Science 205, 160-166 (1979).
19. T. Glonek, and S. F. Marotta, Hormone and Metabolic Res. 10, 420-424 (1978).
20. J. W. Pettegrew, T. Glonek, F. Baskin, and R. N. Rosenberg, Neurochem. Res., in press.
21. T. Glonek and S. F. Marotta, 1978 Fed. Proc. Abs. 37, abs. #3342.
22. R. L. Hilderbrand, T. O. Henderson, T. Glonek, and T. C. Myers, Biochemistry 12, 4756-4762 (1973).
23. D. F. Hoult, S. J. W. Busby, D. G. Gadian, G. K. Radda, R. E. Richards, and P. J. Seeley, Nature (London) 252, 285-287 (1974).
24. W. E. Jacobus, G. J. Taylor, IV, D. P. Hollis, and R. L. Nunnally, Nature 265, 756-758 (1977).
25. K. S. Zaner and R. Damadian, Science 189, 729-731 (1975).
26. F. E. Evans and N. O. Kaplan, Proc. Natl. Acad. Sci. USA 74, 4909-4913 (1977).
27. J. M. Salhany, T. Yamane, R. G. Shulman, and S. Ogawa, Proc. Natl. Acad. Sci. USA 72, 4966-4970 (1975).
28. S. J. Opella, Science 189, 158-165 (1977).
29. T. Glonek, Biochem. Med. 19, 246-251 (1978).

OLIGODEOXYRIBONUCLEOTIDE HIGH SPEED SOLID PHASE SYNTHESIS AS A MOLECULAR BIOLOGY RESEARCH INSTRUMENT

V. K. Potapov, V. P. Veiko, O. N. Koroleva, S. I. Turkin and Z. A. Shabarova

Chemistry Department and A. N. Belozersky Laboratory of Molecular Biology and Bioorganic Chemistry, Moscow State University, Moscow, U.S.S.R.

Abstract - Investigations have been carried out on oligonucleotide synthesis scheme which is based on a successive internucleotide phosphate group protection by aniline residues with the application of a graft type polymer supports. The efficiency of this method has been demonstrated in the synthesis of 10 oligodeoxyribonucleotides, 8 to 15 residues in chain length. The rate of chain propagation is two nucleotides per twenty four hour period. The possible applications of the polimeric condensing agent made by the sulfonylchlorination of polystyrene grafted on a teflon matrix are discussed. Synthesized oligonucleotides have been used for reverse transcription, investigations of the secondary and tertiary structure of viral RNA, for improving the ligase coupling method, and polynucleotide chain cleavage by restriction enzymes.

INTRODUCTION

The progress achieved in the field of the solid phase synthesis of oligodeoxyribonucleotides (1-3) over the last two or three years affirms that the method under discussion can be successfully used to get preparative quantities (3-5 umol) of oligonucleotides of definite base sequence 8 to 15 in chain length. The standartization of synthesis procedures and the possibility to make the process as a whole automated will result in the further progress of the method. It is necessary to mention that out of the two main problems of the solid phase synthesis, i.e. the choice of an optimum polymer support and the method of internucleotide linkage formation, the former at present comes to a close. In fact, both the polar acrylamide supports (1,2) and graft copolymers of the styrene type (3,4) permitt the execution of the solid phase synthesis with sufficiently high and stable yields. At the same time up to now there is no unanimous opinion concerning the optimum polynucleotide chain growing scheme as used in the solid phase synthesis. Therefore most of the experiments in this field have been made according to the classical diester approach (5). The main drawback of this approach is the possible interaction of the condensing agent excess with other components of the reaction mixture and the presence of a number of side reactions resulting in oligonucleotide chain destruction, sulfonylation of the 3'-hydroxy group of a nucleoside component and the polymerization of the reaction mixture owing to the interaction of arylsulphonylchloride with heterocyclic bases.
As in the solid phase method the isolation of the desired product after each step of the synthesis cannot be achieved because of the destruction of the oligonucleotide chain. This is due to the fact that destruction of longer oligonucleotides on the polymer not only lowers their yield but also increases the low molecular weight component fraction that in turn results in poor product separation, after cleavage from the support, thus reducing the yield of the desired oligonucleotide.

RESULTS AND DISCUSSION

To eleminate the above mentioned drawbacks two approaches in terms of the diester scheme can be used; i.e. successive internucleotide phosphate group blocking and the separation of the nucleotide component activation stage

from the stage of internucleotide linkage formation in the absence of condensing agent.
The former of these approaches has recently been studied by us (3) when we used the aniline residue as a blocking group. This was introduced by means of a Ph_3P-CCl_4 complex (6).

$$ⓅMeOTr-N \xrightarrow[TPS]{pN(Ac)} ⓅMeOTr-NpN(Ac) \longrightarrow$$

$$\xrightarrow[PhNH_2]{Ph_3P \cdot CCl_4} ⓅMeOTr-NpN(Ac) \xrightarrow{OH^-} \ldots$$
$$\hspace{5cm} NHPh$$

To optimize conditions for performing all the stages of the scheme the synthesis of a number of oligonucleotides of known structure is presented in this work (these being further used in application to some problems of molecular biology). The synthesis has been performed on the styrene polymer support of the graft type (4) with the initial load being 50-70 μmol of the nucleoside. A semiautomatic installation with a column type reactor was used for this purpose (3). Investigation of the kinetic of introduction and removal of the anilide protecting group indicated that the first reaction required 3,4 or 5 hours before a plateau of 90 to 95% yield was obtained. The final reaction approached completion after 2 or 3 hours at 37°C. The previously described technological operation scheme for the addition of one nucleotide (3) is lengthened by two steps, the oligonucleotide - polymer being treated by the complex and washed with excess pyridine with a total time increase of 12 hours per cycle.
Thus, the theoretical synthesic speed equals two internucleotide links in twenty four hours if the polynucleotide chain is growing by mononucleotides. The results of the synthesis are presented in Table I.

Table I. Synthesized oligonucleotide yields

NN	Oligonucleotide ($\varepsilon \cdot 10^{-3}$)	Support quantity, mg (nucleoside, μmol)	Oligonucleotide quantity			Yield mol.%
			reaction mixture	isolated oligonucleotide		
				A_{200} o.u.	μmol	
I	T_7CCGG (100)	100 (6)	360	55	0.55	9.2
II	T_9CG (98)	110 (6.6)	500	110	1.1	16
III	T_9GC (98)	150 (9)	600	218	2.2	22
IV	T_{15} (132)	80 (4.8)	270	130	1.0	20
V	CTCATGTT (77)	53 (2.6)	120	18	0.23	8.8
VI	CTCATGTA (83)	80 (4)	160	21	0.25	6.3
VII	CTCATGAT (83)	110 (5.5)	170	32	0.38	7.0
VIII	CAAACTCC (84)	60 (3.0)	180	13	0.16	5.2
IX	AGCTTCAA (90)	200 (8)	210	30	0.35	4.3
X	CCAGCGGAAG (120)	260 (15)	350	60	0.5	3.3
XI	TATAG (60)	200 (66)	540	158	2.7	4.1

It is obvious from the table that the yield of 11-15 membered oligonucleotides containing oligothymidine blocks (I-IV) is sufficiently high. Thus their isolation and purification are facilitated after removal of the mixture of oligonucleotides from the support.
At the same time the yield of the heterogeneous octanucleotides (V-X) decreases to 4-9% due to the increasing fraction of low molecular weight components in the reaction mixture. This result seems to be connected with the presence of the inhibiting admixtures of carboxylate ions in the N-protected deoxynucleoside-5'-phosphates.
According special model experiments a thorough purtification of the completely protected purine nucleotides with gel-chromatography on LH-20 Sephadex in pyridine increases the yield during the synthesis of short oligonucleotides to 85-90% at each stage.

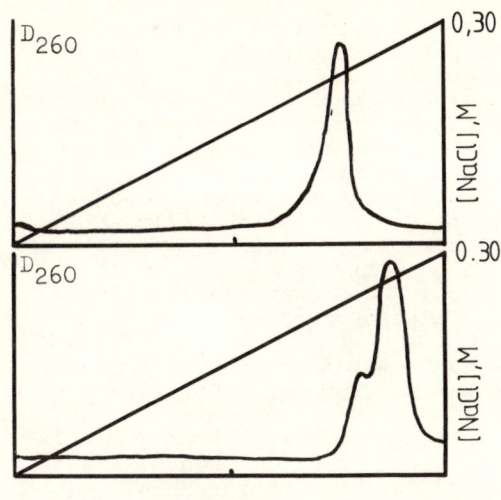

Fig. 1

The microcolumn fractionation of the reaction mixture on DEAE-cellulose in 7 M urea, pH 7.5 in the synthesis 1. T_9, 2. T_9CG

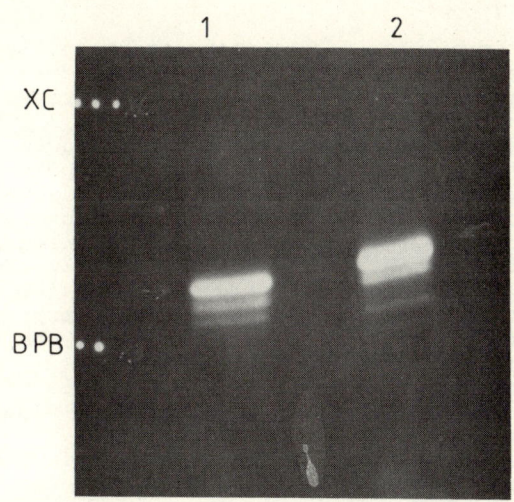

Fig. 2.

Autoradiograph of the reaction mixtures in the syntheses 1. T_9GC, 2. T_9CG, fractionated by polyacrilamide gel electrophoresis after the treatment with γ-^{32}P-ATP and T4-polynucleotide kinase.

A significant criterion of the efficiency of the scheme described is the amount of increment to total nucleotide material per unit weight of a support. While working with the usual diester scheme (4), even at the highest degree of conversion at each stage, such an increment is not more than 2-3 fold in 8-9 membered oligomers as shown by the measurement of the peak areas on the chromatographic profile. Meanwhile the method of internucleotide phosphate group protection, as shown in Table I clearly demonstrates a 4- or 6-fold increace in the amount of nucleotide per gram of polymer support.
Thus the scheme described enables us to carry out the synthesis of 8-11 membered oligodeoxyribonucleotides rapidly and effectively. The final fragment length is limited mostly by the existing technology: i.e. the resolving ability of the separation medium. Evidently the most effective synthetic method involves the block growth of the polynucleotide chain. Application of di- and tri-nucleotide blocks (7) with protected internucleotide phosphate groups made possible the synthesis of the nonanucleotide d (T-A-G-T-A-G-T-A-G) utilizing the (2+3+1+2) scheme. The synthesis required four days and resulted in a 3% product yield in respect to the initial thymidine. Compared to the synthetic scheme involving the sequential addition of mononucleotides the method of "block addition" does result in lower yields (65-70%). However, this drawback is compensated for by several assets: 1. less time is required for the synthesis; 2. the ease of product separation; 3. and the production of the desired oligonucleotide with a larger chain length.
Another approach in terms of the diester scheme which precludes the destruction of the growing chain by the condensing agent is the application of polymeric condensing agent for mononucleotide activation (8). Also as we have indicated previously by the NMR^{31}P method (9), during the reaction the activated nucleotide is merged into the solution and after separation from the support may be used to perform the internucleotide coupling both in the solution and on the polymer support without additional activation.

$$\text{P-SO}_2\text{Cl} + \text{pN(Ac)} \longrightarrow \overset{*}{\text{p}}\text{N(Ac)} \quad \left[\begin{array}{c} \text{O} \diagdown \quad \text{ON(Ac)} \\ \text{P} \\ -\text{O} \diagup \quad \diagdown \text{Py}^+ \end{array} \right]$$

$$\text{P-MeOTr-N} + \overset{*}{\text{p}}\text{N(Ac)} \longrightarrow \text{P-MeOTr-NpN(Ac)}$$

The polymeric condensing agent applied in this work as well as support for
the synthesis was made by modification of the polystyrene grafted in the
amount of 250-300 mg/g on the surface of teflon granules. The sulfonylchlorination of the polymer by a 3-5 fold excess of chlorosulfonic acid
at $0^{o}C$ in chloroform solution proceeds quantitatively and is completed
in 15-20 minutes. The extent of polystyrene sulfonylchlorination is 2 or
2.5 mmol per 1 g of the reagent depending on the amount of graft polystyrene. The existance of active centers on the surface of the support and
the absence of cross-linking between polystyrene chains allows sufficient
improvement or the reagent quality. Thus the rate of mononucleotide activation, measured by means of $NMR^{31}P$, proved to be a little higher than for
TPS due to the high local concentration of reactive conters. The reaction
is complete in 40 or 60 minutes with a 2- fold excess of polymersulfonyl-
chloride.
The reaction takes 6 to 8 hours to proceed with a 6-15 fold excess of the
reagent when using polystyrenesulfonylchloride made from the copolymers
styrene and divynilbenzene (10).
Practically the synthesis is conducted using a column of polystyrenesulfonylchloride which is set up in front of the reactor a column containing
the polymer support. The condensation is carried out by pumping the nucleotide solution through the two columns in succession according to the cycle.
A drawback of this set up is an increasing "dead capacity" because of which
the nucleotide concentration in solution decreases 2 or 2.5 fold thus increasing the reaction time to 10 or 12 hours respectively.
On the other hand, polystyrenesulfonylchloride does not only activate a
nucleotide phosphate group but it also binds the traces of carboxylate ions
and then, what is more important, all the products resulting from the interaction between heterocyclic bases and the excess of sulfonylchloride appear
to be immobilized on the solid phase of a polymer. As a result, the introduction of modified heterocycles into the growing oligonucleotide chain is
minimized. As a matter of fact the nucleotide solution does not change its
colour or even becomes lighter during the condensation while the column with
polystyrenesulfonylchloride on the contrary acquires a distinct brown colour
by the end of the reaction. The synthesis of the T-A-T-A-G (XI) pentanucleotide accomplished according to the scheme investigated demonstrated that the
product isolated by a single chromatography of the reaction mixture on
DEAE-cellulose according to Tomplinson-Tener method does not require additional purification and turns out to be homogeneous according to the chromatographic data with the same method at pH 3.5 and electrophoresis in a
polyacrylamide gel.
The results achieved presently allow us to work out an approach, where polystyrenesulfonylchloride is used for the activation of mononucleotides to
synthesize mixtures of pyro- and tripolyphosphates on a preparative scale.

$$Ⓟ\text{-}SO_2Cl\ +\ pN(Ac)\ \longrightarrow\ \begin{array}{c} p\text{-}N(Ac) \\ | \\ p\text{-}N(Ac) \end{array}\ +\ \begin{array}{c} p\text{-}N(Ac) \\ | \\ p\text{-}N(Ac) \\ | \\ p\text{-}N(Ac) \end{array}$$

Concurrently additional purification of nucleotide components from the mixture is obtained. After washing the polymer with dry pyridine the eluant is evaporated in vacuum and the substance being extracted by precipitation with ether. During the condensation the final activation is accomplished by adding 0.5-1.0 eqv TPS. Reducing the concentration of condensing agent in the solution results in decreasing resining of the reaction mixture which allows an increase of nucleotide concentration in the solution without taking risks that cause salt precipitation (Py·HCl) thus simplifying the synthetic Scheme. In this case the necessity to have a chamber for pre-activation is eliminated.

The structure of all synthesized oligonucleotides was proved by sequencing in accordance with the Maxam-Gilbert method (II) after phosphorylation with γ-^{32}P-ATP and T4-polynucleotide kinase.

The preliminary purification and separation of the desired product from the mixture after removal from the support was obtained by ion-exchange chromatography according to the Tomplinson-Tener method on DEAE-cellulose with neutral and acid pH. The final purification of oligonucleotides was achieved by chromatography on the new support Aminochrom AC-20/100 in the sodium dehydrophosphates.

However, the most direct confirmation that the characteristics of the oligonucleotides obtained are quite feasible to native DNA molecules and do not contain noticible modifications of heterocyclic bases is their biological activity. Therefore, for achieving proof of feasibility the utilization of the oligonucleotides obtained by the solid phase method in molecular biological scientific research all compounds synthesized in this work have been investigated with regard to their ability to form specific complexes between each other and with high molecular weight RNA and interaction of these complexes with appropriate enzymes.

Compounds II-IV containing oligothymidilic blocks have successfully been employed as universal primers of the reverse transcription onkornoviral RNA in doctor L.L. Kiselev's laboratory (IMB AS USSR).V-VIII oligonucleotides, to a certain degree complementary to 16 S ribosomal RNA, were used in the course of investigation of the secondary and tertiary conformation of RNA in ribosome. This work was accomplished in A.N. Belozersky Laboratory of Molecular Biology and Bioorganic Chemistry, under prof. A.A. Bogdanov's leadership.

The most important trend in the work of utilizing synthetic oligonucleotides in molecular biology is the synthesis of genes and their fragments . It is necessary to demonstrate that the properties of "solidphase" nucleotides do not differ from those obtained by various methods in solution. Thus, work done in cooperation with K.G. Skryabin and others (IMB AS USSR) studied ligase coupling of two oligonucleotides (TTCCTTTGA) and (AGCTTCAA) where the former was previously obtained according to the diester scheme (3) and the latter by internucleotide phosphate group protection. After phosphorylation with γ-^{32}P-ATP and T4-polynucleotide kinase, the oligonucleotides under appropriate conditions created a tetramer complex, which upon treatment with DNA-ligase was converted into a 17 membered duplex with cohesive ends containing the recognition of the Hind III (AAGCTT) restriction enzyme.

(5') ^{32}pT-T-C-C-T-T-T-G-A$^{\diagup^{32}p}\diagdown$A-G-C-T-T-C-A-A (3')

(3') A-A-C-T-T-C-G-A$_{\diagdown_{32}p\diagup}$A-G-T-T-T-C-C-T-T$_p$32 (5')

Treatment of the isolated heptadekamer by the enzyme under discussion resulted in the destruction into initial oligonucleotides as shown in fig.3.

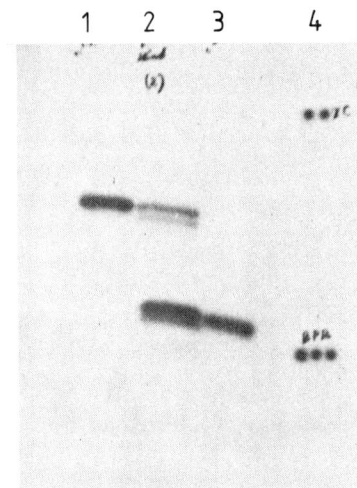

Fig. 3.
Autoradiograph of polyacrylamide gel after electrophoresis.
1. Initial heptadecamer.
2. The mixture after the treatment of oligonucleotide Hind III
3. Nonanucleotide-witness.
4. Dycing markers.

Thus, the results obtained during the investigation of complex formation, ligation and restriction of oligonucleotides obtained by the solid phase method indicate that when standard methods of isolation are employed, they have sufficient activity to conduct further scientific research in the various fields of molecular biology.

REFERENCES

1. M.I. Gait, R.C.Sheppard, Nucleic Acids Research, 4, 4391 (1977)
2. Ch.K. Narang, K. Brunfeldt, K.E. Norris, Tetrahedron Letts., 21, 1819 (1977)
3. V.K. Potapov, V.P. Veiko, O.N.Koroleva, Z.A. Shabarova, Nucleic Acids Research, 6, 2041 (1979)
4. V.K.Potapov, S.I. Turkin, V.P. Veiko, Z.A. Shabarova, Dokl. Akad. Nauk SSSR, 241, 1352 (1978)
5. K.Z. Agarwal, A. Yamazaki, P.I. Cashion, H.G. Khorana, I. Biol. Chem., 250, 5563 (1975)
6. V.K. Potapov, V.P. Veiko, Z.A.Shabarova, Bioorgan.Khim.,5, 468 (1979)
7. G.F. Mishenina, V.V. Samukov, L.N. Semenova, T.N. Shubina, Bioorgan. Khim., 4, 735 (1978)
8. M. Rubinstein, A.Patchornik, Tetrahedron Letts., 28, 2881 (1972)
9. S.I. Turkin, V.K. Potapov, Z.A. Shabarova, V.F. Zarytova, D.G. Knorre, Bioorgan. Khim., 1, 1430 (1975)
10. S.N. Zagrebelny, V.F. Zarytova, A.S. Levina, L.N. Semenova, E.V. Yarmolinskaya, S.M. Yasnetskaya, Bioorgan, Khim., 4, 729 (1978)
11. W. Haseltine, A. Maxam, W. Gilbert, Proc.Natl.Acad.Sci. USA, 74, 989 (1977)

CHEMICAL SYNTHESIS OF DNA - POSSIBILITIES AND APPLICATIONS

H. Köster

Institut für Organische Chemie und Biochemie der Universität,
Martin-Luther-King-Platz 6, D-2000 Hamburg 13, Federal Republic of Germany

<u>Abstract</u> - Total synthesis of DNA requires chemical and biochemical methods. During synthesis of the first DNA coding for a human peptide hormone (angiotensin II) and some other oligodeoxynucleotides e.g. the recognition sequence for the restriction endonuclease Pst I some chemical improvements using diester and triester method for the chemical synthesis of oligodeoxynucleotides could be developed. Both methods are compared critically. For linking the different oligodeoxynucleotides to give the duplex DNA by enzymatic methods a highly effective multistep-one-flask procedure has been introduced. The possibility to use hexanucleotides forming cohesive ends of three nucleotide units for the construction of bihelical DNA has been explored. It turned out that the minimum length to form stable duplex structures which can be ligated enzymatically is given in an octanucleotide forming sticky ends of four nucleotide units. Further applications of chemically synthesized oligodeoxynucleotides are summarized. One example involving promoter (nucleic acid)-RNA polymerase (protein) interactions is discussed in some more detail.

Most oligodeoxynucleotides have been synthesized by either the diester or triester method. Both methods are used by us for the synthesis of defined oligodeoxynucleotide sequences.

The first total synthesis of a DNA coding for a human peptide hormone (angiotensin II) (Refs. 1,2) has been performed by the diester method. The nucleotide sequence has been deduced by the genetic code from the known peptide sequence (Fig. 1).

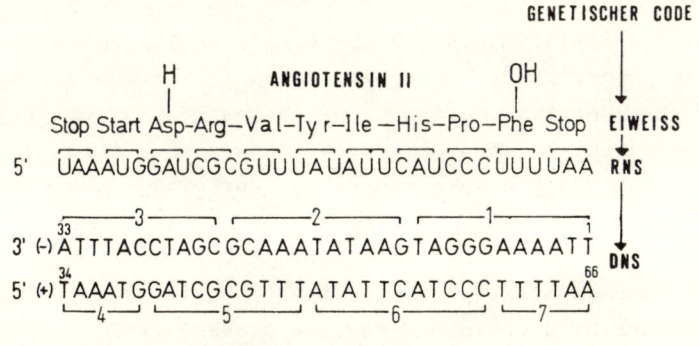

Fig. 1. The angiotensin II DNA

Synthesis of a double helical DNA involves chemical methods for the synthesis of short oligodeoxynucleotides of 6 to 11 nucleotide units and biochemical methods necessary to phosphorylate the oligodeoxynucleotides and to ligate them to form the covalently linked doublehelical structure.

The improvements of chemical methods introduced by us may be summarized:

1) Introduction of a highly effective chromatographic technique (Fig. 2 gives an example), involving anion exchange chromatography with ionic strength and polarity gradients (Refs. 2,4) and stopping the gradients by an automatic device (Ref.5).

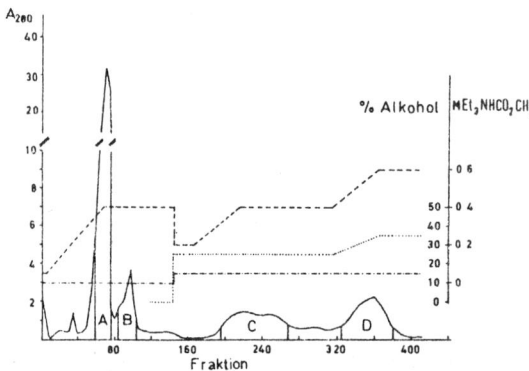

Fig. 2. DEAE-cellulose (acetate form) chromatography of a condensation between $(MeOTr)T_d-A_d^{bz}-A_d^{bz}-A_d^{bz}-T_d-C_d^{an}$ and $pA_d^{bz}-A_d^{bz}(Ac)$ after 3'-O-deacylation (Ref. 3). A = $pA_d^{bz}-A_d^{bz}$, B = symmetrical pyrophosphate of $pA_d^{bz}-A_d^{bz}$, C = $(MeOTr)T_d-A_d^{bz}-A_d^{bz}-A_d^{bz}-T_d-C_d^{an}$ and D = $(MeOTr)T_d-A_d^{bz}-A_d^{bz}-A_d^{bz}-T_d-C_d^{an}-A_d^{bz}-A_d^{bz}$.
——— : UV-absorption at 280nm (A_{280}), ----- : Concentration of triethylammonium acetate (pH 7.0), -·-·-· : Concentration of methanol, ······· : Concentration of propanol-(2).

This technique allows to separate complicated condensation mixtures by one chromatographic step. Furthermore it enables the evaluation of an optimal synthetic strategy for an oligodeoxynucleotide avoiding purification problems of the condensation mixtures at each step (Ref. 2).

2) Introduction of a simplified synthesis of di- and trinucleotide blocks with 5'-phosphate group. The 5'-phosphate group is introduced in high yield by a 10-minutes phosphorylation step using phosphorous oxychloride, thus avoiding the involvement of phosphate protecting groups (Fig. 3) (Refs. 6,7). This procedure is quick and simple and can be performed with cheap reagents.

3) Usage of key intermediates for the synthesis of the 5'-terminal oligodeoxynucleotides of a chain and for the blocks with 5'-phosphate group (Fig. 4) (Refs. 2,7). The degeneracy of the genetic code makes possible the selection of that structural gene sequence which allows the most frequent use of the same key intermediates at different positions in

the DNA sequence to be synthesized (Ref. 2).

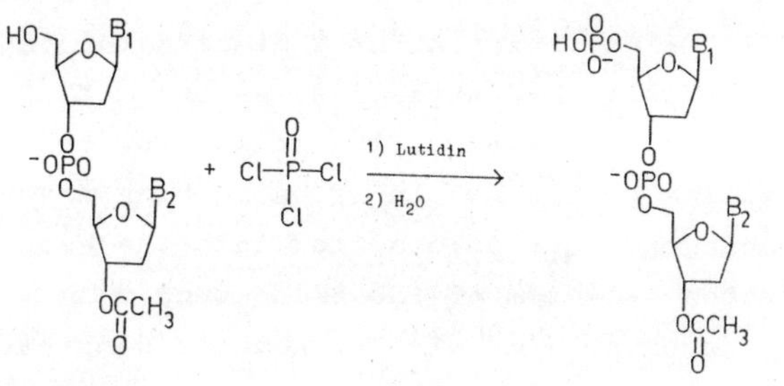

Fig. 3. Synthesis of dinucleotide blocks by direct phosphorylation with phosphorous oxychloride.

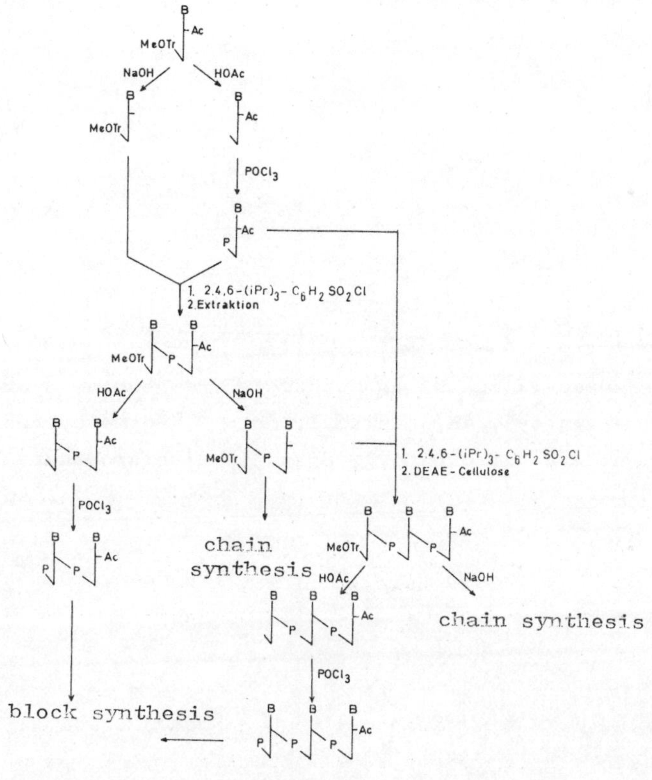

Fig. 4. Schematic representation showing the usage of key intermediates for the synthesis of the 5'-terminal chain segments and for the synthesis of the di- and trinucleotide blocks with the 5'-phosphate groups, the latter being introduced by the direct phosphorylation procedure (see Fig. 3).

After each condensation following the diester approach an alkaline deprotection of the 3'-O-acetyl group was necessary. When using the N-protecting groups and conditions introduced by Khorana and coworkers (Ref. 8), we observed partial N-deprotection at the N^4-anisoyl-deoxycytidine residues, in some cases the isobutyryl group had been cleaved off the guanine ring. We therefore started a quantitative study to determine the stability of various N-protecting groups at the adenine, guanine and the cytosine ring (Ref. 9). It turned out that the influence of the heterocyclic base on the stability of the amide bond towards alkali is tremendous (Fig. 5; compare e.g. the N-benzoylated derivatives of C_d, A_d and G_d). The half lives of C_d^{bz}, A_d^{bz} and G_d^{bz} e.g. have relative values of 1:50:100, being about 7 minutes for the C_d^{bz} using the new developed very effective N-deacylation reagent methanol/0.2 N sodium hydroxide (1:1, v/v) at 22°C (Ref.9). The reason for the particular lability of the N-acyl protecting groups at the cytosine ring might be that in this case the negative charge can be delocalized into the heterocyclic ring, thus leaving the carbonyl-C-atom of the amide bond much more electropositive than in the case of comparable adenine or guanine derivatives (Fig. 6).

Fig. 5. Rate of N-deacylation using methanol/0.2 N sodium hydroxide (1:1, v/v) at 22°C. bz = benzoyl, an = anisoyl, db = 2,4-dimethylbenzoyl, ib = isobutyryl, nt = 4-nitrobenzoyl, dc = 3,4-dichlorobenzoyl, cl = 4-chlorobenzoyl.

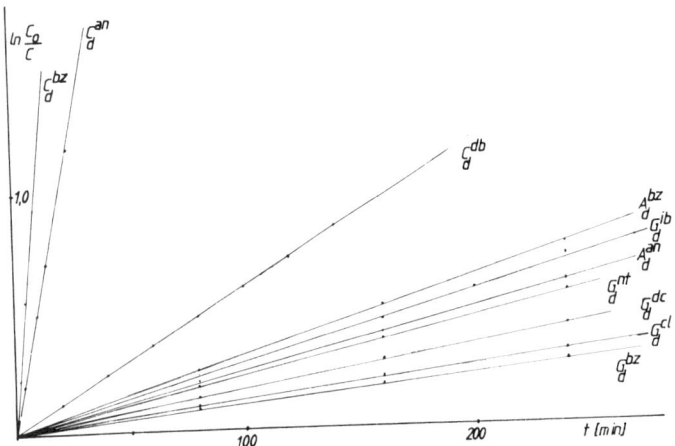

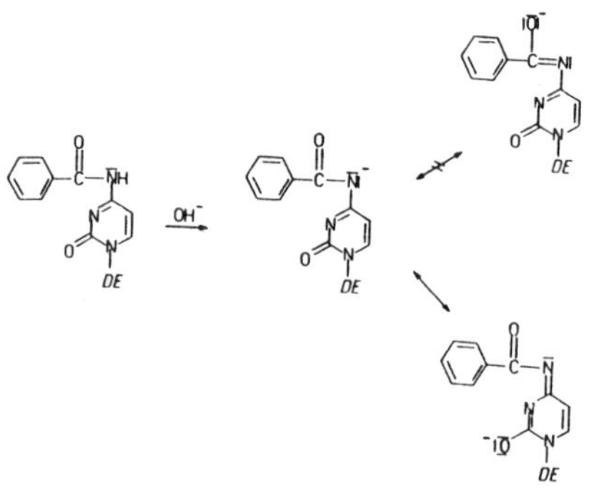

Fig. 6. Suppression of resonance stabilization at the amide bond by the influence of the cytosine ring. DE = deoxyribose.

The most stable protecting group for the cytosine ring we could find so far is the 2,4-dimethylbenzoyl group having a half life of about 116 min. using the methanol/sodium hydroxide reagent (for comparison: under the same conditions the anisoyl group widely used has a half life of 14 minutes). This protecting group can be highly recommended for the syntheses according to the diester and triester method as well. For the latter it is advantageous to have protecting groups with enhanced lipophilicity. The appropriate

protected deoxycytidine derivatives can easily be prepared and are
crystalline compounds. This protecting group has been extensively used
by us for synthesis of various oligodeoxynucleotides (see below).

With the very effective MeOH/NaOH reagent for the guanine ring the
very stable benzoyl group can be used when oligodeoxynucleotide synthesis
is performed according to the diester approach. For the synthesis of
an Pst I-octanucleotide (Fig. 7) (Ref. 10) by the triester method we used
the more labile 3,4-dichlorobenzoyl group. But it turned out that its
use is not recommendable due to an increased sensitivity of the guanine
ring during condensations and strong retention during silica gel column
chromatography. So the benzoyl group should be used for diester approach
using the MeOH/NaOH reagent and the isobutyryl group seems to be at the
moment the best choice for the triester approach.

Introduction of the anisoyl group for protection of the adenine
ring resulted in an 25% increase of stability. This only should be of
importance when the diester method is used.

The disadvantages of the diester approach are (Ref. 11) :
1) Formation of symmetrical pyrophosphates.
2) Formation of side products by activation of the internucleotidic linkage.
3) Necessity of a high excess of phosphate component over OH-component
in a later stage of oligonucleotide synthesis.
4) Relatively much time is needed for anion exchange chromatography.

The last point is no more valid since reversed phase high pressure
liquid chromatography can be used (Ref. 12). Moreover it should be possible
to use our chromatographic technique (see above) at higher pressure.

To be able to compare the two methods we used the triester approach
for the synthesis of some oligodeoxynucleotides. The approach introduced
by Catlin and Cramer (Ref. 13) was adopted by us with some modifications.
As permanent phosphate protecting group we used the 2-chloro- or 4-chloro-
benzoyl group (Ref. 14). As temporary phosphate protecting group the
β-cyanoethyl group is widely used (e.g. Refs. 15,16), but according to
our experience this group is too labile during condensations and silica
gel chromatography. Similar observations have been published recently
by Narang (Ref. 16). We found it superior to use the 2,2,2-trichloroethyl
group (Refs. 13,17) as temporary protecting group which could be removed
very effectively using Zn/acetonylacetone (Ref. 18). As zinc ions do very
effectively inhibit condensations reactions a very careful work-up is
necessary. As condensing agent we used amongst others 2,4,6-triisopropyl-
or 2,4,6-trimethylbenzenesulfonyl tetrazolide (Ref. 19). However, in our
hands this reagent has the disadvantage of being rather unstable and
difficult to prepare in reproducible quality. Moreover it is very expensive.
Better condensing agents with the same effectivity but having not the
described disadvantages have to be developed. Work towards this goal is

in progress in our laboratory.

Phosphorylation of 3'-hydroxyl functions of a suitably protected deoxynucleoside by a monofunctional phosphorylating agent serves as a useful method for the preparation of phosphotriester intermediates. When using 2,2,2-trichloroethyl 2-chlorophenyl phosphorochloridate (Ref. 20) for the phosphorylation of a 5'-dimethoxytritylated N-protected deoxynucleoside we observed considerable amount of cleavage of the acid sensitive dimethoxytrityl group, presumeably due to generation of hydrogen chloride in the reaction mixture. This resulted in some 5'-phosphorylation, a very undesirable side product. On increasing the amount of base (1-methylimidazol) the removal of the 5'-dimethoxytrityl group was reduced but then some N-phosphorylation was observed. To solve this problem we use now powdered molecular sieves (4 Å) which trap hydrogen chloride very effectively and hence protect the acid-sensitive functions in the deoxynucleosides during phosphorylation. The reaction proceeds very quickly (10 minutes) without the formation of any side products. The yields after preparative purification on silica gel columns is in the range of 83 to 96% for the four fully protected phosphotriester intermediates (Ref. 21).

As 3'-terminal protecting group we used the 4-tert.butylbenzoyl (Ref. 21) and more recently the tert. butyldiphenylsilyl group (Ref. 22).

With the methods mentioned some oligodeoxynucleotides have been synthesized e.g. an octanucleotide representing the Pst I recognition sequence (Fig. 7) (Ref. 10) and a nonamer and a tetramer (Fig. 8) necessary for the synthesis of a modified angiotensin II-DNA (Ref. 23).

Having now considerable experience in both diester and triester method a critical comparison is possible. Some disadvantages should be summarized. First of all the use of an additional protecting group increases principally the expenditure of work. This becomes particularly evident during the analytical investigations necessary after each condensation step. Furthermore it will be principally very difficult to find a permanent phosphate protecting group which is stable throughout the oligonucleotide synthesis and which can be cleaved off with 100 % regioselectivity in quantitative yield. This has been also stated by Reese and van Boom (Refs. 24,25). A yield of 95% for the deprotection of each triester function within an oligonucleotide is not sufficient as the residual triester functions can be distributed statistically among all internucleotidic linkages present in the molecule. In the case of synthesis of doublehelical DNA some of the impurities might be purified away during ligation and subsequent cloning. But from chemical point of view this problem is far from being solved although considerable improvements have been achieved (Refs. 26,27). Another drawback is the lack of discriminating power of silica gel chromatography when higher molecular weight material is to be purified (Ref. 16). In this respect it is notable that despite the presence of many diastereomers only one peak is observed in the case of longer oligonucleotides. Furthermore considerable losses of material due to adsorption or degradation can be observed during silica gel chromatography (Ref. 16) the amount depending on the base composition of the sequence. This effect is particularly pronounced if the sequence is rich in guanine residues. Here the diester approach has the advantage that each nucleotide unit within an oligonucleotide chain bears a full negative charge thus serving as a purification handle. Adsorption and degradation can be neglected during DEAE-cellulose chromatography. In addition it has been reported that the triester function (Ref. 16) and the dimethoxytrityl group (Ref. 28) are not stable under condensation conditions and that considerable sulfonylation of the primary 5'-OH group takes place (Ref. 16) amongst other side reactions (Ref. 29). Condensation yields are not regularly better than compared to the diester method and do also depend on the nucleotides involved in the coupling step (Ref. 16).

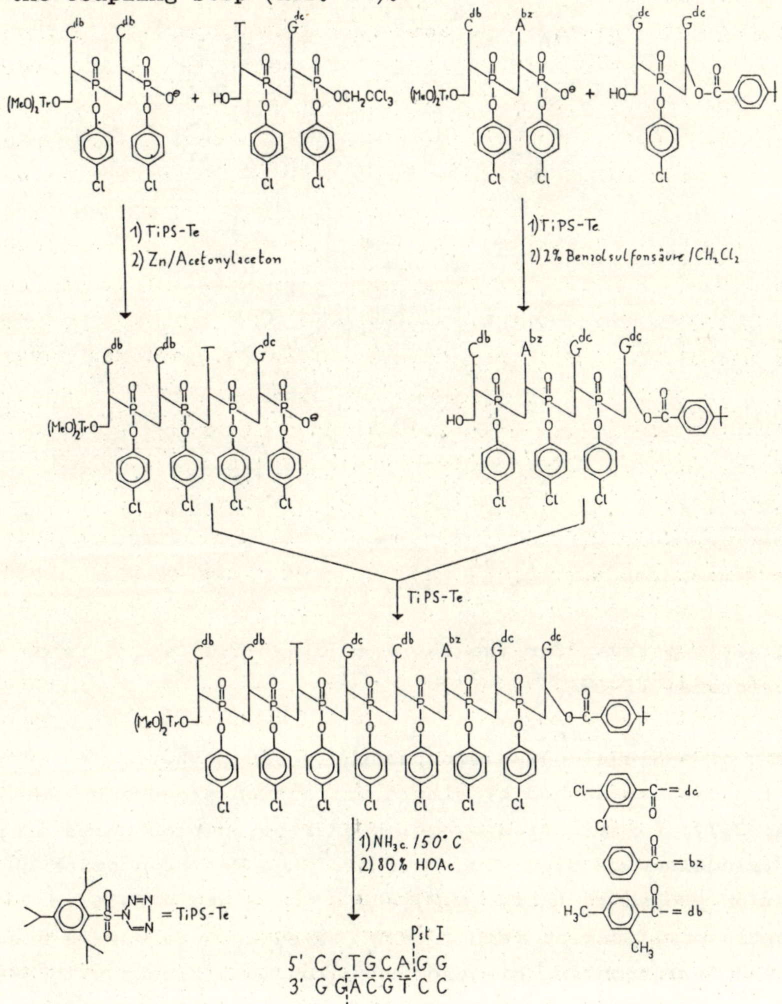

Fig. 7. Synthetic pathway for the Pst I recognition sequence.

It has been pointed out that the synthesis of the building blocks is quicker done when using the triester approach. This is not the case when considering all reaction steps. The additional phosphate protecting group makes some more reaction steps necessary for synthesis, deprotection and analytical investigations.

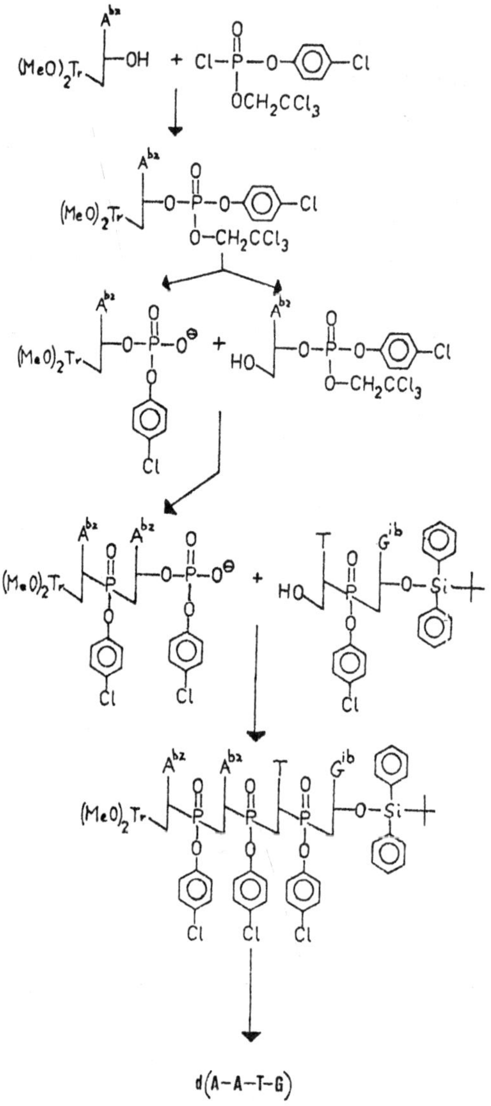

Fig. 8. Synthetic pathway for the tetramer d(A-A-T-G) which is part of a modified angiotensin II-DNA.

The synthesis of a dinucleoside monophosphate can be performed very easily when using the diester method by simple extraction procedures for purification (Refs. 2,7). Purification by quick extraction procedures is possible even with trinucleoside diphosphates (Ref. 30). Having a quick and simple phosphorylation procedure (Refs. 6,7) and a quick and effective chromatographic purification method (see above) at hand one is also able to synthesize the corresponding di- and trinucleotides blocks with 5'-phosphate groups in a comparable short time. To perform the analytical investigations after each condensation step the complications in connection with deprotection at the triester function are omitted. What makes triester approach for the synthesis of long oligonucleotides more effective in the moment is the fact that the fully protected dimers and trimers have been synthesized in very large quantities and that only a low excess of phosphate component needs to be used for condensation reactions. Thus one was able to build up a library of di- and trinucleotides (Ref. 15) with which many different long oligodeoxynucleotides could be synthesized. Having this library of phosphotriester intermediates at hand it is at the moment quicker to get different oligodeoxynucleotides synthesized by the triester method, but it is more difficult to get them as pure as when using the diester approach.

It may be concluded that both methods can be further improved. It is an open question which method will be superior at the end.

A method which should have a great future is the solid phase synthesis. Some time ago we were the first to synthesize longer oligodeoxynucleotides with defined sequence using the diester method and preformed blocks and a macroporous non-swellable polystyrene resin (Fig. 9) (Ref. 31).

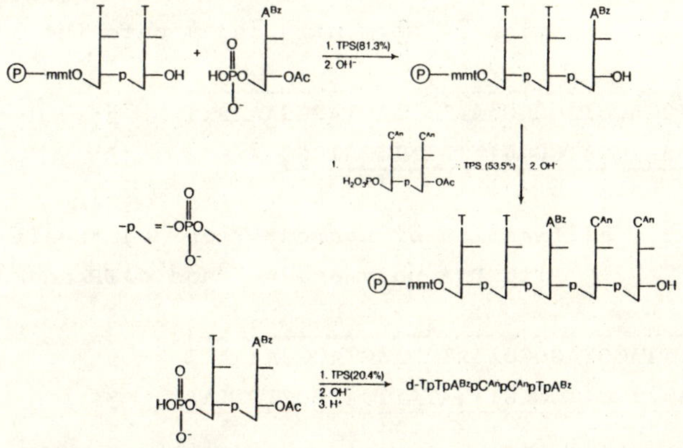

Fig. 9. Polymer support synthesis of d(T-T-A-C-C-T-A) using diester method and preformed blocks on a macroporous highly crosslinked polystyrene matrix (Ref. 31).

In the meantime hydrophilic carriers with N,N-dimethylacrylamide backbone (Ref. 32) (first used by us for oligodeoxynucleotide synthesis using a new developed bead polymer (Ref. 33)) and polyacrylmorpholidate support (Ref. 34) have been introduced successfully. The advantage of this approach can be seen in the simplified techniques of operation which enables the standardization and automatization of oligonucleotide synthesis (Refs. 35, 36,37).

Apart from the fact that both diester and triester method have to be further improved long bihelical nucleic acids have been synthesized (Ref. 38) (and references cited therein).

For linking together the seven segments of the angiotensin II DNA (see Fig. 1) we introduced a procedure which makes possible a multistep-one-flask ligation procedure starting with the unphosphorylated oligodeoxynucleotides. Furthermore an analytical system has been developed which allows to follow each step from phosphorylation of oligodeoxynucleotides up to the last ligation step quantitatively without any

intermediate purification. The main features of this procedure are:
1) Modified Norit assay (Ref. 38).
2) 'Single-strand ligation' (Ref. 38).
3) Polyacrylamidegel electrophoresis using a system of chain length standards which allow the exact calculation of the mobility of any given sequence and to predict whether oligodeoxynucleotides of the same length but different sequence can be separated (Ref. 39). To avoid wrong duplex formation the synthesis of the angiotensin II DNA has been divided into the following steps:

1. Synthetic step :   Ligation of segments 1, 2, 5 and 6 (see Fig. 1)

```
                         2         1
        3'        GCAAATATAAGTAGGGAAAATT         5'
        5'        GATCGCGTTTATATTCATCCC          3'
                     5         6
```

2. Synthetic step :   Ligation of segment 3 and 4 onto the core

```
              3           2         1
        3'  ATTTACCTAGCGCAAATATAAGTAGGGAAAATT    5'
        5'  TAAATGGATCGCGTTTATATTCATCCC          3'
              4         5         6
```

3. Synthetic step :   a) Ligation of segment 7 or   b) repair synthesis
                         with DNA polymerase I and dATP and dTTP

```
              3           2         1
        3'  ATTTACCTAGCGCAAATATAAGTAGGGAAAATT        5'
        5'  TAAATGGATCGCGTTTATATTCATCCCTTTTAA        3'
              4         5         6       7
```

For insertion into a cloning vehicle sticky ends for a restriction endonuclease e.g. Eco RI have to be fused to the DNA by further two steps:

4. Synthetic step :   Blunt end ligation of Eco RI octanucleotide

```
        3'  CCTTAAGGATTTACCTAGCGCAAATATAAGTAGGGAAAATTCCTTAAGG    5'
        5'  GGAATTCCTAAATGGATCGCGTTTATATTCATCCCTTTTAAGGAATTCC    3'
```

5. Synthetic step :   Incubation with Eco RI restriction endonuclease

```
        3'      GGATTTACCTAGCGCAAATATAAGTAGGGAAAATTCCTTAA    5'
        5'  AATTCCTAAATGGATCGCGTTTATATTCATCCCTTTTAAGG        3'
```

Quantification of the reaction steps 1 to 3 is shown in Fig. 10 and 11.

Finally the purified product band can be further characterized by Maxam-Gilbert sequencing (Ref. 40).

This multistep-one-flask ligation procedure is so efficient that the synthesis of the angiotensin II DNA can be performed in one week on a nmol-scale starting from the unphosphorylated oligodeoxynucleotides and including a final purification step of the duplex DNA on Sephadex or preparative polyacrylamidegel electrophoresis.

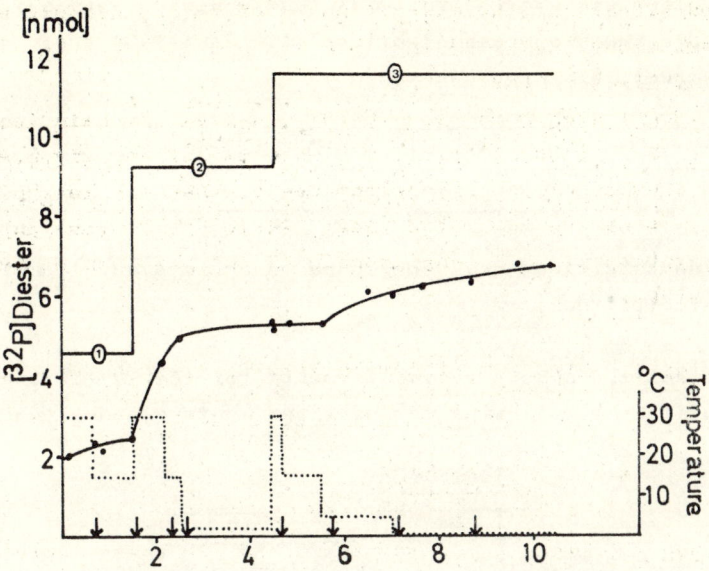

Fig. 10. Three-step ligation to give the 33 base pairs long angiotensin II-DNA duplex. ↓: addition of 0.4 to 2 units T4 DNA ligase. ·····: reaction temperature.

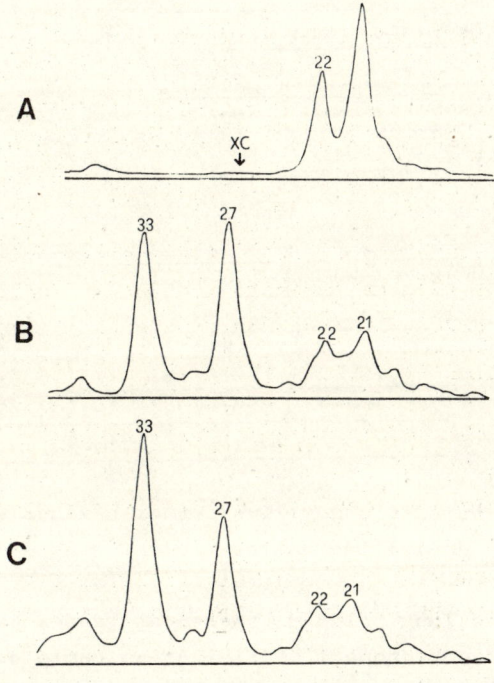

Fig. 11. Product distribution in the three steps of ligation forming the angiotensin II DNA. Polyacrylamidegel electrophoresis was from left to right. XC : xylenecyanol. Densitometric evaluation of an autoradiogramm. A : step 1, B : step 2, C : step 3 (see Fig. 10 and text on previous page).

With regard to the synthesis of structural genes the question is of interest whether hexanucleotides each having the genetic information of two amino acid codons and designed to give sticky ends of three nucleotide units can be used for the construction of bihelical DNA. This would drastically reduce the expenditure of chemical work and with a library of some hexanucleotides every structural gene could be very quickly constructed. To test this the two hexanucleotides d(T-A-A-A-T-G) (Ref. 41) and d(T-A-C-A-T-T) (Ref. 42) have been used :

```
5'|T-A-A-A-T-G|T-A-A-A-T-G|T-A-A-A-T-G|T-A-A-A-T-G|
     |T-A-C-A-T-T|T-A-C-A-T-T|T-A-C-A-T-T|T-A-C-A-T-T| 5'
```

Unfortunately no conditions could be found to achieve any ligation between these two hexanucleotides.

To find out the necessary requirements for the T4 DNA ligase catalyzed ligation the following set of experiments was designed (Fig. 12).

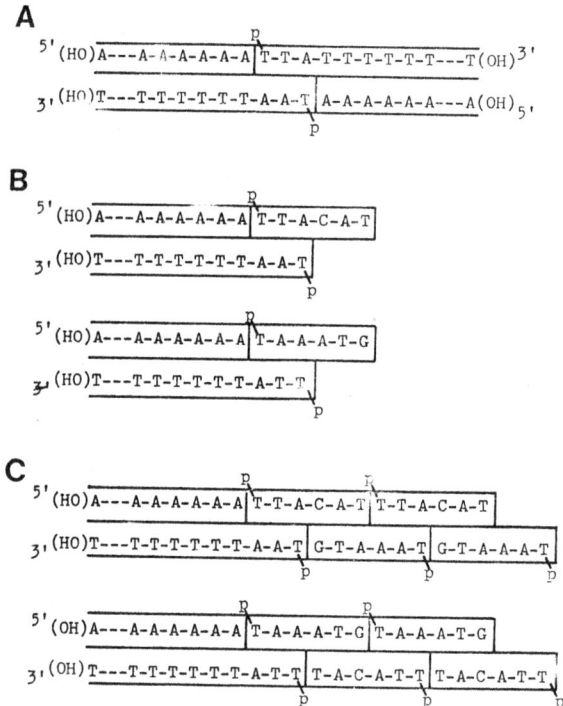

Fig. 12. Model experiments to study ligation of hexanucleotides with cohesive ends of three nucleotide units.

As a consequence of the long polydAdT sequences (model experiment A in Fig. 12) an unwanted two-component hybridization could occur. This is only possible if mispairing takes place. To explore whether this is the case we first performed the following two-component experiments (Table 1) (Ref. 43):

```
                        p α β γ
        5'---T-T-T-T-T-T|T-T-T-T-T-T-----

        3'---A-A-A-A-A-A-A-A-A-A-A-A-----
```

Surprisingly enough there was a very high ligation efficiency. Nearest neighbour analysis demonstrated that the T4 DNA ligase formed a covalent linkage not only when there was an A/A mispair in the γ-position with respect to the ligation point (I, II in Table 1) or in the β- and γ-position (III in Table 1) but also when there was a T/T mispair (IV in Table 1) at the ligation point (α-position) provided at both sides of the ligation point was a long stable doublehelical structure (compare e.g. the ligation yields of experiments I and II in Table 1). Therefore model experiment A (Fig. 12) had to be performed with rather short doublehelices at both sides of the ligation point. It turned out (Table 2) that the doublehelix at both sides of the ligation point had to be at least 7 base pairs long in order to get efficient ligation (experiments II and III in Table 2; compare with experiments IV and V in Table 2). Then the correct ligation takes place via sticky ends of three nucleotide units. No mispairing was tolerated under these conditions (compare experiments II to V with experiment I in Table 2). When a doublehelical structure (even a longer one) was only present at one side of the ligation point no ligation could be observed (model experiment B and C in Fig. 12, results not shown).

TABLE 1. Various ligation experiments according to model A (Fig. 12) with only two components. Incubations were for 15 hours at $4^{\circ}C$.

Reaction No.	Components	$\bar{n}$	Concentration [µM]	Wrong base-pairing	Yield of Ligation	Nearest Neighbour Analysis
I	$d(pT-\underset{+}{T}-A-T_n)$	10	1	γ(A/A)	85%	pT ⟶ Tp
	$d(A-A-A-A_n)$	40	1			
II	as I	9	1	γ(A/A)	3%	pT ⟶ Tp
		7	1			
III	$d(pT-\underset{+}{A}-A-T_n)$	8	5	β+γ(A/A)	69%	pT ⟶ T
	$d(A-A-A-A_n)$	40	5			
IV	$d(pT-\underset{+}{A}-A-A_n)$	25	1	α(T/T)	8%	pT ⟶ Ap
	$d(T-T-T-T_n)$	50	1			

TABLE 2. Ligation experiments according to model A (Fig. 12). Incubations for 15 hours at $4^{\circ}C$. Concentration of component 1, $d(pT-T-A-T_n)$, 10 µM; component 2, $d(pT-A-A-T_n)$, 10 µM; component 3, $d(A-A-A-A_n)$, 20 µM.

Reaction No.	Components $\bar{n}$			Concentration of Mg^{2+}-ions	Yield of Ligation	Nearest Neighbour Analysis
	1	2	3			
I	15	15	20	10 mM	99%	pT ⟶ Tp (90%)
II	9	9	7	10 mM	44%	pT ⟶ Ap (10%)
III	9	9	7	20 mM	50%	pT ⟶ Ap
IV	5	5	4	10 mM	5%	pT ⟶ Ap
V	5	5	4	20 mM	5%	pT ⟶ Ap

Thus the ligase needs essentially a doublehelical structure at both sides of the ligation point; the length of the sticky ends seems to be of less importance. This is supported by the findings that long doublehelical DNAs with sticky ends of only two bases (a Hpa II digest) could be effectively ligated (Ref. 44). Under somewhat different conditions even blunt end ligation can be mediated by this enzyme.

The question now was whether octanucleotides with sticky ends of four nucleotide units can be used. To show that this strategy works well for the construction of bihelical DNA the octanucleotide d(A-A-T-T-C-C-G-G) was chosen as a model system:

```
5'
    | A-A-T-T-C-C-G-G | A-A-T-T-C-C-G-G | A-A-T-T-C-C-G-G |
| G-G-C-C-T-T-A-A | G-G-C-C-T-T-A-A | G-G-C-C-T-T-A-A |
                                                          5'
```

The ligation which involves also the relatively weak interaction of four AT base pairs proceeds very quickly to form high molecular weight DNA (Ref. 43).

Thus the shortest oligodeoxynucleotides which can be used to construct bihelical DNA according to the sticky-end concept are octanucleotides giving rise to cohesive ends of four nucleotide units. It follows that for the construction of doublehelical DNA only octanucleotides have to be synthesized chemically.

Despite the use of synthetic oligodeoxynucleotides for the construction of duplex DNA (structural genes or regulatory signals) the number of further applications is more and more increasing. Some recent applications should be summarized (for references see 45):
1) Specific primers for
    a) Sequencing of DNA and RNA.
    b) Introduction of modifications into DNA.
    c) Introduction of mutations into DNA.
    d) Synthesis of specific cDNA with reverse transcriptase.
2) Gene fishing.
3) Inhibition of virus reproduction.
4) Linkermolecules for genetic engineering.
5) Total synthesis of DNA.

The usefulness of oligodeoxynucleotides of a specially designed sequence is to serve as highly specific probes by selecting a complementary target sequence out of thousands of nucleotides in an interesting DNA.

One example should be given concerning experiments which are in progress in our laboratory.

One way to study the high specific protein-nucleic acid interactions which are involved e.g. in RNA polymerase-promoter recognition is to incorporate specially selected nucleotide analogs into the promoter DNA sequence. As the rate of complex formation between RNA polymerase and promoter DNA is equivalent to the efficiency of in-vitro and in-vivo initiation of transcription we performed a quantitative study with various bacteriophage fd promoters modified by deoxyuridine or 5-bromodeoxyuridine substituting deoxythymidine residues in the minusstrand (codogenic strand).

The results turned out to be very interesting (Ref. 46). In one case one of the strongest promoters of fd (gene II promoter) increased its activity by a factor of about 5 when 5-bromodeoxyuridine is substituting deoxythymidine and lost completely its ability to react with RNA polymerase at all when deoxyuridine replaces deoxythymidine (Ref. 47). To find out which positions are responsible for these dramatic effects the gene II promoter is now being modified sequentially at only single positions. This is only possible by using a synthetic short oligodeoxynucleotide which binds specifically on the about 6400 nucleotide long plusstrand of bacteriophage fd downstream and outside the promoter sequence and which allows to restore the doublestranded specifically modified promoter region using DNA polymerase I (Ref. 48).

We believe that this kind of experiments will contribute considerably to unravel the secrets of protein-nucleic acid interaction.

Acknowledgement - I thank the Deutsche Forschungsgemeinschaft for financial support and my highly motivated coworkers who are cited under references for fruitful collaboration.

REFERENCES

1. H.Köster, H.Blöcker, R.Frank, S.Geussenhainer and W.Kaiser, Hoppe Seyler's Z.Physiol.Chem. 356, 1585(1975).
2. H.Köster, H.Blöcker, R.Frank, S.Geussenhainer and W.Kaiser, Liebigs Ann.Chem. 1978, 839.
3. Synthesis described in H.Köster and H.Blöcker, Liebigs Ann.Chem. 1978, 1007; for abbreviations see reference $^{2)}$.
4. H.Köster and W.Kaiser, Liebigs Ann.Chem. 1974, 336.
5. H.Köster and R.Frank, Chromatographia, 9, 497(1976).
6. H.Köster and W.Heimann, Angew.Chem.Int.Ed.Engl. 12, 859(1973).
7. H.Köster, H.Blöcker, R.Frank, S.Geussenhainer, W.Heidmann, W.Kaiser and D. Skroch, Liebigs Ann.Chem. 1978, 854.
8. H.Weber and H.G.Khorana, J.Mol.Biol. 72, 219(1972).
9. H.Köster, K.Kulikowski, T.Liese, W.Heikens and V.Kohli, in preparation.
10. H.Blöcker, V.Kohli, C.Wolff and H.Köster, Hoppe Seyler's Z. Physiol. Chem. 360, 235(1979).
11. D.G.Knorre, V.F.Zarytova, A.V.Lebedev, L.M.Khalimskaya and E.A.Sheshegova, Nucleid Acids Res. 5, 1253(1978).
12. H.-J.Fritz, R.Belagaje, E.L.Brown, R.H.Fritz, R.A.Jones, R.G.Lees and H.G.Khorana, Biochemistry, 17, 1257(1978).
13. J.C.Catlin and F.Cramer, J.Org.Chem. 38, 245(1973).
14. C.B.Reese, Tetrahedron 34, 3143(1978).
15. R.Crea, T.Hirose, A.Kraszewski and K.Itakura, Proc.Nat.Acad.Sci. (USA) 75, 5765(1978).
16. H.M.Hsiung, R.Brousseau, J.Michniewicz and S.A.Narang, Nucleic Acids Res. 6, 1371(1979).

17. J.H.van Boom, P.M.J.Burgers and P.H. van Deursen, Tetrahedron Letters, 1976, 869.
18. R.W.Adamiak, E.Biala, K.Grzeskowiak, R.Kerzek, A.Kraszewski, W.T.Markiewicz, J.Stawinski and M.Wiewiorowski, Nucleic Acids Res., 4, 2321(1977).
19. J.Stawinski, T.Hozumi, S.A.Narang, C.P.Bahl and R.Wu, Nucleic Acids Res. 4, 353(1977)
20. J.H. van Boom, P.M.J.Burgers, G. van der Marel, C.H.M.Verdegaal and G.Wille, Nucleic Acids Res. 4, 1047(1977).
21. V.Kohli, H.Blöcker and H.Köster, in preparation.
22. R.A.Jones, H.-J.Fritz and H.G.Khorana, Biochemistry, 17, 1268(1978).
23. H.Blöcker, V.Kohli and H.Köster, Hoppe Seyler's Z.Phys.Chem. 360, 1019(1979).
24. J.H. van Boom, P.M.J.Burgers and P.H. van Deursen, Tetrahedron Letters, 1974, 3785.
25. R.W.Adamiak, R.Arentzen and C.B.Reese, Tetrahedron Letters, 1977, 1431.
26. C.B.Reese, R.Titmas and L.Yau, Tetrahedron Letters, 1978, 2727.
27. C.B.Reese and L.Yau, Tetrahedron Letters, 1978, 4443.
28. H.Takaku, M.Yoshida, M.Kato and T.Hata, Chemistry Letters, 1979, 811.
29. Z.F.M.de Rooij, G.Wille-Hazeleger, P.M.Z.Burgers and Z.H. van Boom, Nucleic Acids Res., 6, 2237(1979).
30. V.Kohli and A.Kumar, Bull.Soc.Chim.Belg., 87, 211(1978).
31. H.Köster, A.Pollak and F.Cramer, Liebigs Ann.Chem. 1974, 959.
32. M.J.Gait and R.C.Sheppard, Nucleic Acids Res. 4, 1135(1977).
33. W.Heidmann and H.Köster, Angew.Chem. 88, 577(1976).
34. C.K.Narang, K.Brunfeldt and K.E.Norris, Tetrahedron Letters, 1977, 1819.
35. M.J.Gait and R.C.Sheppard, Nucleic Acids Res. 4, 4391(1977).
36. V.K.Potapov, V.P.Veiko, O.N.Koroleva and Z.A.Shabarova, Nucleic Acids Res. 6, 2041(1979).
37. N.Hoppe and H.Köster, in preparation.
38. H.Blöcker, R.Frank and H.Köster, Liebigs Ann.Chem., 1978, 991.
39. R.Frank and H.Köster, Nucleic Acids Res. 6, 2069(1979).
40. A.M.Maxam and W.Gilbert, Proc.Natl.Acad.Sci.(USA) 74, 560(1977).
41. R.Frank and H.Köster, Liebigs Ann.Chem. 1978, 971.
42. M.Goppelt, A.Pingoud, G.Maas, R.Frank and H.Köster, J.Mol.Biol., submitted for publication.
43. R.Frank and H.Köster, to be published.
44. B.Hofer and H.Köster, to be published.
45. H.Köster, Nachr.Chem.Tech.Lab. 27(11) (1979).
46. B.Hofer and H.Köster, to be published.
47. B.Hofer and H.Köster, FEBS-Letters, 102, 87(1979).
48. M.Kröpelin and H.Köster, to be published.

SILYL- AND STANNYL-ESTERS OF PHOSPHORUS OXYACIDS — INTERMEDIATES FOR THE SYNTHESIS OF PHOSPHATE DERIVATIVES OF BIOLOGICAL INTEREST

T. Hata and M. Sekine

Department of Life Chemistry, Tokyo Institute of Technology, Nagatsuta, Midoriku, Yokohama 227, Japan

<u>Abstract</u> - Among several esters of phosphorus oxyacids, we have recently paid attention to the silyl and stannyl esters to synthesize organophosphorus compounds which require to overcome the synthetic limitation owing to their instability. In this paper we describe a systematic study on the reactions of silyl phosphites compared with that of trialkyl phosphites and also to describe the synthesis of biologically interesting phosphate molecules by making the best use of the differences in reactivity among silyl-, stannyl-, and alkyl-esters of phosphorus oxyacids.

INTRODUCTION

The brilliant development of the chemistry of organophosphorus compounds during the past 40 years with numerous studies on the chemical reactions and their mechanisms has made it possible to introduce heteroatoms into organophosphorus compounds. Consequently, a wide variety of organophosphorus compounds have been registered during the last 20 years. At the same time, the chemistry of organosilicon compounds containing Si-C bonds has been developed rapidly. For example, trimethylsilylation has often been used as a useful technique of gas chromatography because it changes dramatically the solubility and volatility of the organic compounds (Ref. 1). On the other hand, the chemistry of organotin compounds was developed with success of their industrial applications to reagents as stabilizers against photodecomposition for poly(vinyl chloride) and other polymers, and later as fungicides. Since 1965 the rapidly increasing number of groups in many countries have produced various organotin compounds including organophosphorus compounds having stannyl groups. Silicon-containing phosphorus compounds have recently been studied extensively by Pudovik and his coworkers. In consideration of previous reports on the chemistry of silyl esters of phosphorus oxyacids, there still remains the basic framework on the reactivity of silyl and stannyl esters compared with that of known alkyl esters of phosphorus oxyacids because they show several interesting features for the synthesis of biologically important molecules.

PREPARATION OF SILYL PHOSPHITES

Highly reactive silyl phosphites were prepared by the reactions of phosphorous acid, alkyl phosphonates, or dialkyl phosphonates with trimethylsilyl chloride (TMSCl) in the presence of triethylamine, or by metal salts of phosphonates with TMSCl. Bis(trimethylsilyl)acetamide (BSA) is also useful since it is a neutral and powerful silylating agent. Many examples showed that the preparation of silyl phosphites were usually carried out in dry pyridine at room temperature or at a slightly higher temperature since P(O)H form of the phosphonates resists the silylation reactions.

Silyl phosphites having low molecular weights can be purified by distillation. However, non-volatile silyl phosphites such as nucleoside silyl phosphites were used for the successive reactions without any purification because, in most of the cases, the silyl phosphites were practically pure. The purities were estimated by their ^{1}H NMR spectra or monitored by the reaction with 2,2'-dipyridyl disulfide (Ref. 2) before use.

$$(RO)_2\overset{O}{\overset{\|}{P}}-H + ClSiMe_3 + Et_3N \longrightarrow (RO)_2\overset{..}{P}-OSiMe_3 + Et_3NH^+Cl^-$$

The most useful silyl phosphite is tris(trimethylsilyl) phosphite (TMSP) (bp 86.5°/18 mmHg) which serves as versatile agent for the synthesis of organophosphorus compounds. Several reactions indicate that the reactivity of TMSP is higher than that of trimethyl phosphite and triethyl phosphite since Me_3SiO group appears as a strong electron-donating group towards the tervalent phosphorus atom in TMSP compared with RO group of trialkyl phosphites.

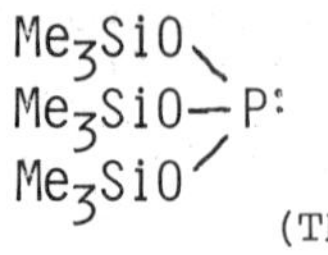

$$Me_3SiO-P(OSiMe_3)_2 \quad (TMSP)$$

A typical feature is represented by deoxygenation reactions and the Arbuzov reaction as described follows.

DEOXYGENATION REACTIONS BY MEANS OF TMSP

Deoxygenation reactions (Ref. 3, 4) using tervalent phosphorus compounds constituted one of important reactions in organic synthesis. TMSP was found to be useful for deoxgenation reactions as a highly reactive agent relative to previously known agents, i.e., tervalent phosphorus compounds substituted with electron-donating groups such as dialkylamino groups (Ref. 5).

The deoxygenation reaction of methyl phenyl sulfoxide by using TMSP was carried out at a higher temperature around 150°C, the reaction proceeded to give methyl phenyl sulfide in 72% yield after 8.5 h. The prolonged heating of the mixture didn't afford a better yield of the sulfide. In this reaction bis(trimethylsilyl) S-phenyl phosphorothioate was formed in 28% yield as the sole phosphorus-containing side-product. From the above facts the deoxygenation might proceed through Path A as shown in the following scheme. The phosphorothioate would be formed with the elimination of a volatile methyl trimethylsilyl ether through Path B.

$$CH_3-\overset{O}{\underset{\|}{S}}-Ph + P(OSiMe_3)_3 \longrightarrow CH_3-S-Ph + O=P(OSiMe_3)_3$$

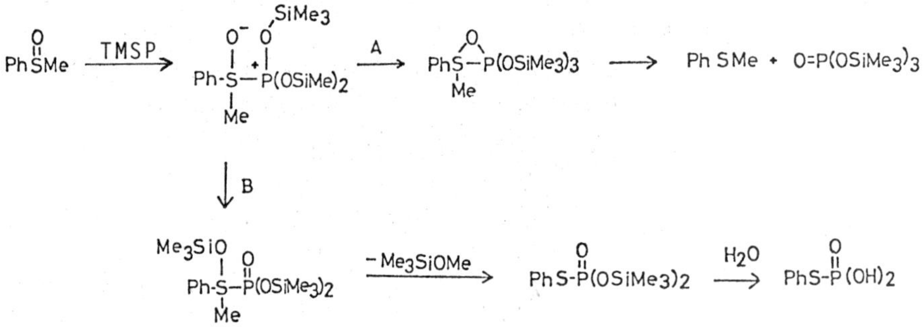

$$Ph\overset{O}{\underset{\|}{S}}Me \xrightarrow{TMSP} \underset{Me}{\overset{O^- \; O^+/SiMe_3}{Ph-S-P(OSiMe)_2}} \xrightarrow{A} \underset{Me}{\overset{O}{Ph\overset{|}{S}-P(OSiMe_3)_3}} \longrightarrow PhSMe + O=P(OSiMe_3)_3$$

$$\downarrow B$$

$$\underset{Me}{\overset{Me_3SiO \; O}{Ph-\overset{|}{S}-\overset{\|}{P}(OSiMe_3)_2}} \xrightarrow{-Me_3SiOMe} Ph\overset{O}{\underset{\|}{S}}-\overset{}{P}(OSiMe_3)_2 \xrightarrow{H_2O} Ph\overset{O}{\underset{\|}{S}}-\overset{}{P}(OH)_2$$

In order to compare the deoxygenating ability of TMSP with those of aminophosphines known as strong deoxygenating agents, the deoxygenation reactions of methyl phenyl sulfoxide with aminophosphines were examined. The results are shown in TABLE 1. Commercially available triethyl phosphite is known to isomerize to diethyl ethylphosphonate at 150°C by the reaction with methyl phenyl sulfoxide losing its deoxygenating ability (Ref. 4, 6).

TABLE 1. Reaction of tervalent phosphorus compounds with methyl phenyl sulfoxide

Compound	Time(h)	Yield of MeSPh (%)
$P(OSiMe_3)_3$	8.5	72
	20	65
$P(NMe_2)_3$	20	39
$P(NEt_2)_3$	20	31
$P(Bu-n)_3$	20	no reaction

Nitrobenzene reacted with TMSP at 180°C to afford an unidentified black mass. Nitrosobenzene reacted violently with TMSP even at room temperature. However, the result was similar as in the case of nitrobenzene. On the other hand, when 2-nitrobiphenyl was heated with TMSP (3 equiv.) at 130°C for 5 h, carbazole was obtained in 84% yield. In contrast to the previous reported carbazole synthesis (Ref. 7) the present reaction proceeds much faster even at lower temperatures and the product was separated easily as a precipitate by simple addition of water after the reaction was completed.

$$\text{2-NO}_2\text{-biphenyl} \xrightarrow{\text{TMSP}} \xrightarrow{\text{H}_2\text{O}} \text{carbazole} + \text{O=P(OH)}_3$$

REACTION OF TMSP WITH ALKYL HALIDES (THE ARBUZOV REACTION)

The reactions of trialkyl phosphites with alkyl halides are well established and the reactivity of various phosphites has been investigated in detail (Ref. 8). TMSP was also employed in the Arbuzov reaction and it afforded the corresponding bis(trimethylsilyl) alkylphosphonates (Ref. 9).

$$RX + P(OSiMe_3)_3 \longrightarrow (Me_3SiO)_2\overset{O}{\underset{\|}{P}}-R + Me_3SiX$$

The reactions were carried out without use of solvent at higher temperatures as in the case of the reactions with trialkyl phosphites.

When n-butyl bromide was used, the reaction was performed at 110°C for 7 h. Benzyl halides (X =Cl, Br, and I) reacted with TMSP at 160°C for 6 h to give bis(trimethylsilyl) benzylphosphonate in 80-91% yields. Benzyl iodide, however, could be employed in this reaction, but side reactions were observed. Probably, it was due to trimethylsilyl iodide formed by the reaction.

It is noteworthy that TMSP is substantially in the stable tervalent form and even if it isomerizes to the phosphonate form, the tervalent form can be regenerated due to the strong affinity between the trimethylsilyl group and the neighbouring oxygen atom of the O=P bond. Therefore, TMSP functions effectively for deoxygenation reactions and also for the Arbuzov reaction where the product is not contaminated with undesirable phosphonates as in the case of trialkyl phosphites.

$$:P(OSiMe_3)_3 \;\rightleftarrows\; Me_3Si\overset{O}{\underset{\|}{P}}(OSiMe_3)_2$$

REACTION OF TMSP WITH ACYL HALIDES —— SYNTHESIS OF UNESTERIFIED ACYLPHOSPHONIC ACIDS

Trimethylsilyl esters are particularly suitable for the synthesis of the compounds having weak bonds since the silyl group can be cleaved off readily under very mild conditions.

Reactions of trialkyl phosphites with acyl halides giving dialkyl acylphosphonates are well known (Ref. 10). However, it has been difficult so far to convert esters of acylphosphonic acids into the corresponding unestrified acids except in the case of diethyl ethoxycarbonylphosphonate reported by Warren (Ref. 11). This is because acid or alkaline treatment of dialkyl acylphosphonates usually leads to cleavage of the P-C bond prior to hydrolysis of the ester-linkage. Orlov and his coworkers (Ref. 12) reported the reaction of tris(tristhylsilyl) phosphite with some acyl chlorides to give bis(triethylsilyl) esters of acylphosphonic acids. We also examined (Ref. 13) that trimethylsilyl groups of the silyl esters of phosphonic and phosphoric acids can be removed simply by addition of water or alcohols. Therefore, it should be possible to prepare unesterified acylphosphonic acids by solvolysis of the trimethylsilyl esters of acylphosphonic acids obtained by the reaction of acyl chlorides with TMSP. For example, when TMSP was allowed to react with 1 equiv. of benzoyl chloride in dry benzene at room temperature for 3 h, bis-(trimethylsilyl) benzoylphosphonate (R=Ph) was obtained in 72% yield. In a similar manner, several acylphosphonate derivatives were obtained in 50-91% yields (Ref. 13). The unesterified acylphosphonic acids were isolated as the monoanilinium salts by addition of aniline-containing methanol or ethanol in high yields as listed in TABLE 2.

On the other hand, when 2 equiv. of TMSP was employed in the above reaction, tetrakis(trimethylsilyl) ester of 1-trimethylsilyloxybenzylphosphonic acid was successfully obtained in 76% yield as the major product. These reactions usually give rise to carbonyl adducts formed as a result of the successive addition of TMSP to bis(trimethylsilyl) acylphosphonates.

Di-anilinium salt of the bisphosphonic acids were successfully obtained in high yields by the same procedure as described in the case of unesterified acylphosphonates.

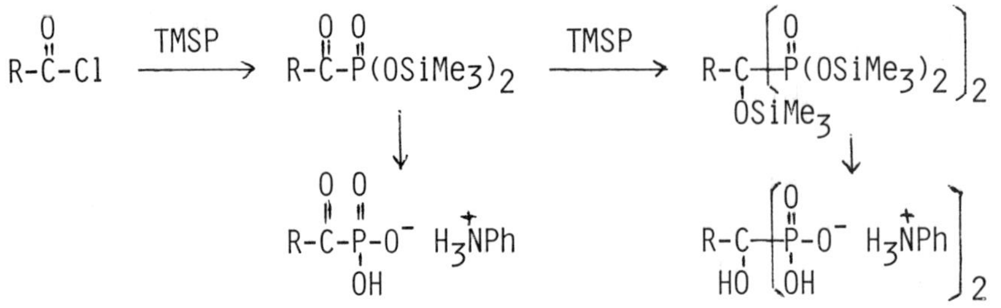

$$R-\overset{O}{\underset{}{C}}-Cl \xrightarrow{TMSP} R-\overset{O}{\underset{}{C}}-\overset{O}{\underset{}{P}}(OSiMe_3)_2 \xrightarrow{TMSP} R-C\left[\overset{O}{\underset{OSiMe_3}{P(OSiMe_3)_2}}\right]_2$$

$$\downarrow \qquad\qquad\qquad\qquad\qquad \downarrow$$

$$R-\overset{O}{\underset{}{C}}-\overset{O}{\underset{OH}{P}}-O^- \ H_3\overset{+}{N}Ph \qquad R-\underset{HO}{C}\left[\overset{O}{\underset{OH}{P}}-O^- \ H_3\overset{+}{N}Ph\right]_2$$

TABLE 2. Reactions of tris(trimethylsilyl) phosphite with acyl chlorides

RCOCl	Yield of TMS deriv. (%)	Time (h)	Yield of unesterified deriv. (%)
C_6H_5COCl	72	2	93
$4-Cl-C_6H_4COCl$	77	2	94
$4-MeO-C_6H_4COCl$	72	2	99
CH_3COCl	50	2	89
C_2H_5COCl	63	2	92
C_2H_5OCOCl	92	5	98
Cl_3CCH_2OCOCl	91	2	96

All the reactions were carried out in dry benzene.
For the preparation of unesterified derivatives, 2 equiv. of aniline in alcohols were used.

REACTIONS OF TMSP WITH CARBONYL COMPOUNDS

The formation of "carbonyl adducts" by the use of phosphites is not unprecedented since trialkyl phosphites react sluggishly with aldehydes, i.e., the reaction requires rather drastic conditions or much long reaction time and gives several addition compounds such as 1:1 adduct (Ref. 14), 1:2 adduct (Ref. 15), and 1:3 adduct (Ref. 15a) in relatively low yields (20-60%). Pudovik and his coworkers (Ref. 16) reported that the 1:1 adducts were also obtained by the reaction of diethyl trimethylsilyl phosphite with carbonyl compounds.

We have found that TMSP reacted smoothly with carbonyl compounds to give the 1:1 carbonyl addition products (Ref. 17).

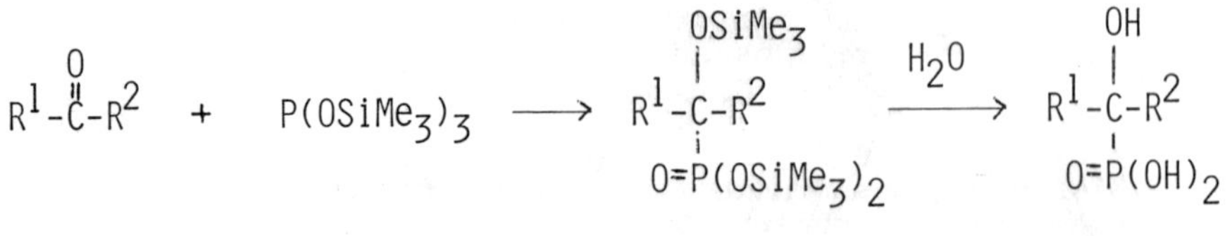

$$R^1-\overset{O}{\underset{}{C}}-R^2 + P(OSiMe_3)_3 \longrightarrow R^1-\underset{O=P(OSiMe_3)_2}{\overset{OSiMe_3}{C}}-R^2 \xrightarrow{H_2O} R^1-\underset{O=P(OH)_2}{\overset{OH}{C}}-R^2$$

According to this reaction, the 1:1 adducts were obtained as the sole product in excellent yields and all the trimethylsilyl groups can be easily removed from the adducts by simple solvolysis using alcohols or water under very mild conditions. Therefore, the reaction by employing TMSP provides a useful and very convenient method for the synthesis of α-hydroxyalkylphosphonic acids.
When the adducts prepared in the above experiments were treated with methanol, the corresponding α-hydroxylakylphosphonic acids were obtained in almost quantitative as shown in TABLE 3.

TABLE 3. Reactions of aldehydes or ketones with TMSP[a]

R^1	R^2	Time (h)	Yield of 1:1 adduct (%)	bp(°C/mmHg)	Yield of unesterified acid (%)	mp(°C)
Ph	H	4	88	126-130/2	98	179-180[b]
$4-ClC_6H_4$	H	4	91	139-142/0.35	93	106-108[b]
$4-MeOC_6H_4$	H	4	83	150-152/0.4	97	124-125[b]
Et	H	2	85	91-95/0.2	quant.	159[b]
i-Pr	H	4	82	106-108/0.8	quant.	166-168[b]
Ph	CH_3	8	75	126-129/0.3	97	180-181[c]
CH_3	CH_3	4	90	86-90/0.7	96	176-178[c]

a) The reaction was carried out at room temperature except the case of acetone(under reflux). b) Free acid. c) Monoanilinium salt.

Kamai (Ref. 18) reported that the reaction of α,β-unsaturated aldehydes with triethyl phosphite at 120°C gave the addition products of the Michael's type, namely, the 1,4-addition products, in ca. 20% yields.
It was found that the reaction of TMSP with α,β-unsaturated aldehydes afforded the 1,2-addition products selectively in excellent yields as listed in TABLE 4.

TABLE 4. Reaction of α,β-unsaturated aldehydes with TMSP[a]

R^1	R^2	Reaction temp.	time (h)	Yield of 1,2-addition prod. (%)	bp(°C/mmHg)	Yield of 1,4-addition prod. (%)	mp(°C/mmHg)
H	H	r.t.	4	94	97-103/1.1	—	—
CH_3	H	r.t.	4	89	93/0.23	—	—
Ph	H	r.t.	4	84	125-135/0.11	—	—
H	CH_3	r.t.	4	—	—	84	119-121/2.1
Ph	Ph	reflux	4	—	—	75	169-171/0.35
CH_3	OEt	reflux	7	—	—	45(71)[b]	113-115/0.4

a) The reaction was carried out in dry benzene. b) Dioxane was used as solvent.

Both 1,2- and 1,4-adducts were converted successfully to monoanilinium or cyclohexylammonium salts (or free acids in some cases) of the corresponding unesterified α-hydroxyalkylphosphonic acids by treatment with aqueous tetrahydrofuran followed by addition of aniline or cyclohexylamine in nearly quantitative yields.

Recently, Evans has reported similar results using dimethyl trialkylsilyl phosphites (Ref. 19). Our efforts were focused on unsymmetical ketones synthesis from aldehydes by means of alkylation of relatively acidic α-hydrogen of carbonyl adducts obtained by the reaction of TMSP with aldehydes followed by elimination of phosphonate.

The Si-O bond of trimethylsilyl-esters of phosphonic acids was sensitive toward metallating agents such as n-BuLi (Ref. 20) and was easily hydrolyzed even when the esters were exposed to air (Ref. 17). Dimethyl esters of phosphonic acids seem to be also unstable toward nucleophiles such as iodide ion produced during the alkylation reaction to decompose partially giving demethylated by-products. In consideration of the above facts, the ethyl ester was chosen of the phosphonic acids and the reactions of diethyl trimethylsilyl phosphite with various carbonyl compounds were examined. The addition products were found to be sufficiently stable during work-up of extraction and on silicagel column (Ref. 21).

$$R^1CH=CH-\overset{O}{\underset{}{C}}-H \xrightarrow{TMSP} R^1CH=CH-\underset{O=P(OSiMe_3)_2}{\overset{OSiMe_3}{CH}} \xrightarrow{H_2O} R^1CH=CH-\underset{}{\overset{OH}{CH}}-\overset{O}{P}(OH)_2$$

$$R^1CH=CH-\overset{O}{\underset{}{C}}-R^2 \xrightarrow{TMSP} R^1CH-CH=\underset{O=P(OSiMe_3)_2}{\overset{OSiMe_3}{C}}-R^2 \xrightarrow{H_2O} R^1CH-CH_2-\underset{O=P(OH)_2}{\overset{O}{C}}-R^2$$

The results were essentially almost the same as those described previously in the case of the reactions with TMSP (Ref. 17) indicating the high regio-selectivities and high yields (Ref. 21).

$$R^1-\underset{O=P(OEt)_2}{\overset{OSiMe_3}{C}}-R^2 \qquad R^1CH=CH-\underset{O=P(OEt)_2}{\overset{OSiMe_3}{CH}} \qquad R^1CH-CH=\underset{O=P(OEt)_2}{\overset{OSiMe_3}{C}}-R^2$$

The adducts could be converted to the lithio-derivatives by means of lithium diisopropylamide in THF. The lithio-intermediates were successively alkylated or acylated by the corresponding electrophiles.

The further studies in alkylation and acylation have been done in this laboratory and the results are described in the literatures (Ref. 21, 22, 23).

REACTIONS OF TMSP WITH ISOCYANATES OR THIOISOCYANATES

Mukaiyama and his coworkers (Ref. 24) reported that trialkyl phosphites reacted with isocyanates or isothiocyanates at elevated temperatures (150-180°C) to give the corresponding isonitriles.

$$R-N=C=X + P(OR)_3 \xrightarrow{\Delta} R-N=C + X=P(OR)_3 \qquad X=O \text{ or } S$$

When TMSP was allowed to react with i-butyl isothiocyanate (one equiv.), an alkyl isothiocyanate, an exothermic reaction took place even at room temperature giving a thiocarbonyl adduct. The reaction mixture was further heated at 100°C for 3 h to afford i-butyl isonitrile in 90% yield by distillation. The adduct could not be isolated. However, formation of the adduct was strongly suggested because a monoanilinium salt of i-butylthiocarbamoylphosphonic acid was isolated in 60% yield by treatment of the reaction mixture with aniline-containing methanol after the reaction was completed.

$$i\text{-BuN=C=S} + \text{P(OSiMe}_3)_3 \rightleftharpoons i\text{-BuN=}\overset{\overset{S^-}{|}}{\underset{}{C}}\text{-}\overset{+}{\text{P}}\text{(OSiMe}_3)_3 \rightleftharpoons \left[i\text{-BuN=}\overset{\overset{S\text{-SiMe}_3}{|}}{\underset{\overset{\|}{O}}{C}}\text{-P(OSiMe}_3)_2 \right]$$

$$i\text{-BuN=C:} + \text{S=P(OSiMe}_3)_3 \xleftarrow{\Delta} i\text{-BuN=}\overset{\overset{S}{\diagdown}}{\underset{}{C}}\text{—P(OSiMe}_3)_2$$

$$\downarrow \text{PhNH}_2/\text{MeOH}$$

$$i\text{-BuN-}\overset{\overset{H}{|}}{\underset{}{C}}\text{-}\overset{\overset{S}{\|}}{\underset{\overset{|}{OH}}{P}}\text{-O}^- \;\; \text{H}_3\overset{+}{\text{N}}\text{Ph}$$

This result suggests the trapping of the hypothetical intermediate dicussed by Mukaiyama (Ref. 24) in this type of desulfurization. Attempts to isolate the adduct by distillation even under highly reduced pressures caused the partial decomposition to i-butyl isonitrile and tris(trimethylsilyl) thionophosphate.

On the other hand, the reaction of n-butyl isocyanate with TMSP (1 equiv.) at room temperature for 1.5 h gave the carbonyl adduct in 93% yield by rapid distillation under reduced pressure at lower temperatures than 100°C. Even when the distillation was conducted at higher temperatures, no deoxygenation

$$\left[\text{RN=}\overset{\overset{X}{|}}{\underset{}{C}}\text{-}\overset{+}{\text{P}}\text{(OR)}_3 \right]$$

reaction took place but the retro-reaction occured to give TMSP releasing volatile n-butyl isocyanate. Unesterified n-butylcarbamoylphosphonic acid was isolated in 96% yield as the monoanilinium salt.

$$\text{BuN=C=O} + \text{P(OSiMe}_3)_3 \underset{\Delta}{\rightleftharpoons} \text{BuN=}\overset{\overset{O^-}{|}}{\underset{}{C}}\text{-}\overset{+}{\text{P}}\text{(OSiMe}_3)_3 \underset{\Delta}{\rightleftharpoons} \text{BuN=}\overset{\overset{OSiMe_3}{|}}{\underset{\overset{\|}{O}}{C}}\text{-P(OSiMe}_3)_2 \longrightarrow \text{BuN-}\overset{\overset{H}{|}}{\underset{}{C}}\overset{\overset{O}{\|}}{\underset{}{}}\text{-}\overset{\overset{O}{\|}}{\underset{O^-}{P}}\text{-OH} + \text{H}_3\overset{+}{\text{N}}\text{Ph}$$

Phenyl isocyanate reacted with TMSP exothermically and the adduct was obtained in 94% yield. It was found to be quite stable at least to 200°C and could be converted to a monoanilinium salt of phenylcarbamoylphosphonic acid in 98% yield. In the case of phenyl isothiocyanate the yield of the adduct was almost quantitative. Distillation of the adduct caused the retro-reaction to the starting materials. Unesterified monoanilinium salt of phenylthiocarbamoylphosphonic acid was obtained in 95% yield.

$$\text{PhN=C=X} + \text{P(OSiMe}_3)_3 \underset{\Delta}{\rightleftharpoons} \text{PhN=}\overset{\overset{XSiMe_3}{|}}{\underset{\overset{\|}{O}}{C}}\text{-P(OSiMe)}_2 \xrightarrow[\text{MeOH}]{\text{PhNH}_2} \text{PhN-}\overset{\overset{H}{|}}{\underset{}{C}}\overset{\overset{X}{\|}}{\underset{}{}}\text{-}\overset{\overset{O}{\|}}{\underset{O^-}{P}}\text{-OH} + \text{H}_3\overset{+}{\text{N}}\text{Ph}$$

$$X = O \text{ or } S$$

From these results, the thermal stability of the carbonyl and thiocarbonyl adducts of TMSP would be ranked in the following order:

$$\text{Ar-N=C(OSiMe}_3)\text{-P(O)(OSiMe}_3)_2 > \text{Alkyl-N=C(OSiMe}_3)\text{-P(O)(OSiMe}_3)_2 >$$
$$\text{Ar-N=C(SSiMe}_3)\text{-P(O)(OSiMe}_3)_2 > \text{Alkyl-N=C(SSiMe}_3)\text{-P(O)(OSiMe}_3)_2$$

The adducts were in equilibrium with the corresponding starting materials due to the instability of the S-Si bonds and/or the nonconjugated N=C bonds. The adduct (R=Ph) was stabilized by the conjugation between the N=C bond and the aromatic ring. The isomer having Si-N bond may also exist, but their structures have not yet been assigned clearly. The thermal desulfurization might be due to the C-S bond cleavage through a three-membered ring transition state. Desulfurization of isothiocyanates is known to be easier than deoxygenetion of isocyanates.

In conclusion, it is noted that the present reactions provide the general and convenient method for the synthesis of a new class of organophosphorus compounds, unesterified carbamoyl- and thiocarbamoyl-phosphonic acids (Ref.25).

REACTION OF TMSP WITH α-HALOCARBONYL COMPOUNDS

The reaction of TMSP with α-halocarbonyl compounds which have not been described to date except for an analogous reaction of bis(trimethylsilyl) hypophosphite with chloroacetone reported by Pudovik (Ref. 26). It is well recognized that trialkyl phosphites react with α-halocarbonyl compounds to give enol-phosphates (the Perkow reaction) and/or phosphonates (the Arbuzov reaction) depending on the nature of the compounds and the reaction conditions employed. A typical example appeared in the literature reported by Kirby (Ref. 27). In contrast to the results, however, when TMSP was allowed to react with chloroacetone in dry THF, an oily product (bp 102-103°/0.45 mmHg) distilled was neither the Perkow product nor the Arbuzov product but bis(trimethylsilyl) 1-methyl-1-trimethylsilyloxy-2-chloroethylphosphonate was obtained in an excellent yield of 92%. The structure of the carbonyl adduct was confirmed by its ^{1}H-NMR and IR spectra. The use of bromoacetone or mesyloxyacetone, substituted with a stronger leaving group, also gave the corresponding carbonyl adduct in 92% yield each, while the reactions of trialkyl phosphites or triphenylphosphines with these reactive ketones were known to give β-oxophosphonates predominantly. Next, TMSP reacted with a disubstituted α-chloro ketone and with two kinds of α-chloroaldehydes. These α-halocabonyl compounds usually gave the exclusive formation of the Prekow reaction products. However, even in the case of the reactions of TMSP with α-chlorocarbonyl compounds [R_1=CH$_3$, R_2=H, CH$_3$, or R_2-R_3=(CH$_2$)$_4$], the carbonyl adducts were obtained in excellent yields. On the other hand, when TMSP was employed to react with phenacyl chloride, a Perkow reaction product was obtained in 74% yield under the same conditions as mentioned in the above experiments. When phenacyl bromide was used in place of phenacyl chloride, Perkow and Arbuzov reaction products were obtained as a mixture in 61 and 14% yields, respectively. Next, the reactions of TMSP with ethyl chloroacetate and with ethyl bromoacetate were examined. It was found that ethyl chloroacetate did not react sufficiently with TMSP at room temperature, but, when it was allowed to react with TMSP under reflux, the reaction proceeded to give an Arbuzov reaction product in 66% yield, while ethyl bromoacetate reacted even at room temperature with TMSP to afford only an Arbuzov product in 84% yield.

Borowitz (Ref. 28) has recently proposed a mechanism of the Perkow reaction involving intial attack of a trialkyl phosphite on the carbonyl carbon. The isolation of several carbonyl adducts described here indicates the trapping of the initially formed carbonyl insertion intermediate, discussed by Borowitz, due to the rapid intramolecular rearrangement of a trimethylsilyl group of the intermediate to its oxygen anion via five-membered transition state as described in the following scheme.

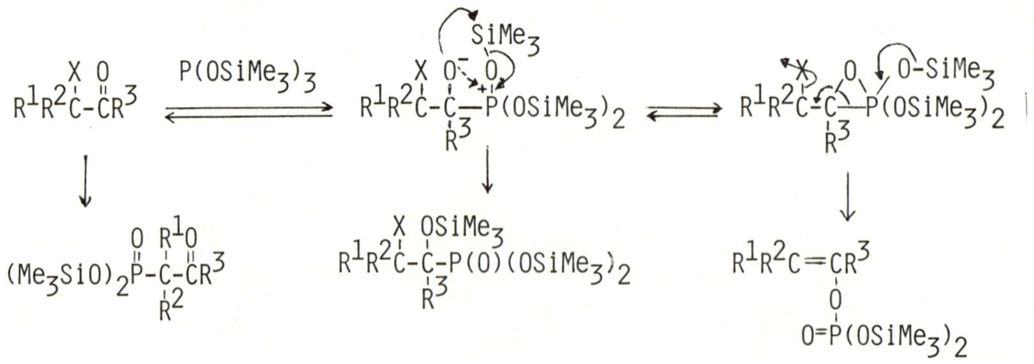

$$R^1R^2\overset{X}{C}-\overset{O}{C}R^3 \xrightleftharpoons{P(OSiMe_3)_3} R^1R^2\overset{X}{C}-\overset{O-SiMe_3}{\underset{R^3}{C}}-P(OSiMe_3)_2 \rightleftharpoons R^1R^2\overset{X}{C}-\overset{O-SiMe_3}{\underset{R^3}{C}}-P(OSiMe_3)_2$$

$$\downarrow \qquad\qquad \downarrow \qquad\qquad\qquad \downarrow$$

$$(Me_3SiO)_2\overset{O}{\overset{\|}{P}}-\overset{R^1}{\underset{R^2}{C}}-\overset{O}{\overset{\|}{C}}R^3 \qquad R^1R^2\overset{X}{C}-\overset{OSiMe_3}{\underset{R^3}{C}}-P(O)(OSiMe_3)_2 \qquad R^1R^2C=CR^3 \atop \underset{O=P(OSiMe_3)_2}{\overset{O}{|}}$$

It is plausible that the ordinary Perkow and/or Arbuzov reactions proceed as a result of the predominant cleavage of P-C bond of the three-membered ring intermediate in equilibrium with carbonyl attacking intermediate. The equilibria are affected by electron-withdrawing groups such as phenyl and ethoxycarbonyl groups (the Perkow reaction) or relatively reactive halogen atoms (bromine) which make the direct diaplacement reaction possible, compared with those of alkyl-substituted α-halocarbonyl compounds (the Arbuzov reaction). The effect of a substituent adjacent to the carbonyl group was clearly elucidated from the fact that, when p-bromophenacyl chloride, which has a stronger electron-withdrawing group, was used instead of phenacyl bromide, ratio of the Perkow to the Arbuzov reaction products increased from 4.4 to 10.0.

The following results illustrate features of the reactions described here. When chloroacetone-TMSP adduct was treated with ethanol (3 equiv.) and aniline (2 equiv.) in ether, a monoanilinium salt of 1-hydroxy-1-methyl-2-chloroethyl-phosphonic acid was obtained in 99% yield. In a similar manner, bromoacetone-TMSP adduct and α-chloropropionaldehyde-TMSP adduct were obtained in 98 and 87% yields, respectively. Therefore, the reactions of TMSP with alkyl-substituted α-halocarbonyl compounds provide the first method for the synthesis of unesterified 1-hydroxy-2-haloalkylphosphonic acids. From ethyl chloroacetate, a newly classified half ester, a monoanilinium salt of ethoxycarbonylmethylphosphonic acid was obtained in 94% yield.

Furthermore, it was of particular interest that a quite unstable enol-phosphate, unesterified phenacyl enol-phosphate was for the first time isolated as the crystalline form from the trimethylsilylated procursor in 81% yield using aniline-containing ethanol at room temperature for 2.5 h. Finally, it was shown that the carbonyl adducts could be converted to bis(trimethylsilyl) esters of 1,2-epoxyalkylphosphonic acids by treatment with sodium methoxide in methanol followed by retrimethylsilylation with trimethylsilyl chloride in dry THF. Thus, for instance, bis(trimethylsilyl) 1-methyl-1,2-epoxyethylphosphonate (bp 68-77°/0.10-0.12 mmHg) was obtained from the corresponding carbonyl adduct in 58% yield. The epoxyphosphonate was readily converted to the corresponding unesterified epoxyphosphonate.

This facile conversion of 1-silyloxy-2-halo-alkylphosphonates to 1,2-epoxyalkylphosphonates reported here is noteworthy in the context of interest in the synthesis of such unesterified epoxyalkylphosphonates following discovery of a naturally occuring unesterified epoxyalkylphosphonate, phosphonomycin (Ref. 29), which is a wide-spectrum antibiotic.

PREPARATION OF ENOL-PHOSPHATES

Phosphoenolpyruvate (PEP) is well known as one of the high-energy phosphate compounds and plays a vital role in the ATP synthesis with phosphoenolpyruvate kinase (Ref. 31). The large free energy change for the hydrolysis (-14.8 kcal/mole) is due to the characteristic enol phosphate. From a mechanistical point of view, the reactivity of the C=C-O-P system seems to be also due to the fact that PEP is an unesterified enol-phosphate with the two dissociated anions of the phosphate which facilitates the elimination of metaphosphate compared with the corresponding dialkyl esters of PEP.

In this section we describe a convenient method for the synthesis of unesterified enol-phosphates and a facile synthesis of PEP.

There have appeared a number of studies on the preparation of PEP (Ref.32). However, even simple unesterified enol-phosphates other than PEP have been described in only a few papers. Most of the synthesis of enol-phosphates have dealt with the corresponding dialkyl esters which were easily accessible through the Perkow reaction of trialkyl phosphites with α-halocarbonyl compounds (Ref. 33). The dialkyl esters could not be converted by acid- or alkaline-treatment to the corresponding unesterified enol-phosphates since the cleavage of the O-P bond in C=C-O-P system occured predominantly. It has recently been demonstrated in this laboratory that the trimethylsilyl esters could be readily converted to monanilinium salts of the corresponding unesterified acids simply by addition of aniline-containing alcohols (Ref. 17). The procedure involves alcoholysis of the trimethylsilyl esters followed by neutralization of the resulting free acids with aniline. This method has proved quite useful and general for isolation of relatively unstable phosphoryl compounds such as unesterified enol-phosphates, α-hydroxyalkylphosphinates, and acylphosphonates (Ref. 34, 35, 13). On the other hand, McKenna (Ref. 36) have recently reported the facile conversion of dialkyl alkylphosphonates to bis(trimethylsilyl) alkylphosphonates by means of TMSBr. More recently, the scope and limitation of this type of reactions have been described by Sakurai and his coworkers (Ref.25). In connection with this type of dealkylation, Jung and Olah (Ref. 37) have developed a procedure for the dealkylation of carboxylic esters and their derivatives by the use of trimethylsilyl iodide (TMSI). Our particular interest was focused on the application of this new type of dealkylation and our procedure involving the alcoholysis of unesterified enol-phosphates.

Several dimethyl enol-phosphates were prepared from trimethyl phosphite and α-halocarbonyl compounds according to the literature procedures (Ref. 26) and the enol-phosphates obtained were successively employed as described below:

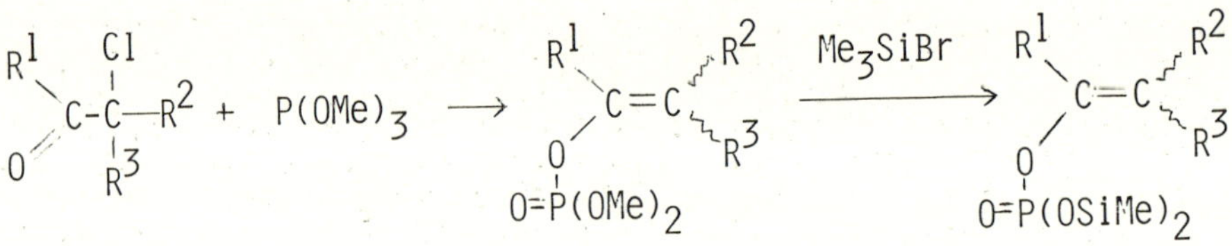

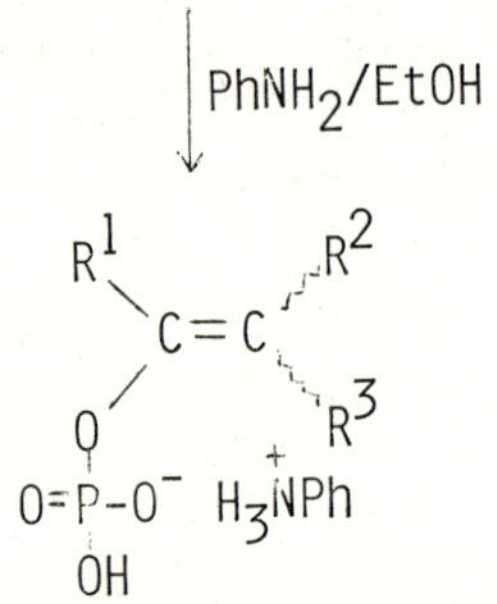

For example, when a dialkyl enol-phosphate, dimethyl 1-phenylvinyl phosphate, was treated with TMSBr bis(trimethylsilyl) 1-phenylvinyl phosphate was obtained in 78% yield after distillation. In a similar manner, three kinds of vinyl phosphates substituted with electron-withdrawing or electron-donating substitutents were treated with TMSBr. The vinyl phosphates obtained from the corresponding phenacyl halides substituted with bromine and chlorine at the p-position of the phenyl groups were successfully converted to the corresponding bis(trimethylsilyl) esters in 82% and 90% yields, respectively. However, the reaction of dimethyl 1-(4-methoxyphenyl)-vinyl phosphate with TMSBr did not give the desired product. Next, the dimethyl enol-phosphates prepared by the reaction of α-chloroaldehydes or α-chloro ketones with trimethyl phosphite were similar treated with TMSBr. It was found that all of these dimethyl esters were successfully converted to the bis(trimethylsilyl) esters with TMSBr in high yields without any side reactions. The structures of the silylated enol-phosphates obtained in the above experiments were chracterized by their NMR and IR spectra (TABLE 5). All of the silylated enol-phosphates were successfully converted to the monoanilinium salts in high yields. These monoanilinium salts had sharp melting points and charateristic, strong absorbances for $\nu_{C=C}$ in the IR spectra (TABLE 6). In the above experiments, unesterified isopropenyl phosphate was found to be extremrly unstable so that the corresponding monoanilinium salt could not be isolated. The NMR spectra of the isolated monoanilinium salts of the enol-phosphates were measured in CDCl$_3$/DMSO-d$_6$ (1:1, v/v).

TABLE 5. Preparation of bis(trimethylsilyl) enol-phosphates from dimethyl enol-phosphates.

Product			Yield	bp
R^1	R^2	R^3	(%)	(°C/mmHg)
C_6H_5	H	H	78	122-127/2.0
4-Cl-C_6H_4	H	H	90	166-168/1.4
4-Br-C_6H_4	H	H	82	140-151/2.5
CH_3	H	H	95	75-77/1.25
CH_3	H	CH_3	88	99-102/1.3
H	H	CH_3	96	94-96/2.0

TABLE 6. Preparation of monoanilinium salts of unesterified enol-phosphates from bis(trimethylsilyl) enol-phosphates.

Product			Yield	mp
R^1	R^2	R^3	(%)	(°C)
C_6H_5	H	H	66	94-95
4-Cl-C_6H_4	H	H	80	98-99
4-Cl-C_6H_4	H	H	78	101-102
CH_3	H	CH_3	78	94-96
H	H	CH_3	99	123-125

SYNTHESIS OF PHOSPHOENOLPYRUVATE

The concept of 'high energy phosphate bond' was originally proposed by Lipman in 1941 (Ref. 38). Several types of high energy phosphates, especially, phosphoenolpyruvate (PEP), phosphocreatine, and acetyl phosphate are well known. They make a role of phosphate-transfer reactions involving ATP synthesis in biological systems. Cramer (Ref. 39) has studied 'Zur Chemie der energiereichen Phosphate' and proposed several types of compounds as phosphorylating agents. One of the interesting high energy phosphates is PEP since it has the highest free energy difference of hydrolysis ($\triangle G° = -14.8$ kcal/mole). While several synthetic routes to the esters of PEP have been developed (Ref. 26), the synthesis of unesterified naturally occuring PEP has been troublesome for removal of the protecting groups and the isolated yields are relatively low (Ref. 26, 39). The most important subject of the synthesis of PEP by the Perkow reaction is selection of the starting materials, both α-halopyruvic acid derivatives and phosphites, since the reaction is known to afford generally an enol-phosphate (the Perkow reaction product) and/or a phosphonate (the Arbuzov reaction product). Tris(trimethylsilyl) phosphite seems to be useful but it can not be employed in this reaction because it affords not only the desired PEP silyl ester but also 1,2-carbonyl addition products as described previously (Ref. 30). After several screenings, it was found that pyruvic acid and dimethyl trimethylsilyl phosphite were the most suitable starting materials for this purpose.

When pyruvic acid was allowed to react with trimethylsilyl chloride (TMSCl) (2 equiv.) and triethylamine (2 equiv.) in the presence of a catalytic amount of 4-dimethylaminopyridine (Ref. 40) in dry benzene under reflux for 2 h, trimethylsilyl 2-trimethylsilyloxy-2-propenoate was obtained in 91% yield and purified by distillation (bp 88-92°C). It was treated with one equiv. of bromine in methylene chloride at -78°C. The reaction mixture was successively treated with dimethyl trimethylsilyl phosphite (0.9 equiv.) at 0°C overnight. After removal of the solvent, tris(trimethylsilyl) ester of PEP (bp 112-116°C/2.2 mmHg) was obtained in 90% yield. The structure was confirmed by its IR and ^{1}H-NMR spectra. It was treated with 3 equiv. of sodium ethoxide in dry ether. The precipitate of trisodium salt of PEP was collected and dried over P_4O_{10}. The yield was almost quantative based on the silyl ester. It was homogeneous on paper chromatogram and the structure of PEP was supported by its IR and ^{1}H-NMR spectra and also by elemental analysis.

The one-flask reaction seems to proceed smoothly and efficiently without addition of trimethylsilyl bromide (TMSBr) as shown in the following scheme.

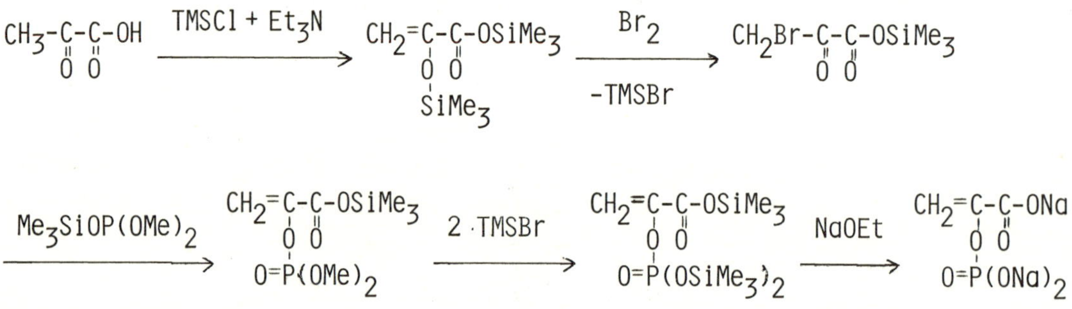

$$CH_3-\underset{O}{\overset{\|}{C}}-\underset{O}{\overset{\|}{C}}-OH \xrightarrow{TMSCl + Et_3N} CH_2=\underset{\underset{SiMe_3}{|}}{C}-\underset{O}{\overset{\|}{C}}-OSiMe_3 \xrightarrow[-TMSBr]{Br_2} CH_2Br-\underset{O}{\overset{\|}{C}}-\underset{O}{\overset{\|}{C}}-OSiMe_3$$

$$\xrightarrow{Me_3SiOP(OMe)_2} CH_2=\underset{\underset{O=P(OMe)_2}{|}}{C}-\underset{O}{\overset{\|}{C}}-OSiMe_3 \xrightarrow{2 \cdot TMSBr} CH_2=\underset{\underset{O=P(OSiMe_3)_2}{|}}{C}-\underset{O}{\overset{\|}{C}}-OSiMe_3 \xrightarrow{NaOEt} CH_2=\underset{\underset{O=P(ONa)_2}{|}}{C}-\underset{O}{\overset{\|}{C}}-ONa$$

It is noteworthy that two equiv. of TMSBr formed from the reaction sequence is very efficiently used for the transformation of the dimethyl ester to the bis(trimethylsilyl) ester according to the McKenna reaction (Ref. 30).

According to this method, the silylated PEP was obtained in one-flask reaction starting from the bis-silylated pyruvic acid and the yield of PEP is very high. In addition, the PEP is the pure sodium salt available for biochemical studies.

SYNTHESIS OF ACYL PHOSPHATES

Development of the synthetic methods of acetyl phosphate is of importance since it is also a high energy phosphate like PEP described in the above section.

In this section a general and convenient method for the synthesis of unesterified acyl phosphates by employing silyl and stannyl esters of phosphoric acid is described.

The reaction of tris(trimethylsilyl) phosphate with acetyl chloride seems to afford acetyl bis(trimethylsilyl) phosphate, however, a mixture of the mono-, di-, and tri-acetyl phosphates were formed even when one equiv. of acetyl chloride was used. On the contrary, tris(tri-n-butylstannyl) phosphate was found to be more reactive and selective with acetyl chloride. Moreover, it was found that diethyl tri-n-butylstannyl phosphate reacted smoothly with trimethylsilyl chloride to give diethyl trimethylsilyl phosphate in very high yield.

$$Bu_3SnO-\overset{\overset{O}{\|}}{P}(OEt)_2 + Me_3SiCl \longrightarrow Me_3SiO-\overset{\overset{O}{\|}}{P}(OEt)_2 + Bu_3SnCl$$

According to the above fact, the stannyl esters can be converted to the corresponding unesterified acids via the silyl ester intermediates which are known to be easily solvolyzed. Therefore, the reaction of tris(tri-n-butyl-stannyl) phosphate (Ref. 41) with acyl chloride was examined in order to obtain unesterified acyl phosphates.

Treatment of tris(tri-n-butylstannyl) phosphate prepared from phosphoric acid and bis(tri-n-butyltin) oxide with one equiv. of benzoyl chloride in carbon tetrachloride under reflux for 2 h afforded benzoyl bis(tri-n-butyl-stannyl) phosphate. It was successively treated with trimethylsilyl chloride (TMSCl) and the resulting benzoyl bis(trimethylsilyl) phosphate was transformed into monoanilinium salt of unesterified benzoyl phosphate by treatment with aniline-containing ethanol.

$$PhCOCl + (Bu_3SnO)_3P=O \longrightarrow Ph-\overset{O}{\underset{\|}{C}}-O-\overset{O}{\underset{\|}{P}}(OSnBu_3)_2 \xrightarrow{2\ TMSCl} Ph-\overset{O}{\underset{\|}{C}}-O-\overset{O}{\underset{\|}{P}}(OSiMe_3)_2$$

$$\downarrow PhNH_2/EtOH$$

$$Ph-\overset{O}{\underset{\|}{C}}-O-\overset{O}{\underset{\underset{OH}{|}}{P}}-O^-\ H_3\overset{+}{N}Ph$$

However, acetyl phosphate could not be obtained in the same manner. After several experiments, it was found that tri-n-butylstannyl bis(trimethylsilyl) phosphate was the most widely applicable to the synthesis of acyl phosphates. This reagent was obtained by the reaction of tris(trimethylsilyl) phosphate (one equiv.) with tri-n-butyltin methoxide (2 equiv.) in carbon tetrachloride at room temperature. The structure of the later compound was confirmed by its ^{1}H and ^{31}P NMR spectra. Bis(trimethylsilyl) tri-n-butylstannyl phosphate is sticky oily substance. When it was treated with acetyl chloride (one equiv.) in carbon tetrachloride under reflux for 1 h, acetyl bis(trimethylsilyl) phosphate was formed. It was successively treated with aniline-containg ethanol. Monoanilinium salt of acetyl phosphate was obtained in 80% yield.

In a similar manner, some of unesterified acyl phosphates were obtained as listed in TABLE 7.

$$Bu_3SnO-\overset{O}{\underset{\|}{P}}(OSiMe_3)_2 + RCOCl \xrightarrow{-Bu_3SnCl} R\overset{O}{\underset{\|}{C}}-O-\overset{O}{\underset{\|}{P}}(OSiMe_3)_2$$

$$\xrightarrow{ArNH_2\ /\ EtOH} R\overset{O}{\underset{\|}{C}}-O-\overset{O}{\underset{\underset{OH}{|}}{P}}-O^-\ H_3\overset{+}{N}Ar$$

TABLE 7. Synthesis of unesterified acyl phosphates.

RCOCl	mp* (°C)	Yield (%)
CH_3COCl	93-94	80
CH_3CH_2COCl	107-108	54
$CH_3(CH_2)_6COCl$	103-104	49
$(CH_3)_3CCOCl$	110-112	54

* Monoanilinium salt

REACTIONS OF SILYL PHOSPHITES WITH AZIDES

The reactions of azides with several tervalent phosphorus compounds were already investigated by Staudinger in the early of this century.(Ref. 42). He obtained several monophosphazenes represented by $X_3P=NR$. They are nitrogen analogs of the phosphorus ylides. Tris(trimethylsilyl) phosphite (TMSP) also reacted with azides but the reactions afforded the corresponding phosphoroamidates were very unstable.

In this section, mechanism of the reaction using TMSP and the properties of the phosphoroamidates are discribed briefly.

The reaction proceeds through a triazo intermediate which in turn decomposes to nitrogen and the corresponding monophosphazene. It further reacts with water to give the phosphoroamidate. Formation of the triazo intermediate was suggested by the precedent example of triphenylphosphine with 2,4,6-trinitrobenzene azide (Ref. 43).

$$(Me_3SiO)_3P: \overset{+}{N}\!\!\equiv\!\!\overset{-}{N}\!-\!NR \longrightarrow \left[(Me_3SiO)_3\overset{+}{P}\!-\!N\!=\!\overset{-}{N}\!-\!NR \leftrightarrow \underset{N=N}{(Me_3SiO)_3P\!-\!NR} \right] \xrightarrow[-N_2]{H_2O} \underset{OH}{\overset{O\ H}{HO\!-\!\underset{|}{\overset{\|}{P}}\!-\!NR}}$$

The phosphoroamidate was detected on tlc but it could not be isolated because of its sensitivity towards moisture. It was readily hydrolyzed to the corresponding amine and phosphoric acid. This shows that the transformation of azides to amines by means of TMSP under very mild conditions make possible.

$$\underset{OH}{\overset{O\ H}{HO\!-\!\underset{|}{\overset{\|}{P}}\!-\!NR}} \longrightarrow \underset{OH}{\overset{O}{HO\!-\!\underset{|}{\overset{\|}{P}}\!-\!OH}} + H_2NR$$

On the other hand, it was found that azide-nucleosides were synthesized from nucleosides in one-flask reaction using carbon tetrabromide, lithium azide, and triphenylphosphine (Ref. 44). Therefore, by the combination of the above two kinds of reactions, nucleosides were successfully converted to the corresponding amino-nucleosides in high yields (Ref. 44, 45). The procedure is widely applicable to the synthesis of amino-nucleosides and amino-sugars.

$$R\!-\!OH \xrightarrow{CBr_4/LiN_3/Ph_3P} R\!-\!N_3 \xrightarrow{(Me_3SiO)_3P} R\!-\!NH_2$$

APPLICATION OF THE PHOSPHITE CHEMISTRY TO THE NUCLEOTIDES SYNTHESIS

Most of the above sections, the chemistry of silyl phosphites are described. The reactions of silyl phosphites can be applied effectively to the nucleotide chemistry. For example, dinucleoside phosphoroamidates having a P-N internucleotidic bond and nucleoside 3',5'-cyclic phosphoroamidates were synthesized via nucleoside silyl phosphite intermediates (Ref. 45). The reactions of nucleoside silyl phosphites with diaryl disulfides are of importance because arylthio groups of phosphorothioates in nucleotides are promised as useful phosphate protecting groups for both internucleotidic phosphate (Ref. 46, 47) and 5'-terminal phosphate (Ref. 48). Recently, three convenient methods for the preparation of S, S-diaryl(dialkyl) phosphorodithioates are established starting from hydrophosphorous acid (Ref. 48), dimethyl phosphorodichloridate (Ref. 49) and trimethylsilyl phosphorodichloridate (Ref. 50). On the other hand, nucleoside phosphonates were oxidized very smoothly by means of 2,2'-dipyridyl disulfide via highly reactive nucleoside silyl phosphite intermediates to the corresponding nucleotides without any side reactions (Ref. 51). Moreover, a new type of selenium containing nucleotide, thymidine 5'-Se-ethyl phosphoroselenoate could also be obtained (Ref. 52).

REFERENCES

1) L. Birkofer and A. Ritter, Newer Methods of Preparative Organic Chemistry, Vol. 5, p. 211, Ed. W. Foerst, Verlag Chemie GmbH, Weinheim/Bergstr.
2) T. Hata and M. Sekine, Tetrahedron Lett., 3943(1974).
3) E. H. Amonoo-Neizer, S. K. Ray, R. A. Shaw, and B. C. Smith, J. Chem. Soc., 4296(1965).
4) S. Oae, A. Nakanishi, and S. Kozuka, Tetrahedron, 28, 549(1972), and references cited therein.
5) M. Sekine, H. Yamagata, and T. Hata, Tetrahedron Lett., 375(1979).
6) In the case of the reaction of triethyl phosphite with DMSO, a rather reactive sulfoxide, triethyl phosphite is known to behave as a good deoxygenating agent to afford dimethyl sulfide.
7) J. I. G. Cadogan, M. Caeron-wood, R. K. Mackie, and R. J. G. Searle, J. Chem. Soc., 4831(1965).
8) A. J. Kirby and S. G. Warren, The Organic Chemistry of Phosphorus, p. 37 Elsevier Pub., Amsterdam (1967).
9) T. Hata, M. Sekine, and N. Kagawa, Chemistry Lett., 635(1975).
10) G. M. Kosolapoff and L. Maier, Organic Phosphorus Compounds, Vol. 5, Wiley, New York (1973).
11) S. Warren and M. R. Williams, J. Chem. Soc., (B), 618(1971).
12) N. F. Orlov, B.K. Kaufman, L.Sukhi, L. N. Slesar, and E. V. Sudakova, Khim. Prakt. Primen. Kreiiorg. Soedin, Tr . Sovesch., 111(1966); Chem. Abstr., 72, 21738g(1970).
13) M. Sekine and T. Hata, J. Chem. Soc., Chem. Commun., 286(1978).
14) V. S. Abramov, Kokl. Akad. Nauk. SSSR, 95, 991(1954); Chem. Abstr., 49, 6084(1955).
15) (a) F. Ramirez, A. V. Patwardhan, and S. R. Heller, J. Am. Chem. Soc., 86, 514(1964); (b) V. A. Ginsburg and A. Y. Yakubovich, Zh. Obshch. Khim., 30, 3979(1960); Chem. Abstr., 55, 22099(1961).
16) (a) I. V. Konovalova, L. A. Burnaeva, N. Sh. Novikova, and A. N. Pudovik, Zh. Obshch. Khim., 46, 18(1976); (b) Z. S. Novikova, S. N. Mashoshina, T. A. Sapozhnikova, and I. F. Lutsenko, ibid., 41, 2662(1971); (c) Z. S. Novikova and I. F. Lutsenko, ibid., 40, 2129(1970); (d) L.V. Nesterov, N. E. Krepysheva, R. A. Sabirova, and G. N. Romanova, ibid., 41, 2449 (1971).
17) M. Sekine, I. Yamamoto, A. Hashizume, and T. Hata, Chemistry Lett., 485 (1977).
18) G. Kamai and V. A. Kukhtin, Zh. Obshch. Khim., 27, 2376(1956).
19) D. A. Evans, K. M. Hurst, L. K. Truesdale, and J. M. Takacs, Tetrahedron Lett., 2495(1977); J. Am. Chem. Soc., 100, 3467(1978).
20) M. Sekine and T. Hata unpublished results.
21) T. Hata, A. Hashizume, M. Nakajima, and M. Sekine, Tetrahedron Lett., 363 (1978).
22) T. Hata, N. Nakajima, and M. Sekine, Tetrahedron Lett., 2047(1979).
23) T. Hata, A. Hashizume, and M. Sekine, Chemistry Lett., 519(1979).
24) T. Mukaiyama, H. Nambu, and M. Okamoto, J. Org. Chem., 27, 3651(1962).
25) Unesterified N,N-disubstituted carbamoylphosphonic acids were recently reported; T. Morita, Y. Okamoto, anf H. Sakurai, Tetrahedron Lett., 2523(1978).
26) A. N. Pudovic, G. V. Romanov, and R. Ya. Nazmutdinov, Zh. Obshch. Khim., 44, 221(1974).
27) P. A. Chpard, V. M. Clark, R. F. Hudson, and A. J. Kirby, Tetrahedron, 21, 1961(1965).
28) I. J. Borowitz, S. F. Firstenberg, E. W. R. Casper, and R. K. Crouch, J. Org. Chem., 36, 3282(1971); I. J. Borowitz, S. Firstenberg, G. B. Borowitz, and D. Schuessler, J. Am. Chem. Soc., 94, 1623(1972); E. M. Gaydov and J. P. Bianchini, Can. J. Chem., 54, 3626(1976).
29) D. Hendlin, E. O. Stapley, M.Jackson, H. Wallick, A. K. Miller, F. J. Wolf, T. W. Muller, L. Chaiet, F. M. Kahan, E. L. Foltz. H. B. Woodruff, J. M. Mata, S. Hernandcz, and S. Mochales, Science, 166, 122(1969); B. G. Christensen, W. J. Leanza, T. R. Beattie, A. A. Patchett, B. H. Arison, R. E. Ormond, F. A. Kuehl, Jr., G. A. Schonberg, and O. Jardetzky, ibid., 166, 123(1969).
30) M. Sekine, K. Okimoto, and T. Hata, J. Am. Chem. Soc., 100, 1001(1978).
31) A. L. Lehninger, Biochemistry, Chap. 11, Worth Pub., New York(1970).
32) (a) W. Kiessling, Chem. Ber.,68, 597(1935); (b) F. Cramer and R. Wittman, Chem. Ber., 94, 328(1961), and references cited therein.
33) F. W. Lichtenthaler, Chem. Rev., 61, 607(1961), and refernces cited therein.
34) T. Hata, K. Yamada, T. Futatsugi, and M. Sekine, Synthesis, 189(1979).
35) T. Hata, H. Mori, and M. Sekine, Chemistry Lett., 1431(1977).

36) C. E. McKenna, M. T. Higa, N. H. Cheung, and M. C. McKenna, Tetrahedron Lett., 155(1977).
37) M. E. Jung, M.A. Lyster, J. Am. Chem. Soc., 99, 869(1977); M. E. Jung, W. A. Andrus, and P..L. Ornstein, Tetrahedron Lett., 4175(1977); T-L. Ho and G. A. Olah, Proc. Natl. Acad. Sci., U. S. A., 75, 4(1978); Synthesis, 417(1977).
38) F. Lipman, Advan. in Enzymol., 1, 99(1941).
39) F. Cramer and D. Voges, Chem. Ber., 92, 952(1959); V. M. Clark and A. J. Kirby, Biochim. Biophys. Acta, 78, 732(1963).
40) H. Vorbrüggen, Angew. Chem., Intern. Ed., 17, 569(1978).
41) J. M.Church, U. S. Patent, 2630436; Chem. Abstr., 448, 1420e(1954).
42) H. Staudinger and J. Meyer, Helv. Chim. Acta, 2, 635(1919); H. Staudinger and E. Hauser, ibid.,4, 861(1921).
43) G. Wittig and K. Schwarzenbach, Ann. Chem., 650, 1(1961).
44) T. Hata, I. Yamamoto, and M. Sekine, Chemistry Lett., 977(1975).
45) T. Hata, I. Yamamoto, and M. Sekine, ibid., 601(1976).
46) T. Hata and.M. Sekine, J. Am. Chem. Soc., 96, 7363(1974).
47) M. Sekine and T. Hata, Tetrahedron Lett., 1711(1975).
48) M. Sekine, K. Hamaoki, and T. Hata, J. Org. Chem., 44, 2325(1979).
49) K. Yamaguchi, S. Honda, and T. Hata, Chemistry Lett., 1057(1979).
50) T. Hata, K. Yamaguchi, S. Honda, and I. Nakagawa, ibid., 507(1978).
51) M. S. Poonian, E. F. Nowoswiat, L. Tobios, and A. L. Nussbaum, Bioorg. Chem., 2, 322(1973), and references cited therein.
52) M. Sekine and T. Hata, Chemistry Lett., 801(1979).

THE USE OF DIETHYL AZODICARBOXYLATE-TRIPHENYLPHOSPHINE SYSTEM IN THE SYNTHESIS AND TRANSFORMATION OF NATURAL PRODUCTS

O. Mitsunobu

Department of Chemistry, College of Science and Engineering, Aoyama Gakuin University, Chitosedai, Setagayaku, Tokyo 157, Japan

Abstract - The reagent formed by combining diethyl azodicarboxylate (DEAD) and triphenylphosphine (TPP) could be utilized in the intermolecular dehydration between an alcohol and various acidic components such as carboxylic acids, phosphoric diesters, imides, and active methylene compounds. By the use of DEAD and TPP, diols and hydroxy acids gave cyclic ethers and lactones, respectively. The reaction of nucleosides with DEAD and TPP afforded triphenylphosphoranylnucleosides. Alcohols reacted with 2,6-di-t-butyl-4-nitrophenol in the presence of DEAD and TPP to give aci-nitroesters which converted into the corresponding carbonyl compounds.

INTRODUCTION

The reaction of diethyl azodicarboxylate (DEAD, I) and triphenylphosphine (TPP, II) in the presence of alcohols and either carboxylic acids or phosphoric diesters gave the corresponding esters in good yields (Scheme 1; Refs. 1 & 2, see Note a).

Scheme 1

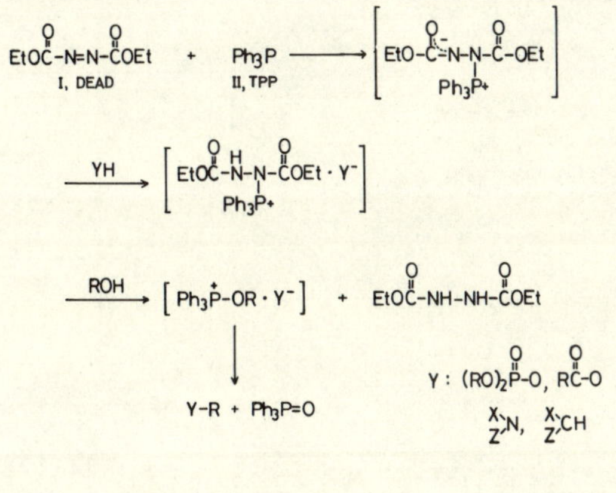

$$\text{EtOC-N=N-COEt} + \text{Ph}_3\text{P} \longrightarrow \left[\text{EtO-C=N-N-C-OEt} \atop \text{Ph}_3\text{P}^+ \right]$$

$$\xrightarrow{\text{YH}} \left[\text{EtOC-N-N-COEt} \cdot \text{Y}^- \atop \text{Ph}_3\text{P}^+ \right]$$

$$\xrightarrow{\text{ROH}} [\text{Ph}_3\overset{+}{\text{P}}\text{-OR} \cdot \text{Y}^-] + \text{EtOC-NH-NH-COEt}$$

$$\downarrow$$

$$\text{Y-R} + \text{Ph}_3\text{P=O} \qquad \text{Y}: (\text{RO})_2\text{P-O, RC-O}$$

$$\overset{X}{\underset{Z}{>}}\text{N}, \quad \overset{X}{\underset{Z}{>}}\text{CH}$$

The intermediacy of an alkoxyphosphonium salt was proved by the reaction of S-(+)-2-octanol with benzoic acid in which R-(-)-2-octyl benzoate was formed (Ref. 2). It was found that no allylic rearrangement took place in this reaction (Ref. 4, see Note b). Since the intermolecular dehydration by the use of DEAD and TPP involves initial activation of alcohols, and not carboxylic acids, other active hydrogen compounds such as imides and active methylene compounds could also be utilized as acidic components.

Note a. The structure of 1 : 1 adduct formed by the reaction of DEAD with TPP was determined by Brunn and Huisgen (Ref. 3).
Note b. Aneja et al. reported that the reaction of cholesterol with benzoic acid, DEAD, and TPP afforded all possible products resulting from homoallylic carbocation. Zamojski et al. found that allylic rearrangement took place in the phthaloylamination of unsaturated carbohydrates (Refs. 5 & 6).

PREPARATION OF NUCLEOSIDE DERIVATIVES

The reaction of unprotected nucleosides with a carboxylic acid, DEAD, and TPP resulted in the selective formation of the corresponding 5'-O-acylnucleosides (Refs. 7 & 8; Table 1). This result could be attributed to the steric hindrance of three phenyl groups attached to the phosphorus cation of the reaction intermediate.

Thymidine and uridine reacted with dibenzyl hydrogenphosphate in the presence of DEAD and TPP to give, after hydrogenolysis, the corresponding nucleoside 5'-phosphates (Scheme 2, B = uracil or thymine; Ref. 9).

Scheme 2

Table 1

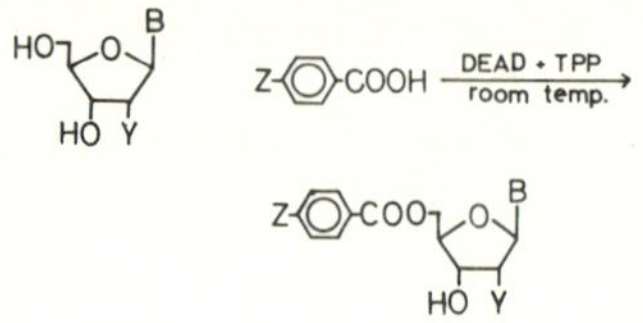

Y	B	Solvent	Z	Yield, %
H	Thymine	HM	NO$_2$	85
			CN	80
			H	74
			OCH$_3$	66
			CH$_3$	65
OH	Uracil	D	H	80 (56)
		HM-D	NO$_2$	83 (50)
OH	Adenine	HM-D	NO$_2$	(43)

HM: HMPA   D: Dioxane
( ): Isolated yield.

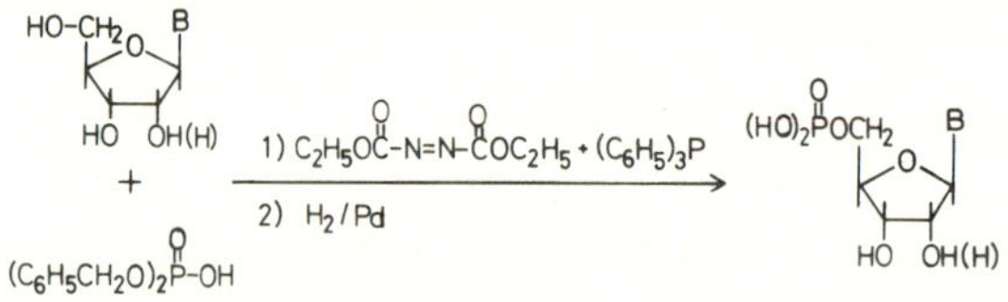

Contrary to the case of pyrimidine nucleosides, the reaction of dibenzyl hydrogenphosphate with adenosine and with guanosine afforded the corresponding cyclonucleosides as main products (Scheme 3; Ref.9).

Scheme 3

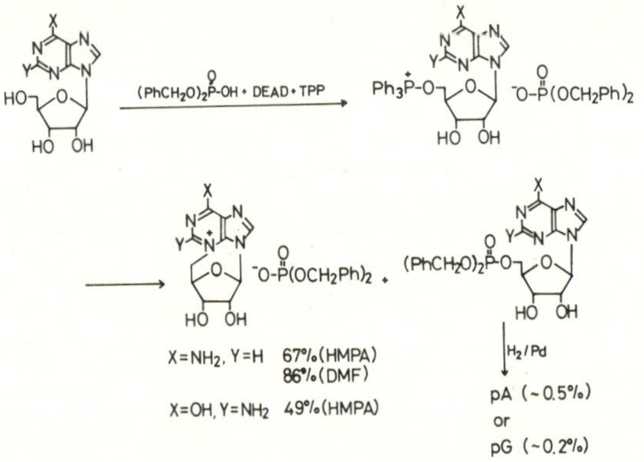

X=NH$_2$, Y=H 67% (HMPA)
             86% (DMF)
X=OH, Y=NH$_2$ 49% (HMPA)

pA (~0.5%)
or
pG (~0.2%)

The reaction of 2,3-isopropylidene-5-O-tritylribose with 2-methyl-4(or 5)-nitroimidazole, DEAD, and TPP resulted in the formation of imidazole nucleosides as shown in Scheme 4.

Scheme 4

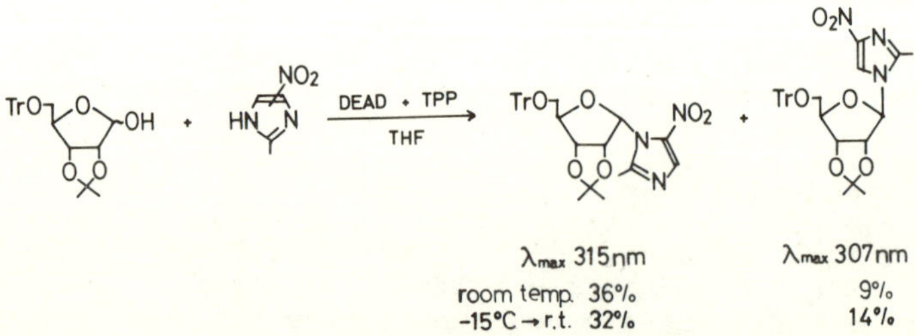

λ_{max} 315 nm
room temp. 36%
−15°C → r.t. 32%

λ_{max} 307 nm
9%
14%

When ribonucleosides having free 2'- and 3'-hydroxyl groups were allowed to react with DEAD and TPP, (triphenyl)phosphoranylnucleosides were isolated (Table 2; Refs. 10 & 11, see Note a).

Table 2

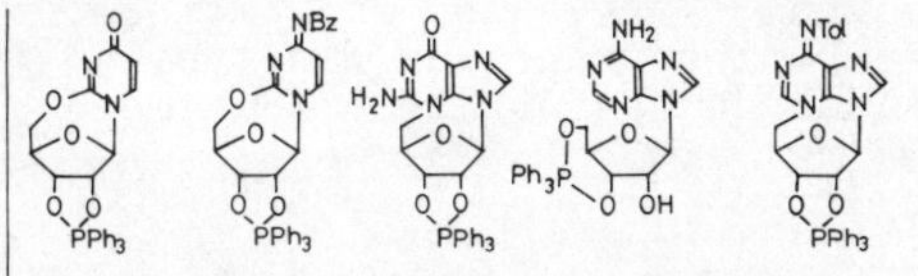

Yield (%)	62	71	67	75	24
Mp (°C)	>250	200-201	>250	171-172	200-204
$\lambda_{max}^{CH_3CN}$ (ε)	232 (26500)	235 sh 318(14600)	259 (12500)	260 (7400)	
^{31}P-NMR* (ppm)	+27.60	+27.54	+26.58	+29.33	

* 36.414 MHz. Solvent: DMF. Ph$_3$P=S (δ -42.07 Ref. 85% H$_3$PO$_4$) was used as internal standard. The reported chemical shifts are relative to 85% H$_3$PO$_4$.

OXIDATION OF ALCOHOLS TO CARBONYL COMPOUNDS VIA ACI-NITROESTERS

Alcohols reacted with 2,6-di-t-butyl-4-nitrophenol, DEAD, and TPP giving aci-nitroesters which, on treatment with a base, converted into aldehydes or ketones (Scheme 5, Table 3; Ref. 13).

Scheme 5

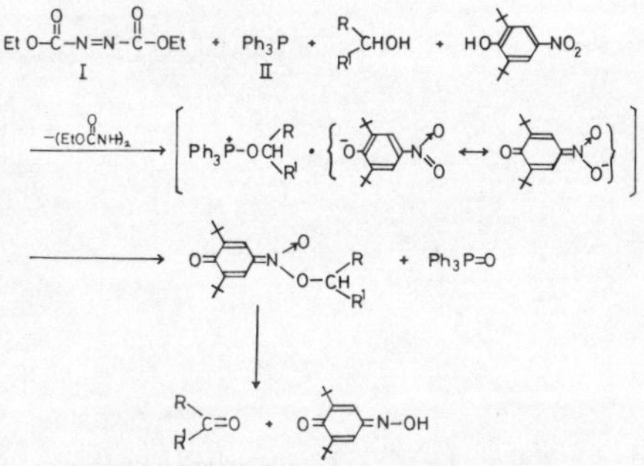

Table 3

Alcohol	aci-Nitroester (%)	Aldehyde or Ketone (%)
β-Phenethyl Alcohol	91	86 [82]
Isopropanol	79	—
2-Octanol	92	63
β-Citronellol	66	55 [48]
3-Hexen-1-ol	73	[27]
1,6-Hexanediol	68(7)	63
2,4-Pentanediol	68(7)	64
1,3-Butanediol	69(8)	
2,2,4-Trimethyl-1,3-pentanediol	No Reaction	—
5-MMTr-2,3-Isop-ribose		68

( ): di-aci-Nitroester.
[ ]: Isolated as 1,3-diphenylimidazolidine deriv.

By the use of this reaction, 5α-cholestan-3β-ol and 5α-androstan-3β, 17β-diol were converted into 5α-cholestan-3-one and 5α-androstan-17β-ol-3-one (Scheme 6).

Scheme 6

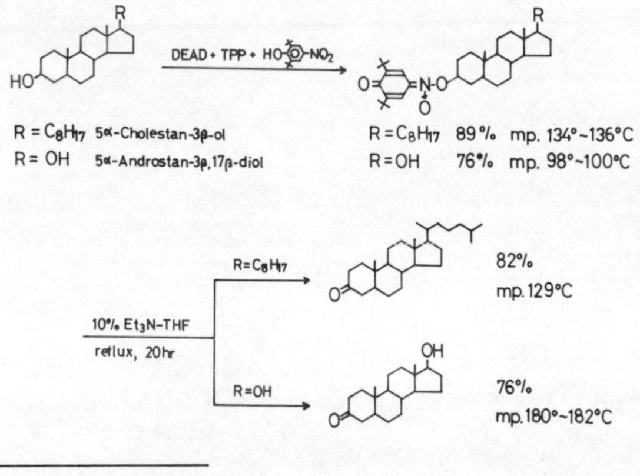

R = C$_8$H$_{17}$ 5α-Cholestan-3β-ol
R = OH 5α-Androstan-3β,17β-diol

R = C$_8$H$_{17}$ 89% mp. 134°~136°C
R = OH 76% mp. 98°~100°C

10% Et$_3$N-THF
reflux, 20hr

R = C$_8$H$_{17}$ 82% mp. 129°C
R = OH 76% mp. 180°~182°C

Note a. Mengel and Bartke demonstrated that the product obtained by the reaction of adenosine with DEAD and TPP is 2',3'-O-(triphenyl)phosphoranyl-adenosine (Ref. 12).

O^2-Methyl-2',3'-O-isopropylideneuridine afforded 2',3'-O-isopropylidene-uridine-5' aldehyde which was isolated as 1,3-diphenylimidazolidine derivative. Carbohydrates having a free anomeric hydroxyl group were, on treatment with 2,6-di-t-butyl-4-nitrophenol, DEAD, and TPP, directly converted into lactones. It was recently found that a base is not essential for the transformation of aci-nitroester into carbonyl compounds. Thus, when aci-nitroester prepared from methyl 2,3-O-isopropylidene-β-D-ribofuranoside was treated in methanol under reflux, the corresponding aldehyde was obtained (Scheme 7).

Scheme 7

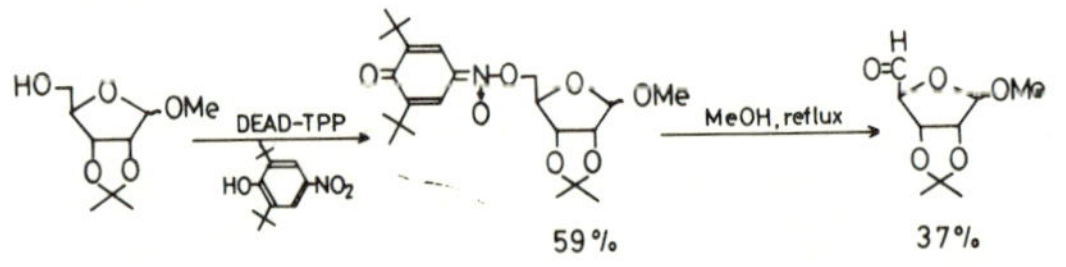

The reaction of aci-nitroesters with Wittig reagent resulted in the formation of the corresponding olefins (Table 4).

Table 4

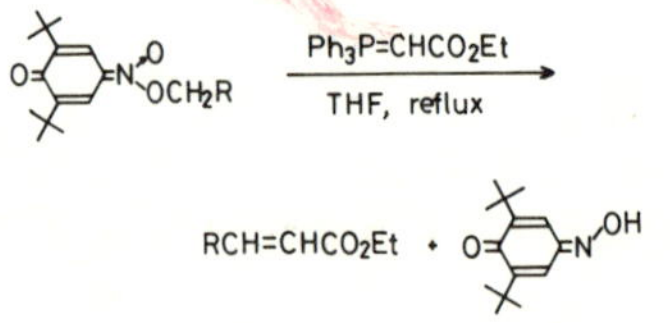

Alcohol	Time (hr)	Yield (%) Olefin	Oxime
1-Octanol	8	50	75
2-Phenylethanol	10	72	92
cis 3-Hexen-1-ol	31	66	80
3-Octyn-1-ol	8	50	75
	16	50	—
Benzene	8	61	87

PREPARATION OF CYCLIC COMPOUNDS

If hydroxyl groups of a diol are located at the favourable position to cyclization, it gave, on treatment with DEAD and TPP, the corresponding cyclic ethers (Scheme 8; Refs. 14 & 15).

Scheme 8

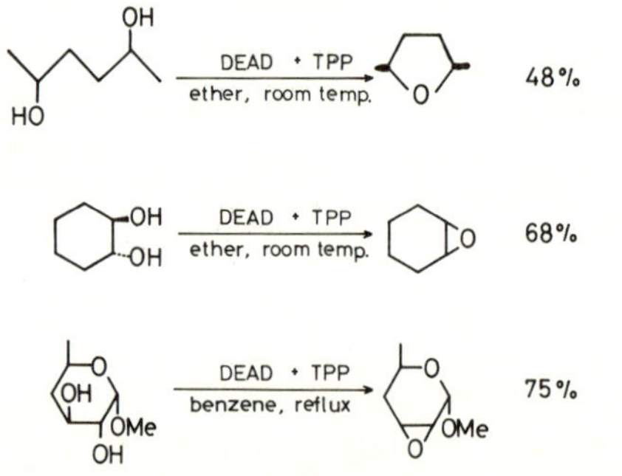

Similarly, ω-hydroxy acids afforded lactones and dilides in varying amounts. Although 13-membered lactone was formed in good yield, 9-membered lactone was scarcely formed (Table 5; Ref. 16).

Table 5

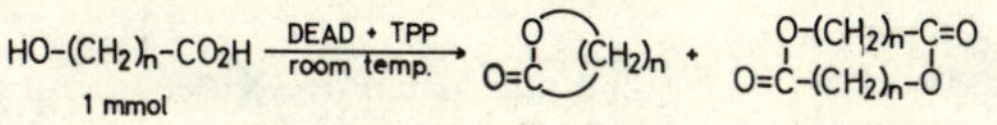

n	Solvent (ml)	Reaction Time day	Isolated Yield, % Lactone	Dilide
5	THF (25)	2	[8]	
	B (25)	2	40	53
7	B (25)	2	0 [0.8]	70
	B (25)	2	60	7
11	B (25)	1(80°C)	[14]	
	B (200)	1	63	32
	Toluene(200)	2(-15°C→r.t.)	[82]	
	B(25)+THF(4)	1	[5]	

B: Benzene. [ ]: Glc yield.

When 2-(10-hydroxyundecyl)benzoic acid prepared by known method (Ref. 17) was allowed to react with DEAD and TPP, dideoxyzearalane was isolated (Scheme 9).

Scheme 9

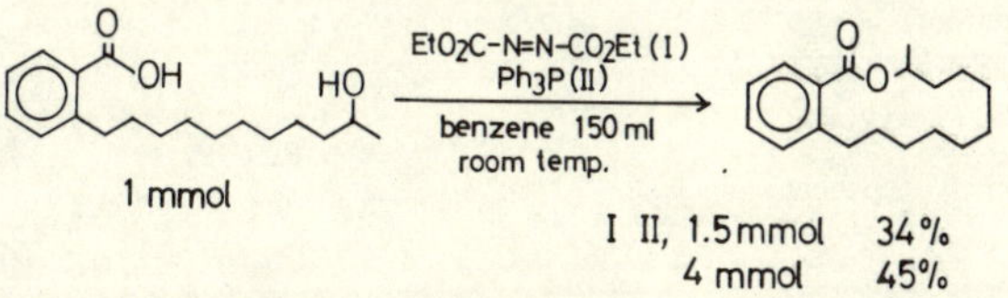

The reaction of 7-hydroxy-4-(1,3-dithianyl)-2-octenoic acid with DEAD and TPP resulted in the formation of protected pyrenophorin (Scheme 10; Ref. 18).

Scheme 10

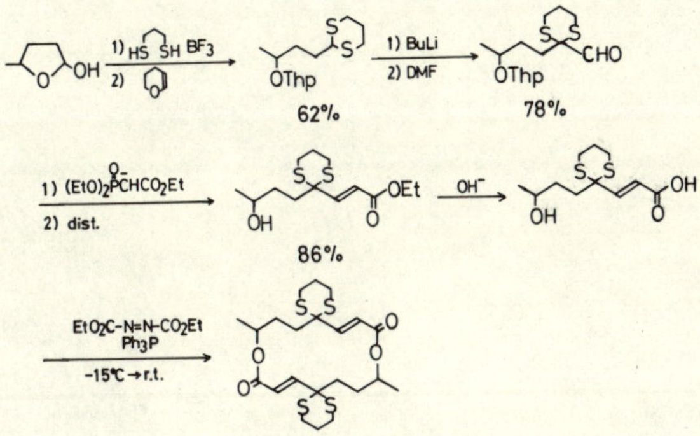

Acknowledgement - I express my appreciation to my coworker - Junji Kimura - and all of the students of my laboratory. This work was supported, in part, by a grant from the Ministry of Education, Japan. Financial support from Yoshida Foundation for Science and Technology is also gratefully acknowleged.

REFERENCES

1. O. Mitsunobu and M. Yamada, Bull. Chem. Soc. Jpn., 40, 2380 (1967).
2. O. Mitsunobu and M. Eguchi, ibid., 44, 3427 (1971).
3. E. Brunn and R. Huisgen, Angew. Chem., 81, 534 (1969).
4. G. Grynkiewicz and H. Burzynska, Tetrahedron, 32, 2109 (1976).
5. R. Aneja, A. P. Davies, and J. A. Knaggs, Tetrahedron Lett., 1033 (1975).
6. A. Banaszek, B. Szechner, J. Mieczkowski, and A. Zamojski, Rocz. Chem., 50, 105 (1976).

7. O. Mitsunobu, J. Kimura, and Y. Fujisawa, Bull. Chem. Soc. Jpn., 45, 245 (1972).
8. S. Shimokawa, J. Kimura, and O. Mitsunobu, ibid., 49, 3357 (1976).
9. J. Kimura, Y. Fujisawa, T. Yoshizawa, K. Fukuda, and O. Mitsunobu, ibid., 52, 1191 (1979).
10. J. Kimura, Y. Fujisawa, T. Sawada, and O. Mitsunobu, Chem. Lett., 691 (1974).
11. J. Kimura, Y. Hashimoto, and O. Mitsunobu, ibid., 1473 (1974).
12. R. Mengel and M. Bartke, Angew. Chem. Int. Ed. Engl., 17, 679 (1978).
13. J. Kimura, A. Kawashima, M. Sugizaki, N. Nemoto, and O. Mitsunobu, J. C. S. Chem. Comm., 303 (1979).
14. O. Mitsunobu, J. Kimura, K. Iiizumi, and N. Yanagida, Bull. Chem. Soc. Jpn., 49, 510 (1976).
15. J. T. Carlock and M. P. Mack, Tetrahedron Lett., 5153 (1978). See also ref. 12 and H. Loibner and E. Zbiral, Helv. Chim. Acta, 60, 417 (1977).
16. T. Kurihara, Y. Nakajima, and O. Mitsunobu, Tetrahedron Lett., 2455 (1976).
17. H. L. Wehrmeister and D. E. Robertson, J. Org. Chem., 33, 4173 (1968).
18. For the synthesis of pyrenophorin, see E. W. Colvin, T. A. Purcell, and R. A. Raphael, J. C. S. Chem. Comm., 1031 (1972). H. Gerlach, K. Oertle, and A. Thalmann, Helv. Chim. Acta, 60, 2860 (1977). Recently Seebach et al. reported the total synthesis of (R, R)-pyrenophorin; D. Seebach, S. Seuring, H.-O. Kalinowski, W. Lubosch, and B. Renger, Angew. Chem. Int. Ed. Engl., 16, 264 (1977). B. Seuring and D. Seebach, Liebigs Ann. Chem., 2044 (1978).

NUCLEOSIDE PHOSPHATES AS PROBES OF THE MECHANISM OF PROTEIN BIOSYNTHESIS

S. M. Hecht

Departments of Chemistry and Biology, University of Virginia, Charlottesville, Virginia 22901, U.S.A.

Abstract - Two types of nucleoside phosphates have been prepared for studies of the mechanism of protein biosynthesis. (1) A variety of aminoacylated P^1,P^2-di(adenosine 5'-)diphosphates were synthesized and shown to be substrates for T4 RNA ligase. When abbreviated tRNA (tRNA-CC_{OH}) was utilized as the acceptor species in the reaction, "chemically" aminoacylated (or misacylated) tRNA's could be prepared from fractionated or unfractionated tRNA-CC_{OH}'s. The potential utility of misacylated tRNA's in the study of peptide bond formation and the elaboration of proteins containing unusual amino acids has been studied quantitatively in a purified, cell-free system. (2) In an effort to facilitate the isolation of proteins involved in the synthesis and recognition of mRNA "caps", an affinity column has been prepared. This column, containing 7-methylguanosine 5'-diphosphate moieties immobilized on AH-Sepharose 4B via a levulinic acid $O^{2'},^{3'}$-acetal, was employed for the purification to homogeneity of a 24,000-dalton polypeptide that stimulated the translation of capped mRNA's.

INTRODUCTION

The central role of nucleic acids in the elaboration of proteins provides the nucleic acid chemist with a unique opportunity for study of the mechanism(s) of protein biosynthesis and alteration of the formed polypeptides via manipulation of the nucleoside phosphates that comprise the nucleic acids. Described herein is a procedure for the "chemical aminoacylation" of tRNA's, a methodology of potential utility for the activation of any tRNA with any amino acid. Since such misacylated tRNA's should, in principle, permit the cell-free synthesis of proteins containing single amino acid substitutions, we have also carried out a quantitative assessment in a purified protein biosynthesizing system of the extent to which expression of the adapter hypothesis (1,2) can be expected to obtain in vitro. Also discussed is the design and preparation of an affinity column for the isolation of proteins that participate in mRNA "cap" (3) biosynthesis and recognition and the use of the column for the facile purification of a factor from rabbit reticulocytes that promoted the translation of capped mRNA's.

PROTEIN BIOSYNTHESIS WITH MISACYLATED tRNA's

According to the adapter hypothesis (1,2) the sequence of amino acids in a polypeptide is specified at the level of codon-anticodon interaction, a process in which the aminoacyl moiety on the tRNA plays only a passive role. Fidelity of protein biosynthesis is therefore maintained in two independent steps—through faithful codon-anticodon interaction and by proper activation of each tRNA with its cognate amino acid. In fact, it has been shown that the use of misacylated tRNA's in in vitro protein biosynthesizing systems resulted in transfer of the misactivated amino acid into protein, although the quantitative aspects of this process have never been verified in a highly purified in vitro system.

The ability to carry out single amino acid substitutions in vitro would clearly depend on the availability of misacylated tRNA's. Although some efforts have been made to develop a methodology that could be used for the preparation of misactivated tRNA's, both by enzymatic (4,5) and chemical (6-8) means, there has been no generally applicable method for the preparation of such species. Moreover, all such studies have focused on the introduction of naturally occurring amino acids onto tRNA's and all but one (6,7) have involved elaboration of the aminoacylated species via bond formation between the aminoacyl and tRNA moieties. An alternate approach has recently been devised using RNA ligase from T4-infected E. coli (9).

Bacteriophage T4 RNA ligase

In addition to circularizing polynucleotides of a certain length (10-13) and resealing the anticodon loop of yeast tRNA^{Phe} (14), RNA ligase has been shown to possess the ability to effect the intermolecular joining of two oligoribonucleotides (13,15,16). The mechanism of the reaction (Scheme 1) has been investigated (14) and may be envisioned to involve initial nucleophilic addition of the 5'-OH group of the "donor" oligonucleotide (in this case pXXX...XXXp) to P$^\alpha$ of ATP, followed by loss of inorganic pyrophosphate. The intermediate adenylylated oligomer is then condensed with the acceptor oligoribonucleotide in a reaction requiring RNA ligase but not ATP (14,17), and resulting in the expulsion of 5'-AMP.

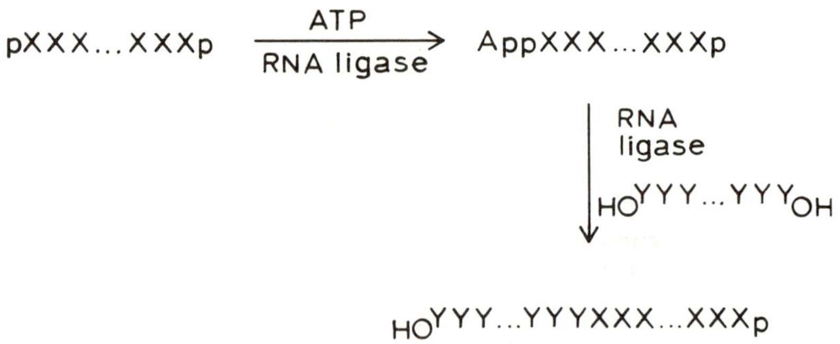

$$pXXX...XXXp \xrightarrow[\text{RNA ligase}]{\text{ATP}} AppXXX...XXXp$$

$$\downarrow \text{RNA ligase}, \; HO^{YYY...YYY}OH$$

$$HO^{YYY...YYYXXX...XXX}p$$

Scheme 1

The minimal structural requirements for the "acceptor" and "donor" oligoribonucleotides have been studied (13,15,16). The efficiency of the acceptor appeared to diminish substantially for very short oligoribonucleotides, but by the use of relatively large amounts of enzyme oligomers 3-6 nucleotides in length gave useful results. More remarkable were the results obtained for the donor; single nucleotides could be added efficiently by the use of an adequate amount of enzyme (18-20). Moreover, P^1,P^2-di(nucleoside 5'-)diphosphates (i.e., the putative adenylylated intermediates) were shown to act as donors in the absence of ATP. Interestingly, while half of the "dinucleotide" had to be 5'-AMP, there seemed to be little or no specificity for the mononucleotide that was actually transferred to the acceptor oligomer. For example, incubation of (Ap)$_3$C in the presence of RNA ligase and nicotinamide adenine dinucleotide (NAD$^+$) resulted in transfer of nicotinamide ribonucleotide to the 3'-terminus of the acceptor oligomer (17).

Transfer RNA has a single stranded acceptor stem, which suggested its possible utility as a substrate for RNA ligase. In particular, incubation of a tRNA lacking the terminal adenosine moiety (tRNA-CC$_{OH}$; abbreviated tRNA) and a preaminoacylated P^1,P^2-di(adenosine 5'-)diphosphate in the presence of RNA ligase might be expected to result in transfer of the aminoacylated nucleotide to tRNA-CC$_{OH}$, thus forming aminoacylated (or misacylated) tRNA.

Preparation of misacylated tRNA's

The preparation of abbreviated tRNA's has been described previously (21). The technique involves initial digestion of fractionated or unfractionated tRNA's with purified venom exonuclease and can be monitored by the loss of amino acid acceptor activity. As shown in Scheme 2, the product of the venom exonuclease digestion consists of tRNA's missing one or more nucleotides; if digestion is carried out carefully, the loss of nucleotides can be confined to the single-stranded portion of the acceptor stem. Although a mixture of tRNA's of varying lengths results from the initial treatment, a uniform population of abbreviated tRNA's can be obtained by further incubation with CTP(ATP):tRNA nucleotidyltransferase in the presence of CTP (7). The structural integrity of the abbreviated tRNA's so derived may be verified by reconstitution of the intact tRNA's with radiolabeled ATP via the agency of the CTP(ATP):tRNA nucleotidyltransferase and, additionally, by activation of the reconstituted tRNA's with their cognate amino acids.

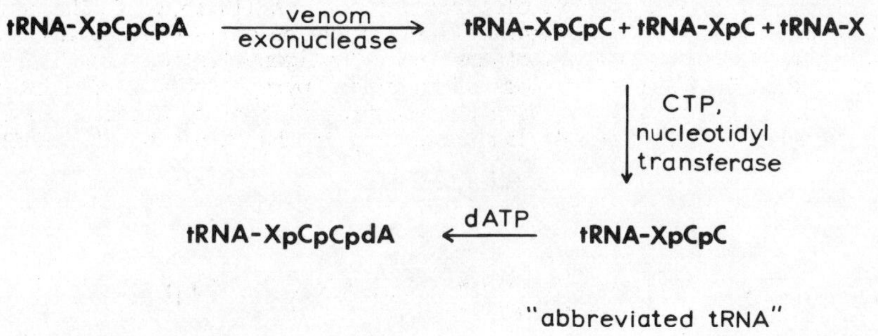

tRNA-XpCpCpA —venom exonuclease→ tRNA-XpCpC + tRNA-XpC + tRNA-X

↓ CTP, nucleotidyl transferase

tRNA-XpCpCpdA ←dATP— tRNA-XpCpC

"abbreviated tRNA"

Scheme 2

Elaboration of the requisite aminoacylated P^1,P^2-bis(5'-adenosyl)diphosphates (3) was effected by coupling P^1,P^2-bis(5'-adenosyl)diphosphate (2) with the appropriate o-nitrophenylsulfenyl amino acids 1 (9). Where the aminoacylated nucleotide to be transferred to tRNA-CC$_{OH}$ was modified in the adenosine moiety (e.g., 3e and 3f), the adenylylated substrate was obtained by condensation of the aminoacylated nucleoside 5'-monophosphate with 5'-AMP. In each case, purification of 3 was achieved by chromatography on DEAE-cellulose.

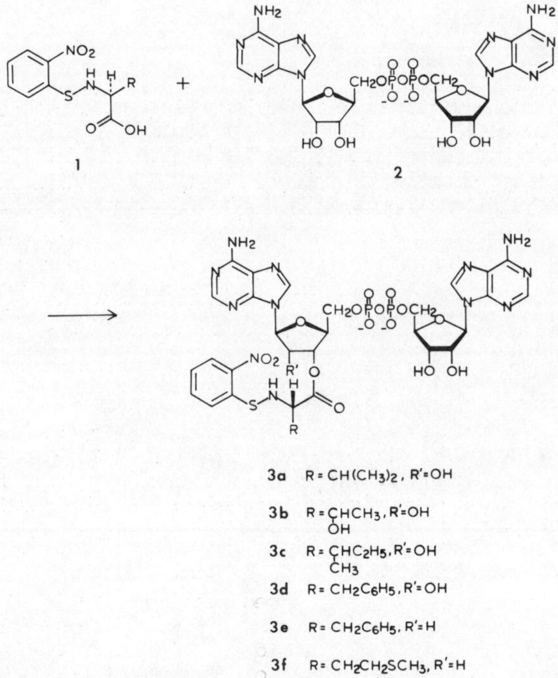

3a R = CH(CH$_3$)$_2$, R' = OH
3b R = CHCH$_3$, R' = OH
      |
      OH
3c R = CHC$_2$H$_5$, R' = OH
      |
      CH$_3$
3d R = CH$_2$C$_6$H$_5$, R' = OH
3e R = CH$_2$C$_6$H$_5$, R' = H
3f R = CH$_2$CH$_2$SCH$_3$, R' = H

Formulas 1 - 3

RNA ligase substrates 3 were prepared with the attached o-nitrophenylsulfenyl groups both to stabilize the aminoacyl moiety (thus permitting the period of incubation to be increased substantially) and also to allow the formed (o-nitrophenylsulfenyl)aminoacyl-tRNA's to be separated from unreacted tRNA-CC$_{OH}$'s by chromatography on BD-cellulose. On the basis of studies with peptidyl-tRNA (22), the o-nitrophenylsulfenyl protecting group was known to be removed with aqueous sodium thiosulfate under conditions compatible with preservation of tRNA structure and function.

Incubation of nucleotides 3 with tRNA-CC$_{OH}$ in the presence of T4 RNA ligase did, in fact,

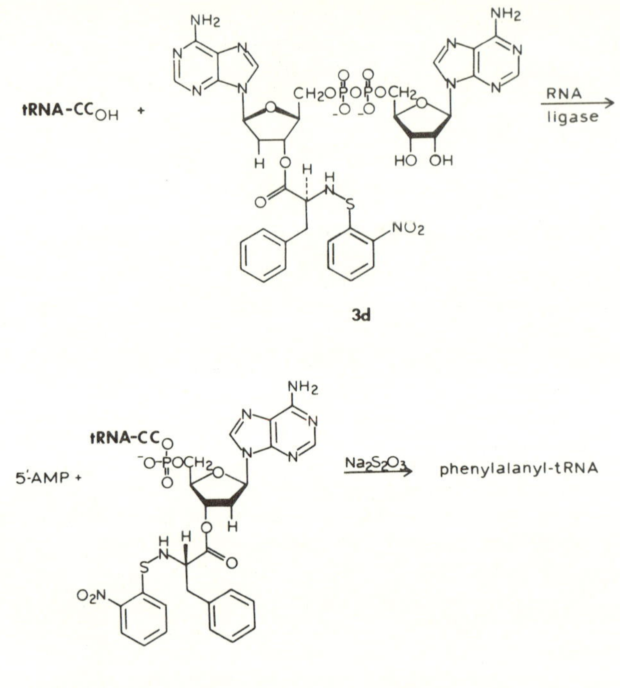

Scheme 3

result in transfer of the o-nitrophenylsulfenylaminoacyladenylate moiety to the 3'-terminus of the abbreviated tRNA (9). As noted above, the products could be purified conveniently by chromatography on BD-cellulose by virtue of the lipophilic o-nitrophenylsulfenyl protecting group, which resulted in selective retention of the product tRNA's. This is illustrated in Figure 1 for the N-(o-nitrophenylsulfenyl)valyl-tRNA's derived from E. coli tRNA-CC$_{OH}$ and

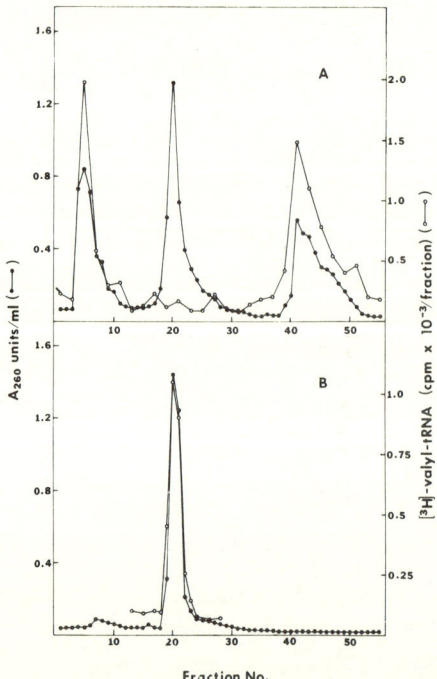

Fig. 1. Chromatography on BD-cellulose of a "chemically aminoacylated" sample of E. coli valyl-tRNA before (A) and after (B) removal of the o-nitrophenylsulfenyl protecting group.

nucleotide 3a. As shown in panel A, the (dialyzed) reaction mixture was applied to a BD-cellulose column in 50 mM NaOAc buffer, pH 4.5, containing 0.4 M NaCl. Washing with the same buffer effected elution of RNA ligase and unreacted 3a; 1.0 M NaCl was then employed for recovery of tRNA-CC$_{OH}$ (fractions 16 to 35). When the column was washed with 50 mM NaOAc buffer containing 1.0 M NaCl and 25% ethanol, the N-(o-nitrophenylsulfenyl)valyl-tRNA's were eluted (fractions 36 to 55). As illustrated in panel B, after removal of the protecting group, the mobility of the "chemically aminoacylated" valyl-tRNA's was that expected of (aminoacyl-)tRNA's.

In addition to the activation of unfractionated E. coli tRNA's with valine, several other activated and misactivated tRNA's were prepared by the use of nucleotides 3a - 3f. The results of some of these experiments are outlined in Table 1. As shown in the table, unfractionated E. coli tRNA was activated with five different aminoacylated nucleotides, and three fractionated tRNA's were successfully activated with cognate or non-cognate amino acids. It is probably worth noting that to date no combination of nucleotide 3 and tRNA-CC$_{OH}$ has failed to give the desired aminoacyl-tRNA.

TABLE 1. Preparation of aminoacyl-tRNA's by the use of RNA ligase (9).

Source of tRNA-CC$_{OH}$	Compound 3		RNA ligase (units/ml)	Yield of isolated aminoacyl-tRNA (%)
	Amino acid	Concentration (A_{256} units/ml)		
E. coli tRNA	Valine (3a)	140	105	39
E. coli tRNA	Threonine (3b)	95	120	38
E. coli tRNA	Isoleucine (3c)	95	75	44
E. coli tRNA	Phenylalanine (3d)	140	180	58
Yeast tRNA^{Trp}	Phenylalanine (3d)	60	75	40
Yeast tRNA^{Phe}	Phenylalanine (3d)	40	75	79
E. coli tRNA	Phenylalanine (3e)	125	375	25
Yeast tRNA^{Phe}	Phenylalanine (3e)	70	70	45
Yeast tRNA$^{Met}_i$	Methionine (3f)	75	75	91

In addition to their chromatographic mobilities on BD-cellulose before and after removal of the o-nitrophenylsulfenyl protecting group, the identity of the "chemically aminoacylated" tRNA's with species obtained by enzymatic activation could be judged in three ways. First, aminoacyl-tRNA's hydrolyze chemically at a characteristic rate; yeast phenylalanyl-tRNA^{Phe}'s prepared both chemically and enzymatically were shown to deacylate at the same rate at pH 8.6 over a period of 90 min. Second, aminoacyl-tRNA's form ternary complexes with elongation factor Tu (EF-Tu) and GTP, and the stabilities of the complexes has been shown to vary considerably according to the specific aminoacyl-tRNA involved (23). Complex formation has been shown to depend on the aminoacyl moieties of the tRNA's, a circumstance reflected in the greatly diminished rate of chemical deacylation of aminoacyl-tRNA's bound to EF-Tu·GTP (23). Therefore, the fact that "chemically" and enzymatically activated yeast phenylalanyl-tRNA^{Phe}'s also hydrolyzed at the same rate in the presence of EF-Tu·GTP was consistent with the interpretation that they were identical species. Finally, the chemical aminoacylation of unfractionated tRNA's with a single amino acid (e.g., valine) ostensibly involves the introduction of both adenosyl and aminoacyl moieties onto the 3'-termini of the various tRNA-CC$_{OH}$'s. Therefore, deacylation of the misacylated species should permit subsequent reacylation with the amino acids corresponding to each of the reconstituted isoacceptors. In fact, deacylation of chemically aminoacylated E. coli phenylalanyl-, threonyl- and valyl-tRNA's afforded species that could be reactivated with arginine, phenylalanine and alanine to about the same extent as unmodified E. coli tRNA.

Fidelity of protein biosynthesis in vitro

Although the validity of the adapter hypothesis in vivo is widely accepted and the misincorporation of several amino acids into protein from misacylated tRNA's has been demonstrated in vitro (24-27), the quantitative nature of the in vitro process has been less fully explored. In those studies that could be evaluated in quantitative terms, the results have often been less than perfect and suggestive of a possible lesser efficiency of incorporation of structurally modified amino acids into protein (see, e.g., ref. 28). Given the additional reports of codon misreading during in vitro translation (29,30), it seemed judicious to study the utilization of a misacylated tRNA in a purified in vitro protein biosynthesizing system (31).

Our well-defined biosynthetic system included E. coli phenylalanyl-tRNA^{Lys} and lysyl-tRNA^{Lys}, purified elongation factors Tu and G, poly A and low salt washed ribosomes from E. coli. As shown in Table 2, in the complete system poly A directed the synthesis of

TABLE 2. Poly A-directed synthesis of polyphenylalanine using phenylalanyl-tRNA^{Lys} [a]

Time (min)	Polypeptide formed (pmol)
5	2.4
10	5.6
20	10.9
30	12.5

[a] Each aliquot contained 16.2 pmol of [^{14}C]-phenylalanine.

polyphenylalanine from phenylalanyl-tRNA^{Lys}. Control reactions omitting each of the components of the system singly (i.e., EF-Tu, EF-G, ribosomes or poly A) failed to produce significant amounts of protein. When [^{14}C]-phenylalanyl-tRNA^{Lys} and [^{3}H]-lysyl-tRNA^{Lys} were admixed in different proportions and the mixture added to the complete protein biosynthesizing system, the resulting polypeptides contained the two amino acids in the same proportion in which they were combined (Figure 2).

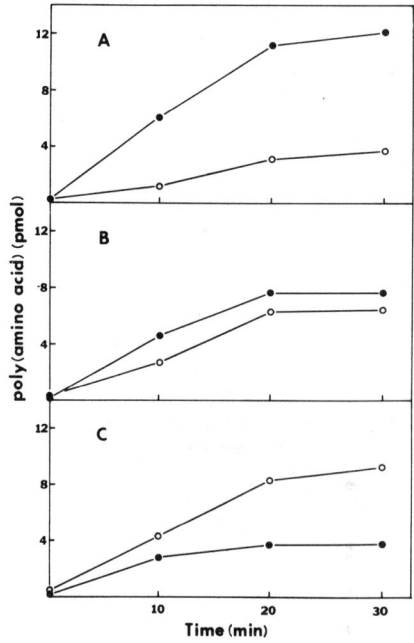

Fig. 2. Polypeptides resulting from admixture of [^{14}C]-phenylalanyl-tRNA^{Lys} (●) and [^{3}H]-lysyl-tRNA^{Lys} (o) to a complete protein biosynthesizing system. The ratios of phenylalanyl/lysyl-tRNA's added (and polyphenylalanine/polylysine observed after 30 min of incubation) were: panel A, 3.32 (3.36); panel B, 1.11 (1.19); panel C, 0.37 (0.40). At the shortest reaction time (10 min) the input and observed ratios agreed less well, ostensibly due to technical difficulties in isolating the small amounts of protein produced.

The ability of other tRNA's to misread the AAA codons utilized for the misincorporation of phenylalanine into protein was assayed by admixture of a large molar excess of unfractionated E. coli tRNA to the incubation mixture or by addition of an excess of E. coli arginyl-tRNA (one species responds to the codon AGA). Neither had any effect on polyphenylalanine synthesis (data not shown). Organic solvents such as methanol and dimethylsulfoxide have been shown to affect the specificity of certain biochemical transformations. When each of these was added (in concentrations up to 20%) to incubation mixtures containing $[^{14}C]$-phenylalanyl-tRNA^{Lys} and $[^{3}H]$-lysyl-tRNA^{Lys}, the ratio of the two amino acids in the formed polypeptides was unaffected, although the absolute amount of protein synthesized was diminished in some experiments.

The composition of the copolymers formed by poly A-directed polymerization of phenylalanyl-tRNA^{Lys} and lysyl-tRNA^{Lys} was also investigated. The polypeptides formed in the incubations described in panels A - C of Fig. 2 were analyzed by solubilization and chromatography on BD-cellulose. The columns were first washed with 0.4 M aqueous sodium acetate to effect elution of unreacted amino acids and polypeptides of low phenylalanine content. Polypeptides containing phenylalanine were then eluted with 0.5 M potassium hydroxide solution. For each of the polypeptides analyzed (Figure 3) the ratio of phenylalanine and lysine in the phenylalanine-containing polypeptides was the same as that utilized in the initial incubation mixture.

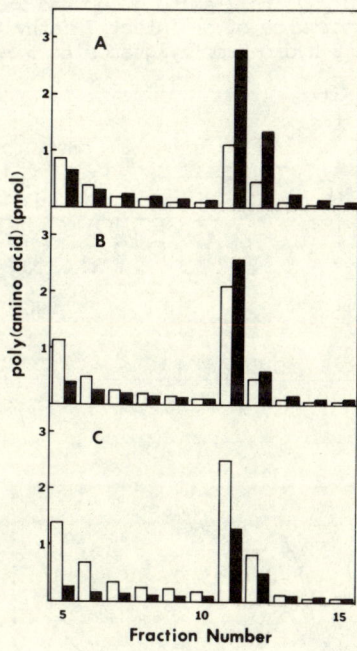

Fig. 3. BD-cellulose chromatography of the co-polypeptides of phenylalanine and lysine formed as described in the corresponding panels in Fig. 2. Analysis for phenylalanine (solid bars) and lysine (open bars) content was carried out as described in the text.

The results of these experiments suggest that a tRNA activated with an amino acid quite distinct structurally from its cognate species can be utilized normally in an in vitro protein biosynthesizing system, i.e., that it may be possible to prepare proteins having predetermined amino acid substitutions at discrete sites. It is less clear at present to what extent aminoacyl-tRNA's can be structurally modified in the aminoacyl moiety without affecting their participation in the peptidyltransferase reaction and the other partial reactions of protein biosynthesis.

PURIFICATION OF A "CAP" BINDING PROTEIN

Most eukaryotic mRNA's have an unusual "cap" structure at their 5'-termini which consists of a 7-methylguanosine moiety attached through a (5' → 5') triphosphate linkage to the polyribonucleotide that is to be translated (m^{7}GpppN...); the first nucleotide of the message is a methylated purine nucleoside (3). Consistent with the decrease in efficiency of translation noted upon removal of the cap (32,33), the cap structure is believed to facilitate translation at the level of initiation complex formation (3). More specifically, it has been suggested (34) that cap recognition may be part of the mechanism by which the ribosome is attached to mRNA. Recently, it was shown that eukaryotic initiation factor preparations contained a polypeptide of molecular weight 24,000 that interacted specifically with the cap structure (35). The ability of these preparations to stimulate mRNA translation (36) has prompted efforts to purify the protein so that its role in protein biosynthesis can be studied. Since the protein is known to bind to the cap structure (35), one attractive approach to its purification might involve affinity chromatography.

Design and preparation of the affinity column

As might be expected of cap analogs, 7-methylguanosine 5'-phosphates have been reported to inhibit translation by limiting the ribosomal binding of mRNA's (37-39). In an effort to obtain insights into the structural features in mRNA caps requisite for expression of cap function, Adams et al. (40) studied the inhibition of reovirus mRNA binding to wheat germ ribosomes by certain analogs of 7-methylguanosine 5'-diphosphate. The analogs prepared included 7-methylguanosine 5'-diphosphates additionally methylated on N-1, N^2 and C-8, as well as 7-methylinosine and xanthosine 5'-diphosphates and 2-amino-6-chloro-7-methyl-9-(β-D-ribofuranosyl)purine 5'-diphosphate. Of these only N^2,7-dimethylguanosine 5'-diphosphate was as effective as 7-methylguanosine 5'-diphosphate in inhibiting mRNA binding, suggesting that cap recognition involved the 2-amino group and was influenced negatively by additional substitution. The importance of an intact 7-methylguanosine moiety was reinforced by the results obtained with 8-hydro-7-methylguanosine 5'-diphosphate (Scheme 4).

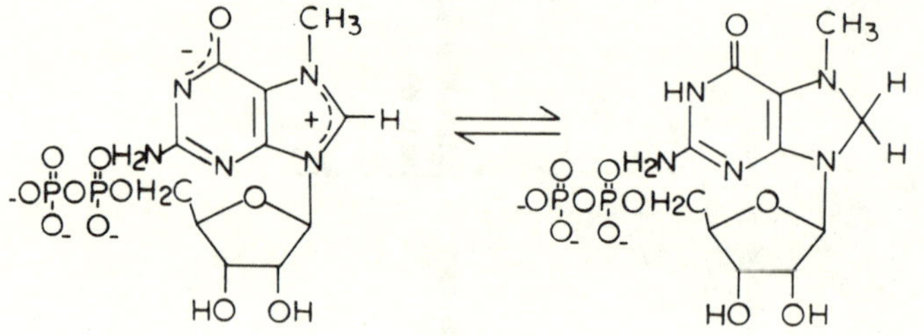

Scheme 4

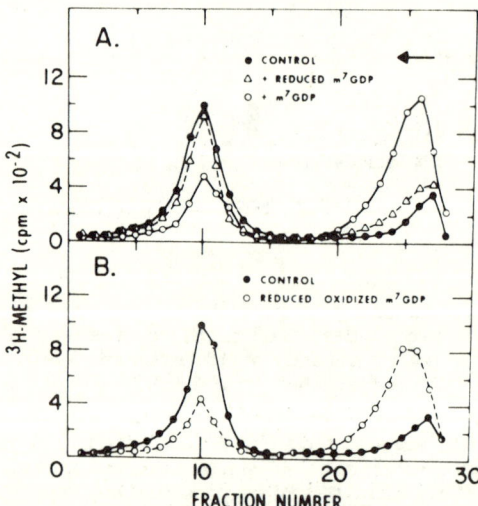

Fig. 4. Inhibition of complex formation between reovirus mRNA and wheat germ ribosomes. [^{3}H]-Methyl mRNA was incubated with wheat germ ribosomes (panel A) in the absence of added nucleotide or in the presence of 100 μM 7-methylguanosine 5'-diphosphate or 8-hydro-7-methylguanosine 5'-diphosphate. Complex formation was assayed by layering the samples onto glycerol gradients (10 - 30% v/v in 20 mM Tris buffer, pH 7.5, containing 70 mM KCl and 3 mM Mg(OAc)$_2$) and centrifugation (SW - 50.1; 90 min; 48,000 rpm), after which 0.2-ml fractions were collected and used for determination of radioactivity. Panel B: repetition of the experiment after reoxidation of 8-hydro-7-methylguanosine 5'-diphosphate.

This species, accessible by borohydride treatment of 7-methylguanosine 5'-diphosphate (41,42), slowly reoxidizes in aqueous media to reafford the parent nucleotide. As shown in Figure 4, when radiolabeled reovirus mRNA was incubated with wheat germ ribosomes (panel A), glycerol gradient centrifugation indicated that most of the mRNA was present as a complex with the ribosomes. Complex formation was inhibited markedly by 100 μM 7-methylguanosine 5'-diphosphate, but little if at all by the same concentration of 8-hydro-7-methylguanosine 5'-diphosphate. As expected (panel B), after reoxidation of the reduced nucleotide inhibition of complex formation was the same as that obtained with authentic 7-methylguanosine 5'-diphosphate.

Perhaps more important in the context of design of an affinity column were those analogs of 7-methylguanosine 5'-diphosphate whose inhibition of mRNA-ribosome complex formation was no different than that of the parent nucleotide. These included 7-ethyl and 7-benzylguanosine 5'-diphosphates as well as 7-methyl-2'-deoxyguanosine 5'-diphosphate. Thus "cap" binding did not seem to involve recognition of a specific alkyl group on the 7-position, nor of an

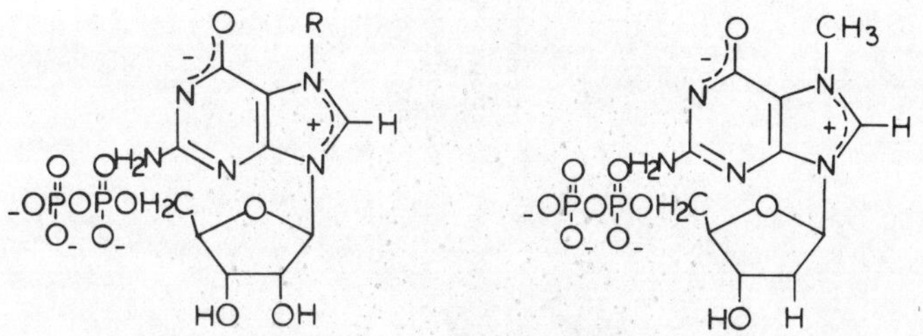

$R = CH_2CH_3$
$R = CH_2C_6H_5$

intact ribofuranose moiety. These results suggested that it might be possible to utilize the 7-position or sugar as points of attachment of the nucleotide to some inert support without affecting its binding properties. Seela and Waldek (43) have described the immobilization of guanosine 5'-diphosphate to aminohexyl-Sepharose-4B via the carboxylate moiety of the corresponding levulinic acid $O^{2'},^{3'}$-acetal.

Synthesis of the desired ligand was accomplished starting from 2',3'-O-[1-(3-ethoxy-3-oxypropyl)ethylidene]guanosine (43), which was phosphenylated with pyrophosphoryl chloride in 71% yield (44) and then converted to the respective 5'-diphosphate via the corresponding phosphorimidazolidate (45,46). After saponification of the ethyl ester, methylation (CH_3I, acidic dimethylsulfoxide) afforded the requisite 2',3'-O-[1-(2-carboxyethyl)ethylidene]-7-methylguanosine 5'-diphosphate. Coupling of the ligand to aminohexyl-Sepharose-4B was effected in aqueous solution, pH 5.5 - 6.0, via the agency of 1-ethyl-3-(3-dimethylaminopropyl)carbodiimide. Analysis of the product revealed that the resin was derivatized with ligand to the extent of 17.4 μmol/g. A control column, containing 2',3'-O-[1-(2-carboxyethyl)ethylidene]guanosine 5'-diphosphate as the ligand was prepared analogously (43) and contained 20.3 μmol of ligand/g of resin.

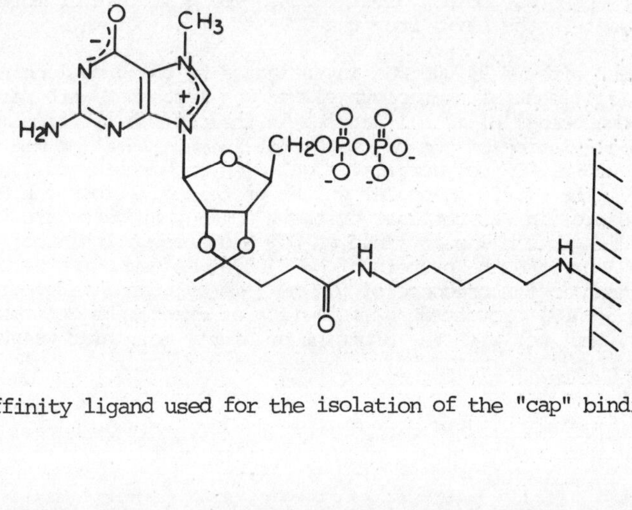

Fig. 5. Affinity ligand used for the isolation of the "cap" binding protein.

Isolation of a "cap" binding protein

A crude initiation factor preparation was obtained by ammonium sulfate fractionation of the high salt wash of rabbit reticulocyte ribosomes; the appropriate (0-40% saturation) fraction was sedimented in a 10-40% glycerol gradient (SW-41; 20 h; 40,000 rpm) and the fractions from the top half of the gradient were used for further purification. Portions of this preparation were applied to the affinity columns containing 7-methylguanosine 5'-diphosphate (m^7GDP) and guanosine 5'-diphosphate (GDP) as ligands. The columns were washed with a Hepes buffer (pH 7.0) containing 0.1 M KCl and then with the same buffer containing 1.0 M KCl. Each of the column elutes was subsequently analyzed by polyacrylamide gel electrophoresis. As shown in Figure 6, the preparation loaded onto the columns consisted of a

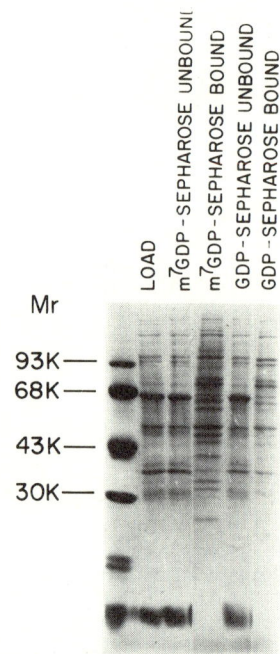

Fig. 6. Analysis by polyacrylamide gel electrophoresis of the proteins resulting from affinity chromatography of an initiation factor preparation. Aliquots of the preparation (38 mg) were applied to each column (equilibrated with 20 mM Hepes, pH 7.0, containing 1 mM dithiothreitol, 0.2 mM EDTA, 10% glycerol and 0.1 M KCl) and the columns were washed with the same buffer containing 1.0 M KCl. The polyacrylamide gels were run in comparison with standards of known molecular weight and visualized by staining with Coomassie blue.

complex mixture of proteins and > 85% of the protein applied was not retained by the columns. The high salt wash from each of the columns also consisted of a mixture of proteins, but the eluate from the m^7GDP-Sepharose column clearly contained a protein of molecular weight 24,000 which was absent in the eluate from the GDP-Sepharose column.

The nature of the band with MW 24,000 was investigated by the crosslinking reaction described previously (35); each of the column eluates was incubated with periodate-oxidized reovirus mRNA that had been [^{3}H]-methyl labeled in the cap. After treatment with NaBH$_3$CN to effect reductive alkylation of the cap with protein, the reaction mixtures were treated with RNAse and analyzed by SDS polyacrylamide gel electrophoresis and fluorography (35,47). As shown in Figure 7, the 24,000 molecular weight protein did crosslink with the oxidized mRNA and all of this protein was retained on the m^7GDP-Sepharose column. In contrast, only a small amount of this protein was retained on GDP-Sepharose. The specific nature of crosslink between the mRNA cap and the protein of MW 24,000 was demonstrated by repeating the crosslinking experiment in the presence of 0.4 mM 7-methylguanosine 5'-diphosphate. This nucleotide (but not 5'-GDP) prevented the formation of crosslinks between the mRNA and the protein of MW 24,000, but not with the proteins of higher molecular weight.

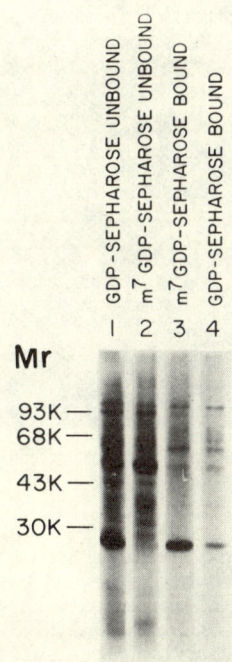

Fig. 7. mRNA crosslinking of the proteins isolated by affinity chromatography. After separation of crude initiation factors as indicated in the legend to Fig. 6, the fractions were crosslinked to [^{3}H]-methylated mRNA, digested with RNAse and analyzed by SDS polyacrylamide gel electrophoresis and fluorography.

In an effort to improve the specificity of the purification process, the elution procedure was modified. Samples were applied to the column in 20 mM Hepes buffer containing 0.1 M KCl as before, but the column was next washed with the same buffer containing 70 μM 7-methylguanosine 5'-diphosphate and finally with 20 mM Hepes buffer containing 1.0 M KCl. As shown in Figure 8, the material applied to the column (lane 1) and the column flow-through (lane 2) contained many proteins, but there was no band at MW 24,000. Elution with buffer containing 70 μM 7-methylguanosine (lanes 3-9) gave exclusively a protein of MW 24,000 and subsequent high salt wash (lane 10) effected the elution of many proteins, but only residual amounts of the 24,000-dalton polypeptide. The eluted protein was dialyzed against 20 mM Hepes buffer containing 0.1 M KCl to remove 7-methylguanosine 5'-diphosphate and was then assayed for its ability to crosslink to reovirus mRNA. The purified protein crosslinked with the mRNA and the interaction could be blocked by admixture of 7-methylguanosine 5'-diphosphate to the incubation mixture. The purity of the isolated material was tested further by one- and two-dimensional gel electrophoresis of the iodinated (48) polypeptide. The protein migrated as a single band when analyzed by SDS polyacrylamide electrophoresis, but was partially resolved by isoelectric focusing (into two species of pI = 5.2 - 5.7). This partial resolution probably reflected differential modification of some of the polypeptide chains in a way that altered net charge.

Functionally, it was found that the purified cap binding protein was capable of stimulating the translation of capped Sindbis virus and rabbit globin mRNA's 2- to 4-fold when protein biosynthesis was carried out using Hela cell extracts. The products elaborated were shown to have the expected composition, as judged by polyacrylamide gel electrophoresis. Significantly, the translation of mouse encephalomyocarditis virus and satellite tobacco necrosis virus (neither of which is capped) were not found to be affected by the cap binding protein.

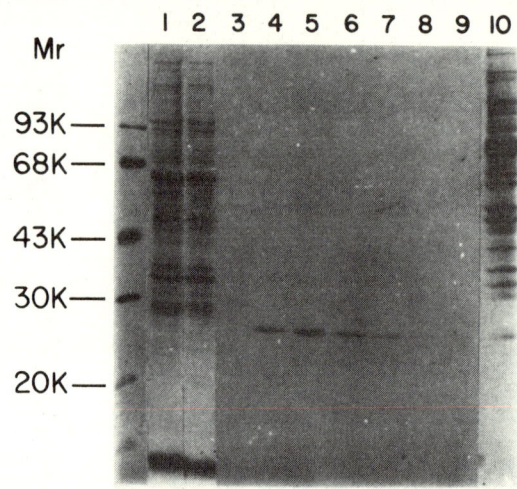

Fig. 8. SDS polyacrylamide gel of the protein eluted from m^{7}GDP-Sepharose by elution with 7-methylguanosine 5'-diphosphate. The column was loaded and washed initially as indicated in the legend to Fig. 6. However, prior to the high salt wash, the column was eluted with 20 mM Hepes buffer containing 0.1 M KCl and 70 μM 7-methylguanosine 5'-diphosphate. All fractions were then analyzed by polyacrylamide gel electrophoresis (and visualized by staining with Coomassie blue). Lane 1 corresponds to the material loaded onto the column, lane 2 to column flow-through, lanes 3-9 to fractions obtained by washing with 7-methylguanosine 5'-diphosphate and lane 10 to material obtained from the high salt wash.

REFERENCES

1. F. H. C. Crick, Symp. Soc. Exp. Biol. 12, 138-163 (1958).
2. M. B. Hoagland, in Brookhaven Symposium in Biology, No. 12: Structure and Function of Genetic Elements, Brookhaven National Laboratory, Upton, New York (1959).
3. A. J. Shatkin, Cell 9, 645-653 (1976).
4. J. P. Ebel, R. Giegé, J. Bonnet, D. Kern, N. Befort, C. Bollack, F. Fasiolo, J. Gangloff and G. Dirheimer, Biochimie 55, 547-557 (1973).
5. R. Giegé, D. Kern, J. P. Ebel, H. Grosjean, S. DeHenau and H. Chantrenne, Eur. J. Biochem. 45, 351-362 (1974).
6. S. M. Hecht, J. W. Kozarich and F. J. Schmidt, Proc. Nat. Acad. Sci. USA 71, 4317-4321 (1974).
7. A. C. Chinault, J. W. Kozarich, S. M. Hecht, F. J. Schmidt and R. M. Bock, Biochemistry 16, 756-765 (1977).
8. J. S. Hwang, P. Bhuta and J. Zemlicka, Biochim. Biophys. Acta 479, 143-151 (1977).
9. S. M. Hecht, B. L. Alford, Y. Kuroda and S. Kitano, J. Biol. Chem. 253, 4517-4520 (1978).
10. R. Silber, V. G. Malathi and J. Hurwitz, Proc. Nat. Acad. Sci. USA 69, 3009-3013 (1972).
11. T. J. Snopek, A. Sugino, K. L. Agarwal and N. R. Cozzarelli, Biochem. Biophys. Res. Commun. 68, 417-424 (1976).
12. G. Kaufmann, T. Klein and U. Z. Littauer, FEBS Lett. 46, 271-275 (1974).
13. G. Kaufmann and N. R. Kallenbach, Nature 254, 452-454 (1975).
14. G. Kaufmann and U. Z. Littauer, Proc. Nat. Acad. Sci. USA 71, 3741-3745 (1974).
15. G. C. Walker, O. C. Uhlenbeck, E. Bedows and R. I. Gumport, Proc. Nat. Acad. Sci. USA 72, 122-126 (1975).
16. T. E. England and O. C. Uhlenbeck, Biochemistry 17, 2069-2076 (1978).
17. T. E. England, R. I. Gumport and O. C. Uhlenbeck, Proc. Nat. Acad. Sci. USA 74, 4839-4842 (1977).
18. J. R. Barrio, M. del C. G. Barrio, N. J. Leonard, T. E. England and O. C. Uhlenbeck, Biochemistry 17, 2077-2081 (1978).

19. Y. Kikuchi, F. Hishinuma and K. Sakaguchi, Proc. Nat. Acad. Sci. USA 75, 1270-1273 (1978).
20. D. M. Hinton, J. A. Baez and R. I. Gumport, Biochemistry 17, 5091-5097 (1978).
21. S. M. Hecht, Tetrahedron 33, 1671-1696 (1977) and references therein.
22. See, e.g., M. Rubenstein, N. DeGroot and Y. Lapidot, Biochim. Biophys. Acta 209, 183-189 (1970).
23. B. L. Alford, J. M. Pezzuto, K. H. Tan and S. M. Hecht, J. Biol. Chem. 254, 6894-6903 (1979).
24. F. Chapeville, F. Lipmann, G. von Ehrenstein, B. Weisblum, W. J. Ray and S. Benzer, Proc. Nat. Acad. Sci. USA 48, 1086-1092 (1962).
25. G. von Ehrenstein, B. Weisblum and S. Benzer, Proc. Nat. Acad. Sci. USA 49, 669-675 (1963).
26. K. B. Jacobson, Cold Spring Harbor Symp. Quant. Biol. 31, 719-722 (1966).
27. M. Yarus, Biochemistry 11, 2352-2361 (1972).
28. A. E. Johnson, W. R. Woodward, E. Herbert and J. R. Menninger, Biochemistry 15, 569-575 (1976).
29. W. M. Holmes, G. W. Hatfield and E. Goldman, J. Biol. Chem. 253, 3482-3486 (1978).
30. P.-C. Tai, B. J. Wallace and B. Davis, Proc. Nat. Acad. Sci. USA 75, 275-279 (1978).
31. J. M. Pezzuto and S. M. Hecht, J. Biol. Chem., in press.
32. S. Muthukrishnan, G. W. Both, Y. Furuichi and A. J. Shatkin, Nature 255, 33-37 (1975).
33. S. Muthukrishnan, W. Filopowicz, J. M. Sierra, G. W. Both, A. J. Shatkin and S. Ochoa, J. Biol. Chem. 250, 9336-9341 (1975).
34. G. W. Both, Y. Furuichi, S. Muthukrishnan and A. J. Shatkin, Cell 6, 185-195 (1975).
35. N. Sonenberg, M. A. Morgan, W. C. Merrick and A. J. Shatkin, Proc. Nat. Acad. Sci. USA 75, 4843-4847 (1978).
36. J. E. Bergmann, H. Trachsel, N. Sonenberg, A. J. Shatkin and H. F. Lodish, J. Biol. Chem. 254, 1440-1443 (1979).
37. E. D. Hickey, L. A. Weber and C. Baglioni, Proc. Nat. Acad. Sci. USA 73, 19-23 (1976).
38. R. Roman, J. D. Brooker, S. N. Seal and A. Marcus, Nature 260, 359-363 (1976).
39. D. Canaani, M. Revel and Y. Groner, FEBS Lett. 64, 326-331 (1976).
40. B. L. Adams, M. Morgan, S. Muthukrishnan, S. M. Hecht and A. J. Shatkin, J. Biol. Chem. 253, 2589-2595 (1978).
41. T. Igo-Kemenes and H. G. Zachau, Eur. J. Biochem. 18, 292-298 (1971).
42. S. M. Hecht, B. L. Adams and J. W. Kozarich, J. Org. Chem. 41, 2303-2311 (1976).
43. F. Seela and S. Waldek, Nucleic Acids Res. 2, 2343-2354 (1975).
44. K. Imai, S. Fujii, K. Takanohashi, Y. Furukawa, T. Masuda and M. Honjo, J. Org. Chem. 34, 1547-1550 (1969).
45. D. E. Hoard and D. G. Ott, J. Am. Chem. Soc. 87, 1785-1788 (1965).
46. J. W. Kozarich, A. C. Chinault and S. M. Hecht, Biochemistry 12, 4458-4463 (1973).
47. N. Sonenberg and A. J. Shatkin, Proc. Nat. Acad. Sci. USA 74, 4288-4292 (1977).
48. A. E. Bolton and W. M. Hunter, Biochem. J. 133, 529-539 (1973).